Student Solutions Manual

Vector Calculus

Miroslav Lovrić

McMaster University

John Wiley & Sons, Inc.

10 9 8 7 6 5 4 3 2

Printed and bound by Lightning Source

FOREWORD

This manual contains solutions to odd-numbered exercises from the textbook *Vector Calculus* by Miroslav Lovrić, published by John Wiley & Sons.

In most cases, all details of a solution are provided. When there is a need, a related theoretical concept or a method (or a formula) are recalled and discussed first. Sometimes, details of evaluation of definite integrals are skipped and a reader is referred to a table of integrals or is advised to use a numeric method or appropriate software. (The objective of this course is not to master a dozen integration techniques but rather to understand what the integral involved is all about.)

Do not read this manual! It is far more beneficial to try to solve an exercise on your own. Start, and see how far you can go. If you get stuck, identify the problem first - try to understand why you are having difficulties, and then look up the solution in the manual. This way, you will learn not only what the problem is (or what you have problems with), but also how that particular problem has been resolved. If you just read a solution, you might not recognize the hard part(s) - or, even worse, you might miss the whole point of the exercise.

It is assumed that you have access to a table of integrals and a calculator. You are strongly encouraged to use a graphing calculator and/or appropriate software (such as Maple or Mathematica).

I accept full responsibility for errors in this text and will be grateful to anybody who brings them to my attention. Your comments and suggestions will be greatly appreciated.

Miroslav Lovrić
Department of Mathematics and Statistics
McMaster University
Hamilton, Ontario, Canada L8S 4K1
e-mail: lovric@mcmaster.ca

November 2006

CONTENTS

1. VECTORS, MATRICES AND APPLICATIONS

1.1. Vectors

1. By the Triangle Inequality, $\|\mathbf{v} + \mathbf{w}\| \le \|\mathbf{v}\| + \|\mathbf{w}\|$ for two vectors $\mathbf{v}$ and $\mathbf{w}$. Assume that $\mathbf{v}$ and $\mathbf{w}$ are parallel and have the same direction, i.e., $\mathbf{w} = t\mathbf{v}$ and $t > 0$. In that case,

$$\|\mathbf{v} + \mathbf{w}\| = \|\mathbf{v} + t\mathbf{v}\| = \|(1 + t)\mathbf{v}\| = |1 + t|\,\|\mathbf{v}\| = (1 + t)\,\|\mathbf{v}\|$$

(since $t > 0$, it follows that $1 + t > 0$ and so $|1 + t| = 1 + t$) and

$$\|\mathbf{v}\| + \|\mathbf{w}\| = \|\mathbf{v}\| + \|t\mathbf{v}\| = \|\mathbf{v}\| + |t|\,\|\mathbf{v}\| = (1 + t)\|\mathbf{v}\|$$

(since $t > 0$, $|t| = t$). Therefore, to illustrate $\|\mathbf{v} + \mathbf{w}\| = \|\mathbf{v}\| + \|\mathbf{w}\|$ we can choose any pair of parallel vectors with the same direction. For example, $\mathbf{v} = \mathbf{i}$ and $\mathbf{w} = 4\mathbf{i}$.

If $\mathbf{v}$ and $\mathbf{w}$ are non-parallel, then they form a triangle with sides equal to $\|\mathbf{v}\|$, $\|\mathbf{w}\|$ and $\|\mathbf{v} + \mathbf{w}\|$, in which case $\|\mathbf{v} + \mathbf{w}\| < \|\mathbf{v}\| + \|\mathbf{w}\|$. As an example, take $\mathbf{v} = \mathbf{i}$ and $\mathbf{w} = \mathbf{i} + \mathbf{j}$.

Notice that as the angle between two vectors (whose magnitudes are fixed) gets larger and larger, the magnitude of their sum gets smaller and smaller. Therefore, to find $\mathbf{w}$, $\mathbf{v}$ such that $\|\mathbf{v}+\mathbf{w}\| < (\|\mathbf{v}\|+\|\mathbf{w}\|)/2$ we can look for vectors that form a larger angle, like $\mathbf{v} = -\mathbf{i}+\mathbf{j}$ and $\mathbf{w} = \mathbf{i}$. In this case, $\|\mathbf{v} + \mathbf{w}\| = \|\mathbf{j}\| = 1$ and $(\|\mathbf{v}\| + \|\mathbf{w}\|)/2 = (\sqrt{2} + 1)/2 > 1$.

3. Assume that $\mathbf{v} = (v_1, v_2) \in \mathbb{R}^2$ (same proof works in any dimension). If $\|\mathbf{v}\| = 0$ then $\sqrt{v_1^2 + v_2^2} = 0$ and $v_1^2 + v_2^2 = 0$. Therefore, $v_1 = v_2 = 0$ and $\mathbf{v} = \mathbf{0}$. This proves it one way. The other implication is immediate: if $\mathbf{v} = \mathbf{0} = (0, 0)$, then $\|\mathbf{v}\| = \sqrt{0^2 + 0^2} = 0$.

5. If $x = y = 2$, then $r = \sqrt{x^2 + y^2} = \sqrt{8}$ and $\tan\theta = 1$. Hence $\theta = \pi/4$ and the point $(2, 2)$ is represented in polar coordinates as $(\sqrt{8}, \pi/4)$. For the remaining three points, $r = \sqrt{8}$ as well. For $(-2, 2)$ we get $\tan\theta = 2/(-2) = -1$, i.e., $\theta = -\pi/4 + k\pi$ (k is an integer). Since the point $(-2, 2)$ is in the second quadrant, $\theta = 3\pi/4$, and $(\sqrt{8}, 3\pi/4)$ are its polar coordinates. Similarly, for the point $(2, -2)$ we get $\tan\theta = -1$ and (the point is in the fourth quadrant) $\theta = 7\pi/4$. Hence $(\sqrt{8}, 7\pi/4)$ are its polar coordinates. Finally, from $x = -2$ and $y = -2$ we get $\tan\theta = 1$ and (we are now in the third quadrant) $\theta = 5\pi/4$. Consequently, $(\sqrt{8}, 5\pi/4)$ are the polar coordinates of $(-2, -2)$.

7. Rewrite the Triangle Inequality as $\|\mathbf{a} + \mathbf{b}\| \le \|\mathbf{a}\| + \|\mathbf{b}\|$ and substitute $\mathbf{a} = \mathbf{v} - \mathbf{w}$ and $\mathbf{b} = \mathbf{w}$. Thus $\|\mathbf{v}-\mathbf{w}+\mathbf{w}\| \le \|\mathbf{v}-\mathbf{w}\|+\|\mathbf{w}\|$, so $\|\mathbf{v}\| \le \|\mathbf{v}-\mathbf{w}\|+\|\mathbf{w}\|$ and $\|\mathbf{v}-\mathbf{w}\| \ge \|\mathbf{v}\|-\|\mathbf{w}\|$. Consider a triangle whose sides are $\mathbf{v}$, $\mathbf{w}$ and $\mathbf{v} - \mathbf{w}$. The above inequality states that the difference of lengths of two sides in a triangle is smaller than the length of the third side. The equality holds in the case when $\mathbf{v}$ and $\mathbf{w}$ are parallel, of the same direction, and $\|\mathbf{v}\| \ge \|\mathbf{w}\|$.

9. If $\mathbf{v}$ and $\mathbf{w}$ are parallel of opposite directions, then $\mathbf{v} + \mathbf{w} = \mathbf{0}$ and hence 0 is the smallest value of $\|\mathbf{v} + \mathbf{w}\|$. By the Triangle Inequality $\|\mathbf{v} + \mathbf{w}\| \le \|\mathbf{v}\| + \|\mathbf{w}\|$, the largest value of $\|\mathbf{v} + \mathbf{w}\|$ is $\|\mathbf{v}\| + \|\mathbf{w}\| = 2$ (and this happens when $\mathbf{v}$ and $\mathbf{w}$ are parallel and of the same direction; i.e., when $\mathbf{v} = \mathbf{w}$, since both vectors are of the same length).

11. By definition, $\|\mathbf{v}\| = \sqrt{0^2 + 2^2 + (-1)^2} = \sqrt{5}$.

13. Since $\|\mathbf{w}\| = \sqrt{9 + 16} = 5$, it follows that

$$\mathbf{v} = (3/5)\mathbf{i} + (4/5)\mathbf{j} \qquad \text{and} \qquad \|\mathbf{v}\| = \sqrt{9/25 + 16/25} = 1.$$

In general, $\mathbf{v} = \mathbf{w}/\|\mathbf{w}\| = (1/\|\mathbf{w}\|)\mathbf{w}$ has the same direction as $\mathbf{w}$. Its length is 1 (it is called the unit vector in the direction of $\mathbf{w}$), since

$$\|\mathbf{v}\| = \left\| \frac{\mathbf{w}}{\|\mathbf{w}\|} \right\| = \frac{1}{\|\mathbf{w}\|} \|\mathbf{w}\| = 1.$$

15. Let $A(a_1, a_2)$, $B(b_1, b_2)$ be points in $\mathbb{R}^2$. If A and B lie on the same horizontal line, then their distance is $|b_1 - a_1|$. Since, in that case, $a_2 = b_2$, we can write (recall that $|x| = \sqrt{x^2}$)

$$|b_1 - a_1| = \sqrt{(b_1 - a_1)^2} = \sqrt{(b_1 - a_1)^2 + (b_2 - a_2)^2}.$$

If A and B lie on the same vertical line, then their distance is $|b_2 - a_2|$; since $a_1 = b_1$, we obtain the same formula. If A and B do not lie on the same horizontal or the same vertical line, then construct the triangle as shown in Figure 15(a). It is a right triangle with sides $|b_1 - a_1|$ and $|b_2 - a_2|$. By the Pythagorean Theorem, its hypotenuse is $d(A, B) = \sqrt{(b_1 - a_1)^2 + (b_2 - a_2)^2}$.

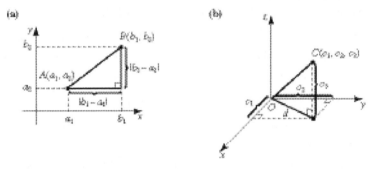

Figure 15

The proof in $\mathbb{R}^3$ is done in a similar way. Here is an alternative: first, compute the distance $d(O, C)$ from a point $C(c_1, c_2, c_3)$ to the origin, see Figure 15(b). By the Pythagorean Theorem, $d^2 = c_1^2 + c_2^2$ and (again by the Pythagorean Theorem) $d(O, C) = \sqrt{d^2 + c_3^2} = \sqrt{c_1^2 + c_2^2 + c_3^2}$. Now take any points $A(a_1, a_2, a_3)$ and $B(b_1, b_2, b_3)$ in $\mathbb{R}^3$. Take a line segment $\overline{AB}$ and translate it by the vector $-(a_1, a_2, a_3)$, thus obtaining the segment $\overline{OC}$, joining the origin O and the point $C(b_1 - a_1, b_2 - a_2, b_3 - a_3)$. By the formula we have just derived, the length of $\overline{AB} = $ distance from the origin to $C = \sqrt{(b_1 - a_1)^2 + (b_2 - a_2)^2 + (b_3 - a_3)^2}$.

17. Recall that the representative of a vector $\mathbf{v} = (v_1, v_2, v_3)$ whose tail is at $A(a_1, a_2, a_3)$ is the directed line segment $\overrightarrow{AB}$, where the point B has coordinates $B(a_1 + v_1, a_2 + v_2, a_3 + v_3)$. Let $\mathbf{v} = (0, 2, -1)$. The directed line segment $\overrightarrow{A_1B_1}$, where $A_1(0, 1, 1)$ and $B_1(0, 3, 0)$ is the representative that starts at $A_1(0, 1, 1)$. Similarly, the directed line segments $\overrightarrow{A_2B_2}$,

$\overrightarrow{A_3B_3}$ and $\overrightarrow{A_4B_4}$, where $A_2(0,3,0)$, $B_2(0,5,-1)$, $A_3(8,9,-4)$, $B_3(8,11,-5)$, $A_4(10,-1,4)$ and $B_4(10,1,3)$, are the representatives of $\mathbf{v}$ that start at A_2, A_3 and A_4 respectively.

19. We use elementary vector operations. First of all, $\mathbf{a}-2\mathbf{b} = (2\mathbf{i}-\mathbf{j}+\mathbf{k})-2(\mathbf{k}-3\mathbf{i}) = 8\mathbf{i}-\mathbf{j}-\mathbf{k}$. Since $\|\mathbf{c}\| = 2$, it follows that $\mathbf{a} - \mathbf{c}/\|\mathbf{c}\| = (2\mathbf{i} - \mathbf{j} + \mathbf{k}) - 2\mathbf{i}/2 = \mathbf{i} - \mathbf{j} + \mathbf{k}$. Similarly, $3\mathbf{a}+\mathbf{c}-\mathbf{j}+\mathbf{k} = 3(2\mathbf{i}-\mathbf{j}+\mathbf{k})+2\mathbf{i}-\mathbf{j}+\mathbf{k} = 8\mathbf{i}-4\mathbf{j}+4\mathbf{k}$. Since $\mathbf{b}+2\mathbf{a} = (\mathbf{k}-3\mathbf{i})+2(2\mathbf{i}-\mathbf{j}+\mathbf{k}) = \mathbf{i} - 2\mathbf{j} + 3\mathbf{k}$ and $\|\mathbf{b} + 2\mathbf{a}\| = \sqrt{14}$, it follows that the unit vector in the direction of $\mathbf{b} + 2\mathbf{a}$ is $(\mathbf{b} + 2\mathbf{a})/\|\mathbf{b} + 2\mathbf{a}\| = (\mathbf{i} - 2\mathbf{j} + 3\mathbf{k})/\sqrt{14}$.

21. The vector $-3\mathbf{i}$ has length 3 and it makes an angle of π radians with respect to the positive x-axis. Consequently, its polar form is $-3\mathbf{i} = 3(\cos\pi\,\mathbf{i}+\sin\pi\,\mathbf{j})$. The length of $\mathbf{i}/2-\mathbf{j}$ is $\sqrt{5}/2 \approx 1.11803$. We need the angle between the positive x-axis and $\mathbf{i}/2-\mathbf{j}$ (notice that $\mathbf{i}/2-\mathbf{j}$ is in the fourth quadrant). From $\tan\theta_1 = -1/(1/2) = -2$ it follows that $\theta_1 = \arctan(-2)$, and the required angle is $2\pi + \arctan(-2) \approx 5.17604$ rad. Hence $\mathbf{i}/2 - \mathbf{j} \approx 1.11803(\cos 5.17604\,\mathbf{i} + \sin 5.17604\,\mathbf{j})$. Finally, let $\mathbf{v} = \mathbf{i} - 4\mathbf{j}$ ($\mathbf{v}$ is in the fourth quadrant). Then $\|\mathbf{v}\| = \sqrt{17} \approx 4.12311$ and $\tan\theta_2 = -4$, so that the angle between the positive x-axis and the vector $\mathbf{v}$ is $2\pi + \arctan(-4) \approx 4.95737$ rad. Therefore, $\mathbf{i} - 4\mathbf{j} \approx 4.12311(\cos 4.95737\,\mathbf{i} + \sin 4.95737\,\mathbf{j})$.

23. Let $\mathbf{v} = (v_1, v_2, v_3)$. Since $(\alpha+\beta)\mathbf{v} = (\alpha+\beta)(v_1, v_2, v_3) = ((\alpha+\beta)v_1, (\alpha+\beta)v_2, (\alpha+\beta)v_3)$ and

$$\alpha\mathbf{v} + \beta\mathbf{v} = \alpha(v_1, v_2, v_3) + \beta(v_1, v_2, v_3) = (\alpha v_1, \alpha v_2, \alpha v_3) + (\beta v_1, \beta v_2, \beta v_3)$$
$$= (\alpha v_1 + \beta v_1, \alpha v_2 + \beta v_2, \alpha v_3 + \beta v_3),$$

it follows (by the distributivity of multiplication of real numbers) that $(\alpha + \beta)\mathbf{v} = \alpha\mathbf{v} + \beta\mathbf{v}$. Similarly, $(\alpha\beta)\mathbf{v} = (\alpha\beta)(v_1, v_2, v_3) = ((\alpha\beta)v_1, (\alpha\beta)v_2, (\alpha\beta)v_3)$ and

$$\alpha(\beta\mathbf{v}) = \alpha(\beta(v_1, v_2, v_3)) = \alpha(\beta v_1, \beta v_2, \beta v_3) = (\alpha(\beta v_1), \alpha(\beta v_2), \alpha(\beta v_3)),$$

together with the associativity of multiplication of real numbers prove the second identity.

1.2. Applications in Geometry and Physics

1. The equation of the line is

$$\ell(t) = (1,3) + t\,(3\mathbf{v}) = (1,3) + t(-3,-15) = (1 - 3t, 3 - 15t),$$

where $t \in \mathbb{R}$. If we replace $3\mathbf{v}$ by a vector $m\mathbf{v}$ (where $m \neq 0$; clearly, $m\mathbf{v}$ is parallel to $\mathbf{v}$), we get infinitely many parametrizations, one for each non-zero value of m:

$$\ell(t) = (1,3) + t\,m(-1,-5) = (1 - tm, 3 - 5tm),$$

where $t \in \mathbb{R}$.

3. The equation of the line is $\ell(t) = \mathbf{a} + t\mathbf{v}$, where $\mathbf{a} = (1,1)$ and $\mathbf{v}$ is the vector from $(1,1)$ to $(-2,4)$; i.e., $\mathbf{v} = (-3,3)$. Hence $\ell(t) = (1,1) + t(-3,3) = (1 - 3t, 1 + 3t)$, where $t \in \mathbb{R}$, parametrizes the given line.

To describe the half-line starting at $(1, 1)$ in the direction towards $(-2, 4)$, we need all vectors that point in the same direction as $\mathbf{v}$. Hence $\ell(t) = (1 - 3t, 1 + 3t)$ with $t \geq 0$ is its parametric representation.

When $t = 0$, $\ell(0) = (1, 1)$; for $t = 1$, we get $\ell(1) = (-2, 4)$. If $0 < t < 1$, then $\ell(t)$ is a point on the line between $(1, 1)$ and $(-2, 4)$. Consequently $\ell(t) = (1 - 3t, 1 + 3t)$, where $0 \leq t \leq 1$, describes the desired line segment.

5. Choose $\mathbf{a} = (2, 1)$ and let $\mathbf{v}$ be the vector from $(2, 1)$ to $(-1, 5)$; i.e., $\mathbf{v} = (-3, 4)$. Then $\ell(t) = (2, 1) + t(-3, 4)$, $t \in \mathbb{R}$, is a parametric equation of the line containing the given points. The half-line ℓ' "contains" all vectors parallel to $\mathbf{v}$ that are of the same direction as $\mathbf{v}$. Hence $\ell'(t) = (2, 1) + t(-3, 4)$ with $t \geq 0$ parametrizes ℓ'.

7. The rectangle in question is spanned by $\mathbf{v} = 3\mathbf{i}$ and $\mathbf{w} = \mathbf{j}$; see Figure 7.

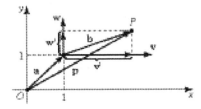

Figure 7

Let $P(x, y)$ be a point in the rectangle, $\mathbf{b}$ the vector from $(1, 1)$ to P and $\mathbf{a}$ the vector from the origin to $(1, 1)$. Then ($\mathbf{p}$ is the position vector of P) $\mathbf{p} = \mathbf{a} + \mathbf{b}$, where $\mathbf{b} = \mathbf{v}' + \mathbf{w}'$. Since P belongs to the rectangle, $\mathbf{v}' = \alpha \mathbf{v}$ and $0 \leq \alpha \leq 1$. Similarly, $\mathbf{w}' = \beta \mathbf{w}$ and $0 \leq \beta \leq 1$. It follows that $\mathbf{b} = \alpha(3\mathbf{i}) + \beta \mathbf{j}$, and therefore $\mathbf{p} = \mathbf{i} + \mathbf{j} + 3\alpha \mathbf{i} + \beta \mathbf{j} = (1 + 3\alpha)\mathbf{i} + (1 + \beta)\mathbf{j}$, $0 \leq \alpha, \beta \leq 1$, is a vector description of the rectangle.

9. The relative position of the two cars is either of the two differences of the displacement vectors, $\mathbf{v} = \mathbf{i} + 3\mathbf{j} - \mathbf{k} - (\mathbf{j} + 2\mathbf{k}) = \mathbf{i} + 2\mathbf{j} - 3\mathbf{k}$ or $-\mathbf{v}$. The distance between the cars is $\|\mathbf{v}\| = \sqrt{14}$.

11. The particle moves along the line

$$\ell(t) = \textbf{initial point} + \textbf{displacement} = (3, 2, 4) + t(3, 0, -1),$$

$t \geq 0$. It will cross the xy-plane when the z-coordinate of $\ell(t)$ is zero; i.e., when $4 - t = 0$ or $t = 4$. The point where it crosses the plane is $\ell(4) = (15, 2, 0)$.

13. The information given describes the vector $\mathbf{F}$ in polar form. Hence

$$\mathbf{F} = \|\mathbf{F}\|(\cos(\pi/6)\mathbf{i} + \sin(\pi/6)\mathbf{j}) = 10\left(\tfrac{\sqrt{3}}{2}\mathbf{i} + \tfrac{1}{2}\mathbf{j}\right) = 5\sqrt{3}\mathbf{i} + 5\mathbf{j}.$$

15. The position vectors of the four masses are $\mathbf{0}$, $\mathbf{i} - 2\mathbf{j}$, $\mathbf{i} + \mathbf{j}$ and $(\mathbf{i} - 2\mathbf{j}) + (\mathbf{i} + \mathbf{j}) = 2\mathbf{i} - \mathbf{j}$. By (1.6), (total mass is $M = 8$)

$$\mathbf{c}_M = \tfrac{1}{8}(2(\mathbf{0}) + 2(\mathbf{i} - 2\mathbf{j}) + 2(\mathbf{i} + \mathbf{j}) + 2(2\mathbf{i} - \mathbf{j})) = \tfrac{1}{8}(8\mathbf{i} - 4\mathbf{j}) = \mathbf{i} - \tfrac{1}{2}\mathbf{j}.$$

Therefore, the center of mass of the system is located at $(1, -1/2)$.

17. By (1.6), $\mathbf{c}_M = \frac{1}{M}(m_1\mathbf{r}_1 + m_2\mathbf{r}_2 + m_3\mathbf{r}_3)$, where $\mathbf{c}_M = (1,0)$, $m_1 = 2$, $\mathbf{r}_1 = (-1,0)$, $m_2 = 2$, $\mathbf{r}_2 = (1,-2)$ and $\mathbf{r}_3 = (2,1)$. We need to find m_3. Note that $M = 2 + 2 + m_3 = 4 + m_3$. From

$$(1,0) = \tfrac{1}{4+m_3}\Big(2(-1,0) + 2(1,-2) + m_3(2,1)\Big),$$

we get

$$1 = \frac{2m_3}{4+m_3} \qquad \text{and} \qquad 0 = \frac{-4+m_3}{4+m_3}.$$

Solving either of the two equations, we get $m_3 = 4$.

19. Place the triangle in the first quadrant (call it the triangle ABC) so that its vertex A is at the origin and the side $\overline{AB}$ lies on the x-axis — thus $A(0,0)$ and $B(2,0)$; see Figure 19. To find the coordinates of $C(c_1, c_2)$ use the fact that the distances from C to A and from C to B must be 2. Since $d(A,C) = 2$, it follows that $d(A,C)^2 = 4$ and $c_1^2 + c_2^2 = 4$. Similarly, $d(B,C) = 2$ implies that $(c_1 - 2)^2 + c_2^2 = 4$. Subtracting the second equation from the first we get $(c_1 - 2)^2 - c_1^2 = 0$ and $c_1 = 1$. Consequently, $c_2 = \sqrt{3}$ and the coordinates of C are $(1, \sqrt{3})$.

We decided to solve this question so as to practice the distance formula. There are other ways of getting C: for example, by symmetry, the x-coordinate of C must be half-way between 0 and 2; i.e., $c_1 = 1$. The y-coordinate is the height of an equilateral triangle of side 2; i.e., $c_2 = \sqrt{2^2 - 1^2} = \sqrt{3}$ by the Pythagorean Theorem. Alternatively, using trigonometry, $c_1 = 2\cos(\pi/3) = 2(1/2) = 1$ and $c_2 = 2\sin(\pi/3) = 2(\sqrt{3}/2) = 1\sqrt{3}$.

All masses are 1 kg. The position vectors are $\mathbf{r}_A = \mathbf{0}$, $\mathbf{r}_B = 2\mathbf{i}$ and $\mathbf{r}_C = \mathbf{i} + \sqrt{3}\mathbf{j}$. Since the total mass is $M = 3$, we get that

$$\mathbf{c}_M = \tfrac{1}{3}(1(\mathbf{0}) + 1(2\mathbf{i}) + 1(\mathbf{i} + \sqrt{3}\mathbf{j})) = \tfrac{1}{3}(3\mathbf{i} + \sqrt{3}\mathbf{j}) = \mathbf{i} + \tfrac{\sqrt{3}}{3}\mathbf{j}.$$

In words, the center of mass $\mathbf{c}_M$ is located on the axis of symmetry of the triangle (one such axis is the line through C perpendicular to $\overline{AB}$), $\sqrt{3}/3$ units away from the side $\overline{AB}$.

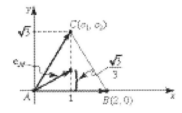

Figure 19

21. The forces are

$$\mathbf{F}_1 = 5(\cos(\pi/10)\mathbf{i} + \sin(\pi/10)\mathbf{j}) \approx 4.755\mathbf{i} + 1.545\mathbf{j},$$
$$\mathbf{F}_2 = 5(\cos(\pi/5)\mathbf{i} + \sin(\pi/5)\mathbf{j}) \approx 4.045\mathbf{i} + 2.939\mathbf{j},$$
$$\mathbf{F}_3 = 5(\cos(\pi/2)\mathbf{i} + \sin(\pi/2)\mathbf{j}) = 5\mathbf{j}$$

and

$$\mathbf{F}_4 = 5(\cos(13\pi/10)\mathbf{i} + \sin(13\pi/10)\mathbf{j}) \approx -2.939\mathbf{i} - 4.045\mathbf{j}.$$

Their resultant is $\mathbf{F} = \mathbf{F}_1 + \mathbf{F}_2 + \mathbf{F}_3 + \mathbf{F}_4 \approx 5.861\mathbf{i} + 5.439\mathbf{j}.$

1.3. The Dot Product

1. Let $\mathbf{v} = (v_1, \ldots, v_n)$ and $\mathbf{w} = (w_1, \ldots, w_n)$. Then $\mathbf{v} \cdot \mathbf{w} = v_1 w_1 + \ldots + v_n w_n$ and $\mathbf{w} \cdot \mathbf{v} = w_1 v_1 + \ldots + w_n v_n$ are equal due to commutativity of multiplication of real numbers.

3. Let $\mathbf{u} = (u_1, u_2, u_3)$, $\mathbf{v} = (v_1, v_2, v_3)$ and $\mathbf{w} = (w_1, w_2, w_3)$. Then $\mathbf{v} + \mathbf{w} = (v_1 + w_1, v_2 + w_2, v_3 + w_3)$ and

$$\mathbf{u} \cdot (\mathbf{v} + \mathbf{w}) = u_1(v_1 + w_1) + u_2(v_2 + w_2) + u_3(v_3 + w_3).$$

Since

$$\mathbf{u} \cdot \mathbf{v} + \mathbf{u} \cdot \mathbf{w} = u_1 v_1 + u_2 v_2 + u_3 v_3 + u_1 w_1 + u_2 w_2 + u_3 w_3,$$

it follows (by distributivity of multiplication of real numbers) that $\mathbf{u} \cdot (\mathbf{v} + \mathbf{w}) = \mathbf{u} \cdot \mathbf{v} + \mathbf{u} \cdot \mathbf{w}$.

The remaining two identities are proven in a similar way. From $\mathbf{u} = (u_1, u_2, u_3)$ we get $\alpha \mathbf{u} = (\alpha u_1, \alpha u_2, \alpha u_3)$ and

$$(\alpha \mathbf{u}) \cdot \mathbf{v} = (\alpha u_1)v_1 + (\alpha u_2)v_2 + (\alpha u_3)v_3 = \alpha u_1 v_1 + \alpha u_2 v_2 + \alpha u_3 v_3.$$

Similarly,

$$\alpha(\mathbf{u} \cdot \mathbf{v}) = \alpha(u_1 v_1 + u_2 v_2 + u_3 v_3)$$

and the identity $(\alpha \mathbf{u}) \cdot \mathbf{v} = \alpha(\mathbf{u} \cdot \mathbf{v})$ follows from the distributivity of multiplication of real numbers.

To prove the third identity we use the commutativity of the dot product (twice) and the identity that we have just proved:

$$\mathbf{u} \cdot (\alpha \mathbf{v}) = (\alpha \mathbf{v}) \cdot \mathbf{u} = \alpha(\mathbf{v} \cdot \mathbf{u}) = \alpha(\mathbf{u} \cdot \mathbf{v}).$$

5. From linear algebra we know that any vector in $\mathbb{R}^3$ can be written as a linear combination of three (non-zero) mutually orthogonal vectors; i.e., $\mathbf{a} = a_{\mathbf{u}}\mathbf{u} + a_{\mathbf{v}}\mathbf{v} + a_{\mathbf{w}}\mathbf{w}$, where $a_{\mathbf{u}}$, $a_{\mathbf{v}}$ and $a_{\mathbf{w}}$ are real numbers. Computing the dot product of $\mathbf{a}$ with $\mathbf{u}$, we get

$$\mathbf{a} \cdot \mathbf{u} = (a_{\mathbf{u}}\mathbf{u} + a_{\mathbf{v}}\mathbf{v} + a_{\mathbf{w}}\mathbf{w}) \cdot \mathbf{u}$$
$$= a_{\mathbf{u}}\mathbf{u} \cdot \mathbf{u} + a_{\mathbf{v}}\mathbf{v} \cdot \mathbf{u} + a_{\mathbf{w}}\mathbf{w} \cdot \mathbf{u} = a_{\mathbf{u}}\mathbf{u} \cdot \mathbf{u},$$

since (by orthogonality) the remaining two terms on the right side vanish. Hence $\mathbf{a} \cdot \mathbf{u} = a_{\mathbf{u}}\|\mathbf{u}\|^2$ and $a_{\mathbf{u}} = \mathbf{a} \cdot \mathbf{u}/\|\mathbf{u}\|^2$. The expressions for $a_{\mathbf{v}}$ and $a_{\mathbf{w}}$ are obtained analogously, by computing the dot product of $\mathbf{a}$ with $\mathbf{v}$ and $\mathbf{w}$.

7. Let θ_1 denote the angle at $(0,3,4)$: it is the angle between the vectors from $(0,3,4)$ to $(0,3,0)$ (call it $\mathbf{u}_1$) and from $(0,3,4)$ to $(12,0,5)$ (call it $\mathbf{v}_1$). It follows that $\mathbf{u}_1 = (0,0,-4)$ and $\mathbf{v}_1 = (12,-3,1)$ and

$$\cos\theta_1 = \frac{\mathbf{u}_1 \cdot \mathbf{v}_1}{\|\mathbf{u}_1\| \, \|\mathbf{v}_1\|} = \frac{-4}{4\sqrt{154}} = \frac{-1}{\sqrt{154}}$$

and therefore $\theta_1 = \arccos(-1/\sqrt{154}) \approx 1.6515$ rad. Denote by θ_2 the angle at $(0,3,0)$; that is, θ_2 is the angle between the vectors $\mathbf{u}_2 = (0,0,4)$ (from $(0,3,0)$ to $(0,3,4)$) and $\mathbf{v}_2 = (12,-3,5)$ (from $(0,3,0)$ to $(12,0,5)$). We get

$$\cos\theta_2 = \frac{\mathbf{u}_2 \cdot \mathbf{v}_2}{\|\mathbf{u}_2\| \, \|\mathbf{v}_2\|} = \frac{20}{4\sqrt{178}}$$

and $\theta_2 = \arccos(5/\sqrt{178}) \approx 1.1867$ rad. Since $\theta_1 + \theta_2 + \theta_3 = \pi$, it follows that $\theta_3 = 0.3034$ rad.

9. Computing the dot product of the given vectors, we get

$$\left(\mathbf{w} - \frac{\mathbf{v} \cdot \mathbf{w}}{\|\mathbf{v}\|^2} \mathbf{v} \right) \cdot \mathbf{v} = \mathbf{w} \cdot \mathbf{v} - \frac{\mathbf{v} \cdot \mathbf{w}}{\|\mathbf{v}\|^2} \mathbf{v} \cdot \mathbf{v} = \mathbf{w} \cdot \mathbf{v} - \frac{\mathbf{v} \cdot \mathbf{w}}{\|\mathbf{v}\|^2} \|\mathbf{v}\|^2 = 0,$$

by the distributivity and the commutativity of the dot product, and the fact that $\mathbf{v} \cdot \mathbf{v} = \|\mathbf{v}\|^2$.

11. Consider the cube spanned by $\mathbf{i}, \mathbf{j}$ and $\mathbf{k}$ with one vertex at the origin (as far as angles are concerned, we are free to choose any cube). In that case, the (direction of the) diagonal can be expressed as $\mathbf{i}+\mathbf{j}+\mathbf{k}$, and hence the angle θ between the diagonal and the side represented by $\mathbf{i}$ is

$$\cos\theta = \frac{\mathbf{i} \cdot (\mathbf{i}+\mathbf{j}+\mathbf{k})}{\|\mathbf{i}\| \, \|\mathbf{i}+\mathbf{j}+\mathbf{k}\|} = \frac{1}{\sqrt{3}},$$

i.e., $\theta = \arccos(1/\sqrt{3}) \approx 0.9553$ rad. The remaining two angles (between the diagonal and the sides represented by $\mathbf{j}$ and $\mathbf{k}$) are computed in the same way, and are both equal to θ.

13. The angle $\theta_{\mathbf{v}}$ between $\mathbf{v}$ and $\mathbf{a}$ is

$$\cos\theta_{\mathbf{v}} = \frac{\mathbf{v} \cdot \mathbf{a}}{\|\mathbf{v}\| \, \|\mathbf{a}\|} = \frac{\mathbf{v} \cdot (\|\mathbf{v}\|\mathbf{w} + \|\mathbf{w}\|\mathbf{v})}{\|\mathbf{v}\| \, \|\mathbf{a}\|} = \frac{\|\mathbf{v}\|\mathbf{v} \cdot \mathbf{w} + \|\mathbf{w}\|\mathbf{v} \cdot \mathbf{v}}{\|\mathbf{v}\| \, \|\mathbf{a}\|}$$
$$= \frac{\|\mathbf{v}\|(\mathbf{v} \cdot \mathbf{w} + \|\mathbf{v}\|^2)}{\|\mathbf{v}\| \, \|\mathbf{a}\|} = \frac{\mathbf{v} \cdot \mathbf{w} + \|\mathbf{v}\|^2}{\|\mathbf{a}\|},$$

since $\|\mathbf{v}\| = \|\mathbf{w}\|$. Similarly, the angle $\theta_{\mathbf{w}}$ between $\mathbf{w}$ and $\mathbf{a}$ is computed to be

$$\cos\theta_{\mathbf{w}} = \frac{\mathbf{w} \cdot \mathbf{a}}{\|\mathbf{w}\| \, \|\mathbf{a}\|} = \frac{\mathbf{w} \cdot (\|\mathbf{v}\|\mathbf{w} + \|\mathbf{w}\|\mathbf{v})}{\|\mathbf{w}\| \, \|\mathbf{a}\|} = \frac{\|\mathbf{v}\|\mathbf{w} \cdot \mathbf{w} + \|\mathbf{w}\|\mathbf{w} \cdot \mathbf{v}}{\|\mathbf{w}\| \, \|\mathbf{a}\|}$$
$$= \frac{\|\mathbf{w}\|(\|\mathbf{w}\|^2 + \mathbf{w} \cdot \mathbf{v})}{\|\mathbf{w}\| \|\mathbf{a}\|} = \frac{\|\mathbf{w}\|^2 + \mathbf{w} \cdot \mathbf{v}}{\|\mathbf{a}\|}.$$

By assumption, $\|\mathbf{v}\| = \|\mathbf{w}\|$, and therefore $\cos\theta_{\mathbf{v}} = \cos\theta_{\mathbf{w}}$, and since (in the definition of the angle between two vectors there is a requirement that $0 \leq \theta_{\mathbf{v}}, \theta_{\mathbf{w}} < \pi$) it follows that $\theta_{\mathbf{v}} = \theta_{\mathbf{w}}$.

15. Let $\mathbf{a} = \mathbf{i} + 2\mathbf{j}$, $\mathbf{v} = \mathbf{i} + \mathbf{j}$ and $\mathbf{w} = \mathbf{i} - \mathbf{j}$. Vectors $\mathbf{v}$ and $\mathbf{w}$ are orthogonal ($\mathbf{v} \cdot \mathbf{w} = 0$) and hence $\mathbf{a} = a_{\mathbf{v}}\mathbf{v} + a_{\mathbf{w}}\mathbf{w}$, where $a_{\mathbf{v}}\mathbf{v} = pr_{\mathbf{v}}\mathbf{a} = (\mathbf{a} \cdot \mathbf{v}/\|\mathbf{v}\|^2)\mathbf{v}$ and $a_{\mathbf{w}}\mathbf{w} = pr_{\mathbf{w}}\mathbf{a} = (\mathbf{a} \cdot \mathbf{w}/\|\mathbf{w}\|^2)\mathbf{w}$

are the vector orthogonal projections of $\mathbf{a}$ onto $\mathbf{v}$ and $\mathbf{w}$; see (1.11) and the text following it. Since

$$a_{\mathbf{v}}\mathbf{v} = \frac{(\mathbf{i} + 2\mathbf{j}) \cdot (\mathbf{i} + \mathbf{j})}{\|\mathbf{i} + \mathbf{j}\|^2}(\mathbf{i} + \mathbf{j}) = \tfrac{3}{2}(\mathbf{i} + \mathbf{j})$$

and

$$a_{\mathbf{v}}\mathbf{w} = \frac{(\mathbf{i} + 2\mathbf{j}) \cdot (\mathbf{i} - \mathbf{j})}{\|\mathbf{i} - \mathbf{j}\|^2}(\mathbf{i} - \mathbf{j}) = -\tfrac{1}{2}(\mathbf{i} - \mathbf{j}),$$

the desired decomposition is

$$\mathbf{a} = \mathbf{i} + 2\mathbf{j} = \tfrac{3}{2}(\mathbf{i} + \mathbf{j}) - \tfrac{1}{2}(\mathbf{i} - \mathbf{j}).$$

17. Using the formula of Example 1.26, we get $3(x - 0) + 0(y + 2) - 1(z - 2) = 0$; i.e., $3x - z + 2 = 0$.

19. The fact that the plane must be perpendicular to the line $\mathbf{l}(t) = (2 + 3t, 1 - t, 7)$, $t \in \mathbb{R}$, implies that the vector $(3, -1, 0)$ (direction vector of the line) is, at the same time, normal vector to the plane. Using (1.12), we get $3(x + 2) - 1(y - 1) + 0(z - 4) = 0$, i.e., $3x - y + 7 = 0$.

21. According to the formula of Example 1.26, we need a point (and we have it — the origin) and a normal vector $\mathbf{n}$. Take $\mathbf{n}$ to be the vector from $(3, 2, -1)$ to $(0, 0, 7)$ (we could have taken the vector in the opposite direction — it would not make any difference). Then $\mathbf{n} = (-3, -2, 8)$ and the equation of the plane is $-3(x - 0) - 2(y - 0) + 8(z - 0) = 0$; i.e., $3x + 2y - 8z = 0$.

23. Consider a constant force $\mathbf{F}$ moving an object around a triangle ABC: first from A to B, then to C and then back to A. Let $\mathbf{a}$, $\mathbf{b}$ and $\mathbf{c}$ be the vectors represented by the directed line segments $\overrightarrow{AB}$, $\overrightarrow{BC}$ and $\overrightarrow{CA}$ respectively. The work along $\overrightarrow{AB}$ is the dot product of the force and the displacement vector; i.e., it is equal to $\mathbf{F} \cdot \mathbf{a}$. Similarly, the work along the remaining two sides is $\mathbf{F} \cdot \mathbf{b}$ and $\mathbf{F} \cdot \mathbf{c}$, and the total work of $\mathbf{F}$ is $W = \mathbf{F} \cdot \mathbf{a} + \mathbf{F} \cdot \mathbf{b} + \mathbf{F} \cdot \mathbf{c} = \mathbf{F} \cdot (\mathbf{a} + \mathbf{b} + \mathbf{c})$. Since $\mathbf{c} = -\mathbf{b} - \mathbf{a}$, it follows that $W = 0$.

An analogous proof works for any polygon. The key fact that is needed is that if the sides of a polygon are oriented so that the terminal point of one side is the initial point of the neighboring one (i.e., all sides are oriented counterclockwise, or all sides are oriented clockwise), then the vector sum of all directed line segments thus obtained is zero. As above, the work of $\mathbf{F}$ is the dot product of $\mathbf{F}$ and the vector sum of all sides, which is zero.

25. From $\mathbf{u} \cdot \mathbf{v} = (\mathbf{i} + \mathbf{j} - 2\mathbf{k}) \cdot (2\mathbf{j} + \mathbf{k}) = 2 - 2 = 0$, $\mathbf{u} \cdot \mathbf{w} = (\mathbf{i} + \mathbf{j} - 2\mathbf{k}) \cdot (5\mathbf{i} - \mathbf{j} + 2\mathbf{k}) = 5 - 1 - 4 = 0$ and $\mathbf{v} \cdot \mathbf{w} = (2\mathbf{j} + \mathbf{k}) \cdot (5\mathbf{i} - \mathbf{j} + 2\mathbf{k}) = -2 + 2 = 0$, it follows that $\mathbf{u}$, $\mathbf{v}$ and $\mathbf{w}$ are mutually orthogonal. From the three-dimensional version of Theorem 1.6 (the formula appears immediately after the proof; see also Exercise 5) we get (replace $\mathbf{a}$ by $\mathbf{i}$) $\mathbf{i} = i_{\mathbf{u}}\mathbf{u} + i_{\mathbf{v}}\mathbf{v} + i_{\mathbf{w}}\mathbf{w}$, where $i_{\mathbf{u}} = \mathbf{i} \cdot \mathbf{u}/\|\mathbf{u}\|^2 = 1/6$, $i_{\mathbf{v}} = \mathbf{i} \cdot \mathbf{v}/\|\mathbf{v}\|^2 = 0$ and $i_{\mathbf{w}} = \mathbf{i} \cdot \mathbf{w}/\|\mathbf{w}\|^2 = 1/6$. Thus $\mathbf{i} = \mathbf{u}/6 + \mathbf{w}/6$.

27. We will show that any vector $\mathbf{a} \in \mathbb{R}^3$ perpendicular to all three of $\mathbf{u}$, $\mathbf{v}$ and $\mathbf{w}$ must be a zero vector. Since $\mathbf{u}$, $\mathbf{v}$ and $\mathbf{w}$ are mutually orthogonal, it follows that (see Theorem 1.6 and the text immediately following it) $\mathbf{a} = a_\mathbf{u}\mathbf{u} + a_\mathbf{v}\mathbf{v} + a_\mathbf{w}\mathbf{w}$, for some real numbers $a_\mathbf{u}$, $a_\mathbf{v}$ and $a_\mathbf{w}$. Computing the dot product of $\mathbf{a}$ with $\mathbf{u}$, we get $\mathbf{a} \cdot \mathbf{u} = a_\mathbf{u}\mathbf{u} \cdot \mathbf{u} + \mathbf{v} \cdot \mathbf{u} + \mathbf{w} \cdot \mathbf{u}$. Since $\mathbf{a}$ must be orthogonal to $\mathbf{u}$ and $\mathbf{u}$, $\mathbf{v}$ and $\mathbf{w}$ are mutually orthogonal, it follows that $\mathbf{a} \cdot \mathbf{u} = 0$, $\mathbf{v} \cdot \mathbf{u} = 0$ and $\mathbf{w} \cdot \mathbf{u} = 0$, and so $0 = a_\mathbf{u}\|\mathbf{u}\|^2$ and $a_\mathbf{u} = 0$, since $\mathbf{u}$ is a non-zero vector by assumption. Computing the dot product of $\mathbf{a}$ with $\mathbf{v}$ and $\mathbf{w}$ and proceeding as above, we will get $a_\mathbf{v} = 0$ and $a_\mathbf{w} = 0$, so that $\mathbf{a} = \mathbf{0}$. Therefore, in $\mathbb{R}^3$, there cannot exist four (or more) mutually orthogonal vectors).

29. Let $\mathbf{p} = (x, y, z)$. Then from $(x, y, z) = (2, 0, -1) + t(0, -1, 1) + s(3, 0, 1)$ we get $x = 2 + 3s$, $y = -t$ and $z = -1 + t + s$. Substituting $s = (x - 2)/3$ and $t = -y$ into the equation for z, we get $z = -1 - y + (x - 2)/3$, i.e., $-x + 3y + 3z + 5 = 0$.

31. We need to solve $Ax + By + Cz + D = 0$ for x, y and z (clearly, at least one of A or B or C must be non-zero). Assuming that $C \neq 0$, we take $x = t$ and $y = s$. Then $At + Bs + Cz + D = 0$ implies that $z = -D/C - At/C - Bs/C$, and

$$(x, y, z) = (t, s, -\tfrac{D}{C} - \tfrac{A}{C}t - \tfrac{B}{C}s) = (0, 0, -\tfrac{D}{C}) + t(1, 0, -\tfrac{A}{C}) + s(0, 1, -\tfrac{B}{C}).$$

Thus, $\mathbf{p} = \mathbf{a} + t\mathbf{v} + s\mathbf{w}$, where $\mathbf{a} = (0, 0, -D/C)$, $\mathbf{v} = (1, 0, -A/C)$, $\mathbf{w} = (0, 1, -B/C)$ and $t, s \in \mathbb{R}$ (clearly, $\mathbf{v}$ and $\mathbf{w}$ are non-parallel). Similar parametric equations are obtained when $A \neq 0$ or $B \neq 0$.

1.4. Matrices and Determinants

1. Using elementary matrix operations,

$$2B - 16I_2 = 2\begin{bmatrix} 0 & 5 \\ 4 & 0 \end{bmatrix} - 16\begin{bmatrix} 1 & 0 \\ 0 & 1 \end{bmatrix} = \begin{bmatrix} 0 & 10 \\ 8 & 0 \end{bmatrix} - \begin{bmatrix} 16 & 0 \\ 0 & 16 \end{bmatrix} = \begin{bmatrix} -16 & 10 \\ 8 & -16 \end{bmatrix}.$$

3. Since C is a 3×2 matrix and B a 2×2 matrix, the product CB is defined and is a 3×2 matrix. The product BA is defined, and is a 2×3 matrix. Since CB and BA are not of the same type, the difference $CB - BA$ is not defined.

5. The product AC is a 2×2 matrix, and consequently, the expression $AC + I_2$ is defined. Its value is

$$AC + I_2 = \begin{bmatrix} 2 & -1 & 1 \\ 0 & -5 & 4 \end{bmatrix}\begin{bmatrix} -1 & 0 \\ 1 & 2 \\ 3 & -1 \end{bmatrix} + \begin{bmatrix} 1 & 0 \\ 0 & 1 \end{bmatrix} = \begin{bmatrix} 0 & -3 \\ 7 & -14 \end{bmatrix} + \begin{bmatrix} 1 & 0 \\ 0 & 1 \end{bmatrix} = \begin{bmatrix} 1 & -3 \\ 7 & -13 \end{bmatrix}.$$

7. The product BA is defined, and is equal to the 2×3 matrix

$$BA = \begin{bmatrix} 0 & 5 \\ 4 & 0 \end{bmatrix}\begin{bmatrix} 2 & -1 & 1 \\ 0 & -5 & 4 \end{bmatrix} = \begin{bmatrix} 0 & -25 & 20 \\ 8 & -4 & 4 \end{bmatrix}.$$

Since C is of type 3×2, the product $C(BA)$ is defined. Its value is

$$C(BA) = \begin{bmatrix} -1 & 0 \\ 1 & 2 \\ 3 & -1 \end{bmatrix} \begin{bmatrix} 0 & -25 & 20 \\ 8 & -4 & 4 \end{bmatrix} = \begin{bmatrix} 0 & 25 & -20 \\ 16 & -33 & 28 \\ -8 & -71 & 56 \end{bmatrix}.$$

9. The product AC is of type 2×2, hence $(AC)^2 = (AC)(AC)$ is also of type 2×2. Since B^2 is a 2×2 matrix, the expression $(AC)^2 + 4B^2$ is defined. From

$$AC = \begin{bmatrix} 2 & -1 & 1 \\ 0 & -5 & 4 \end{bmatrix} \begin{bmatrix} -1 & 0 \\ 1 & 2 \\ 3 & -1 \end{bmatrix} = \begin{bmatrix} 0 & -3 \\ 7 & -14 \end{bmatrix}$$

we get

$$(AC)^2 = \begin{bmatrix} 0 & -3 \\ 7 & -14 \end{bmatrix} \begin{bmatrix} 0 & -3 \\ 7 & -14 \end{bmatrix} = \begin{bmatrix} -21 & 42 \\ -98 & 175 \end{bmatrix}.$$

Since

$$B^2 = \begin{bmatrix} 0 & 5 \\ 4 & 0 \end{bmatrix} \begin{bmatrix} 0 & 5 \\ 4 & 0 \end{bmatrix} = \begin{bmatrix} 20 & 0 \\ 0 & 20 \end{bmatrix},$$

it follows that

$$(AC)^2 + 4B^2 = \begin{bmatrix} -21 & 42 \\ -98 & 175 \end{bmatrix} + \begin{bmatrix} 80 & 0 \\ 0 & 80 \end{bmatrix} = \begin{bmatrix} 59 & 42 \\ -98 & 255 \end{bmatrix}.$$

11. We use properties of matrix operations. From $3A - X = I_2 + 4(C - X)$ it follows that $3X = I_2 + 4C - 3A$ and $X = (I_2 + 4C - 3A)/3$. Hence

$$X = \tfrac{1}{3}\left(\begin{bmatrix} 1 & 0 \\ 0 & 1 \end{bmatrix} + \begin{bmatrix} 0 & 4 \\ 4 & 0 \end{bmatrix} - \begin{bmatrix} 6 & -3 \\ 12 & 0 \end{bmatrix} \right) = \tfrac{1}{3} \begin{bmatrix} -5 & 7 \\ -8 & 1 \end{bmatrix} = \begin{bmatrix} -5/3 & 7/3 \\ -8/3 & 1/3 \end{bmatrix}.$$

13. Let $X = \begin{bmatrix} x_{11} & x_{12} \\ x_{21} & x_{22} \end{bmatrix}$. Then $AX = B$ implies

$$\begin{bmatrix} 2 & -1 \\ 4 & 0 \end{bmatrix} \begin{bmatrix} x_{11} & x_{12} \\ x_{21} & x_{22} \end{bmatrix} = \begin{bmatrix} 10 & 1 \\ 0 & 0 \end{bmatrix}$$

and

$$\begin{bmatrix} 2x_{11} - x_{21} & 2x_{12} - x_{22} \\ 4x_{11} & 4x_{12} \end{bmatrix} = \begin{bmatrix} 10 & 1 \\ 0 & 0 \end{bmatrix}.$$

Comparing corresponding entries, we get $2x_{11} - x_{21} = 10$, $2x_{12} - x_{22} = 1$, $4x_{11} = 0$ and $4x_{12} = 0$. It follows that $x_{11} = x_{12} = 0$, and so $x_{21} = -10$ and $x_{22} = -1$, and hence $X = \begin{bmatrix} 0 & 0 \\ -10 & -1 \end{bmatrix}$.

15. The map $\mathbf{F}_C$ is given by $\mathbf{F}_C(\mathbf{v}) = C \cdot \mathbf{v}$, where $\mathbf{v} \in \mathbb{R}^2$. If $\mathbf{v} = (v_1, v_2)$, then

$$\mathbf{F}_C(\mathbf{v}) = \mathbf{F}_C\left(\begin{bmatrix} v_1 \\ v_2 \end{bmatrix} \right) = \begin{bmatrix} 0 & 1 \\ 1 & 0 \end{bmatrix} \begin{bmatrix} v_1 \\ v_2 \end{bmatrix} = \begin{bmatrix} v_2 \\ v_1 \end{bmatrix}.$$

In words, $\mathbf{F}_C$ switches the x- and y-components of a vector. Geometrically, $\mathbf{F}_C$ is a mapping that assigns to a vector its symmetric image with respect to the line $y = x$. Similarly,

$$\mathbf{F}_{I_2}(\mathbf{v}) = \mathbf{F}_{I_2}\left(\begin{bmatrix} v_1 \\ v_2 \end{bmatrix}\right) = \begin{bmatrix} 1 & 0 \\ 0 & 1 \end{bmatrix}\begin{bmatrix} v_1 \\ v_2 \end{bmatrix} = \begin{bmatrix} v_1 \\ v_2 \end{bmatrix};$$

i.e., $\mathbf{F}_{I_2}$ is the identity mapping: it maps a vector to itself.

17. By definition,

$$\det\begin{bmatrix} 3 & 4 \\ 4 & 3 \end{bmatrix} = 3 \cdot 3 - 4 \cdot 4 = -7.$$

19. By definition,

$$\det\begin{bmatrix} \cos\theta & \sin\theta \\ -r\sin\theta & r\cos\theta \end{bmatrix} = r\cos^2\theta + r\sin^2\theta = r.$$

21. By definition,

$$\begin{vmatrix} 2 & 1 & 4 \\ 0 & 6 & 5 \\ 0 & 2 & 0 \end{vmatrix} = 2\begin{vmatrix} 6 & 5 \\ 2 & 0 \end{vmatrix} - 1\begin{vmatrix} 0 & 5 \\ 0 & 0 \end{vmatrix} + 4\begin{vmatrix} 0 & 6 \\ 0 & 2 \end{vmatrix} = 2(-10) - 1(0) + 4(0) = -20.$$

23. Let $A = \begin{bmatrix} a_{11} & a_{12} \\ a_{21} & a_{22} \end{bmatrix}$ and $B = \begin{bmatrix} b_{11} & b_{12} \\ b_{21} & b_{22} \end{bmatrix}$. Then

$$AB = \begin{bmatrix} a_{11} & a_{12} \\ a_{21} & a_{22} \end{bmatrix}\begin{bmatrix} b_{11} & b_{12} \\ b_{21} & b_{22} \end{bmatrix} = \begin{bmatrix} a_{11}b_{11} + a_{12}b_{21} & a_{11}b_{12} + a_{12}b_{22} \\ a_{21}b_{11} + a_{22}b_{21} & a_{21}b_{12} + a_{22}b_{22} \end{bmatrix}$$

and

$$\det(AB) = (a_{11}b_{11} + a_{12}b_{21})(a_{21}b_{12} + a_{22}b_{22}) - (a_{21}b_{11} + a_{22}b_{21})(a_{11}b_{12} + a_{12}b_{22})$$

$$= a_{11}b_{11}a_{21}b_{12} + a_{11}b_{11}a_{22}b_{22} + a_{12}b_{21}a_{21}b_{12} + a_{12}b_{21}a_{22}b_{22}$$

$$- a_{21}b_{11}a_{11}b_{12} - a_{21}b_{11}a_{12}b_{22} - a_{22}b_{21}a_{11}b_{12} - a_{22}b_{21}a_{12}b_{22}$$

$$= a_{11}b_{11}a_{22}b_{22} + a_{12}b_{21}a_{21}b_{12} - a_{21}b_{11}a_{12}b_{22} - a_{22}b_{21}a_{11}b_{12}.$$

Since

$$\det(A)\det(B) = (a_{11}a_{22} - a_{21}a_{12})(b_{11}b_{22} - b_{21}b_{12})$$

$$= a_{11}a_{22}b_{11}b_{22} - a_{11}a_{22}b_{21}b_{12} - a_{21}a_{12}b_{11}b_{22} + a_{21}a_{12}b_{21}b_{12},$$

it follows that $\det(AB) = \det(A)\det(B)$. In words, the determinant of the product (of two 2×2 matrices) is equal to the product of their determinants (as a matter of fact, the same formula holds for $n \times n$ matrices, $n \geq 2$).

We will now use this fact to prove the second part. Start with $\det(AB) = \det(A)\det(B)$. Since $\det(A)$ and $\det(B)$ are real numbers, it follows that $\det(A)\det(B) = \det(B)\det(A)$, and, again by the above fact, $\det(B)\det(A) = \det(BA)$. Putting it all together, we get $\det(AB) = \det(BA)$.

25. Let $A = \begin{bmatrix} a_{11} & a_{12} \\ a_{21} & a_{22} \end{bmatrix}$. Then $B = \begin{bmatrix} a_{11} & a_{12} \\ \alpha a_{11} & \alpha a_{12} \end{bmatrix}$ and $\det(B) = \alpha a_{11}a_{12} - a_{12}\alpha a_{11} = 0$.

27. Every time we interchange two rows or two columns in a matrix, its determinant changes sign. In order to obtain B from A, we make two interchanges, and therefore $\det(B) = -(-\det(A)) = \det(A)$.

29. Let us look a bit closer at the formula for the determinant of a matrix. Let $A = [a_{ij}]$ be a 3×3 matrix. Then $\det(A)$ is equal to

$$\begin{vmatrix} a_{11} & a_{12} & a_{13} \\ a_{21} & a_{22} & a_{23} \\ a_{31} & a_{32} & a_{33} \end{vmatrix} = a_{11} \begin{vmatrix} a_{22} & a_{23} \\ a_{32} & a_{33} \end{vmatrix} - a_{12} \begin{vmatrix} a_{21} & a_{23} \\ a_{31} & a_{33} \end{vmatrix} + a_{13} \begin{vmatrix} a_{21} & a_{22} \\ a_{31} & a_{32} \end{vmatrix}$$

$$= a_{11}a_{22}a_{33} - a_{11}a_{32}a_{23} - a_{12}a_{21}a_{33} + a_{12}a_{23}a_{31} + a_{13}a_{21}a_{32} - a_{13}a_{31}a_{22}.$$

The expression for $\det(A)$ consists of six terms. Each term contains *exactly one* entry from each row (and exactly one entry from each column) of A. Therefore, if we multiply all elements in one row of A (or in one column of A) by α, each of the six terms will contain the factor α, and therefore $\det(B) = \alpha \det(A)$.

An alternative to this reasoning is a straightforward computation.

31. Let $A = \begin{bmatrix} 1 & 2 \\ 0 & 1 \end{bmatrix}$ and $B = \begin{bmatrix} 1 & 0 \\ 1 & 2 \end{bmatrix}$. Then $AB = \begin{bmatrix} 3 & 4 \\ 1 & 2 \end{bmatrix}$ and $(AB)^t = \begin{bmatrix} 3 & 1 \\ 4 & 2 \end{bmatrix}$. However, since $A^t = \begin{bmatrix} 1 & 0 \\ 2 & 1 \end{bmatrix}$ and $B^t = \begin{bmatrix} 1 & 1 \\ 0 & 2 \end{bmatrix}$, we get $= \begin{bmatrix} 1 & 1 \\ 2 & 4 \end{bmatrix}$.

To prove that $(AB)^t = B^t A^t$, we let $A = \begin{bmatrix} a_{11} & a_{12} \\ a_{21} & a_{22} \end{bmatrix}$ and $B = \begin{bmatrix} b_{11} & b_{12} \\ b_{21} & b_{22} \end{bmatrix}$. Then

$$(AB)^t = \begin{bmatrix} a_{11}b_{11} + a_{12}b_{21} & a_{11}b_{12} + a_{12}b_{22} \\ a_{21}b_{11} + a_{22}b_{21} & a_{21}b_{12} + a_{22}b_{22} \end{bmatrix}^t = \begin{bmatrix} a_{11}b_{11} + a_{12}b_{21} & a_{21}b_{11} + a_{22}b_{21} \\ a_{11}b_{12} + a_{12}b_{22} & a_{21}b_{12} + a_{22}b_{22} \end{bmatrix}.$$

From

$$B^t A^t = \begin{bmatrix} b_{11} & b_{21} \\ b_{12} & b_{22} \end{bmatrix} \begin{bmatrix} a_{11} & a_{21} \\ a_{12} & a_{22} \end{bmatrix} = \begin{bmatrix} b_{11}a_{11} + b_{21}a_{12} & b_{11}a_{21} + b_{21}a_{22} \\ b_{12}a_{11} + b_{22}a_{12} & b_{12}a_{21} + b_{22}a_{22} \end{bmatrix}$$

it follows that $(AB)^t = B^t A^t$.

33. By definition,

$$\mathbf{F}_A(x\mathbf{i} + y\mathbf{j}) = \mathbf{F}_A\left(\begin{bmatrix} x \\ y \end{bmatrix}\right) = \begin{bmatrix} 0 & 0 \\ 0 & 1 \end{bmatrix} \cdot \begin{bmatrix} x \\ y \end{bmatrix} = \begin{bmatrix} 0 \\ y \end{bmatrix} = y\mathbf{j}.$$

In words, $\mathbf{F}_A$ maps $x\mathbf{i} + y\mathbf{j}$ to $y\mathbf{j}$ (i.e., it forgets the $\mathbf{i}$ component of a vector). Thus, $\mathbf{F}_A$ is the orthogonal projection onto the y-axis. From

$$\mathbf{F}_B(x\mathbf{i} + y\mathbf{j}) = \mathbf{F}_B\left(\begin{bmatrix} x \\ y \end{bmatrix}\right) = \begin{bmatrix} 2 & 0 \\ 0 & 2 \end{bmatrix} \cdot \begin{bmatrix} x \\ y \end{bmatrix} = \begin{bmatrix} 2x \\ 2y \end{bmatrix} = 2x\mathbf{i} + 2y\mathbf{j}$$

we see that $\mathbf{F}_B$ magnifies the vector by the factor of 2, i.e., to each non-zero vector it assigns the vector of the same direction and orientation, but twice the magnitude (and $\mathbf{F}_B(\mathbf{0}) = \mathbf{0}$). The linear map $\mathbf{F}_C$ is computed to be

$$\mathbf{F}_C(x\mathbf{i} + y\mathbf{j}) = \mathbf{F}_C\left(\begin{bmatrix} x \\ y \end{bmatrix}\right) = \begin{bmatrix} 1 & 0 \\ 0 & -1 \end{bmatrix} \cdot \begin{bmatrix} x \\ y \end{bmatrix} = x\mathbf{i} - y\mathbf{j}.$$

We see that $\mathbf{F}_C$ changes the sign of the y coordinate of a vector. Thus, $\mathbf{F}_C$ assigns to each vector its mirror image with respect to the x-axis.

35. By the definition of $\mathbf{F}$ and the distributivity of matrix multiplication,

$$\mathbf{F}_{A+B}(\mathbf{v}) = (A+B)\mathbf{v} = A\mathbf{v} + B\mathbf{v} = \mathbf{F}_A(\mathbf{v}) + \mathbf{F}_B(\mathbf{v});$$

i.e., $\mathbf{F}_{A+B} = \mathbf{F}_A + \mathbf{F}_B$. Similarly (using associativity)

$$\mathbf{F}_{2AB}(\mathbf{v}) = (2AB)\mathbf{v} = 2A(B\mathbf{v}) = 2A(\mathbf{F}_B(\mathbf{v})) = 2\mathbf{F}_A(\mathbf{F}_B(\mathbf{v}));$$

i.e., $\mathbf{F}_{2AB} = 2(\mathbf{F}_A \circ \mathbf{F}_B)$, where $\circ$ denotes the composition of functions. Finally,

$$\mathbf{F}_{7B}(\mathbf{v}) = 7B\mathbf{v} = 7\mathbf{F}_B(\mathbf{v}),$$

and so $\mathbf{F}_{7B} = 7\mathbf{F}_B$.

1.5. The Cross Product

1. Using the definition, we get

$$\mathbf{v} \times \mathbf{w} = \begin{vmatrix} \mathbf{i} & \mathbf{j} & \mathbf{k} \\ 2 & 0 & -3 \\ 1 & 4 & 1 \end{vmatrix} = \mathbf{i}\begin{vmatrix} 0 & -3 \\ 4 & 1 \end{vmatrix} - \mathbf{j}\begin{vmatrix} 2 & -3 \\ 1 & 1 \end{vmatrix} + \mathbf{k}\begin{vmatrix} 2 & 0 \\ 1 & 4 \end{vmatrix} = 12\mathbf{i} - 5\mathbf{j} + 8\mathbf{k}.$$

Similarly,

$$(\mathbf{v} + \mathbf{w}) \times \mathbf{k} = \begin{vmatrix} \mathbf{i} & \mathbf{j} & \mathbf{k} \\ 3 & 4 & -2 \\ 0 & 0 & 1 \end{vmatrix} = 4\mathbf{i} - 3\mathbf{j}.$$

By the distributivity of the cross product,

$$(2\mathbf{v} - \mathbf{w}) \times (\mathbf{v} + \mathbf{w}) = (2\mathbf{v}) \times \mathbf{v} - \mathbf{w} \times \mathbf{v} + 2\mathbf{v} \times \mathbf{w} - \mathbf{w} \times \mathbf{w}.$$

Since $(2\mathbf{v}) \times \mathbf{v} = 2\mathbf{v} \times \mathbf{v} = \mathbf{0}$, $\mathbf{w} \times \mathbf{w} = \mathbf{0}$ and $\mathbf{v} \times \mathbf{w} = -\mathbf{w} \times \mathbf{v}$, it follows that

$$(2\mathbf{v} - \mathbf{w}) \times (\mathbf{v} + \mathbf{w}) = 3(\mathbf{v} \times \mathbf{w}) = 3(12\mathbf{i} - 5\mathbf{j} + 8\mathbf{k}) = 36\mathbf{i} - 15\mathbf{j} + 24\mathbf{k}.$$

Finally,

$$\mathbf{i} \times (\mathbf{v} \times \mathbf{w}) = \begin{vmatrix} \mathbf{i} & \mathbf{j} & \mathbf{k} \\ 1 & 0 & 0 \\ 12 & -5 & 8 \end{vmatrix} = -8\mathbf{j} - 5\mathbf{k}.$$

3. Let $\mathbf{u} = (u_1, u_2, u_3)$, $\mathbf{v} = (v_1, v_2, v_3)$ and $\mathbf{w} = (w_1, w_2, w_3)$. Then $\mathbf{u} + \mathbf{v} = (u_1 + v_1, u_2 + v_2, u_3 + v_3)$ and

$$(\mathbf{u} + \mathbf{v}) \times \mathbf{w} = \begin{vmatrix} \mathbf{i} & \mathbf{j} & \mathbf{k} \\ u_1 + v_1 & u_2 + v_2 & u_3 + v_3 \\ w_1 & w_2 & w_3 \end{vmatrix}$$

$$= ((u_2 + v_2)w_3 - (u_3 + v_3)w_2)\mathbf{i} - ((u_1 + v_1)w_3 - (u_3 + v_3)w_1)\mathbf{j}$$

$$+ ((u_1 + v_1)w_2 - (u_2 + v_2)w_1)\mathbf{k}.$$

On the other hand,

$$\mathbf{u} \times \mathbf{w} + \mathbf{v} \times \mathbf{w} = \begin{vmatrix} \mathbf{i} & \mathbf{j} & \mathbf{k} \\ u_1 & u_2 & u_3 \\ w_1 & w_2 & w_3 \end{vmatrix} + \begin{vmatrix} \mathbf{i} & \mathbf{j} & \mathbf{k} \\ v_1 & v_2 & v_3 \\ w_1 & w_2 & w_3 \end{vmatrix}$$

$$= (u_2 w_3 - u_3 w_2)\mathbf{i} - (u_1 w_3 - u_3 w_1)\mathbf{j} + (u_1 w_2 - u_2 w_1)\mathbf{k}$$

$$+ (v_2 w_3 - v_3 w_2)\mathbf{i} - (v_1 w_3 - v_3 w_1)\mathbf{j} + (v_1 w_2 - v_2 w_1)\mathbf{k}.$$

Comparing the components, we conclude that $(\mathbf{u} + \mathbf{v}) \times \mathbf{w}$ and $\mathbf{u} \times \mathbf{w} + \mathbf{v} \times \mathbf{w}$ are equal.

5. All vectors orthogonal to both $\mathbf{i} + \mathbf{j} - 2\mathbf{k}$ and $\mathbf{k}$ must be parallel to their cross product $(\mathbf{i} + \mathbf{j} - 2\mathbf{k}) \times \mathbf{k} = -\mathbf{j} + \mathbf{i}$. Divide $-\mathbf{j} + \mathbf{i}$ by its length to get a unit vector: $(\mathbf{i} - \mathbf{j})/\sqrt{2}$; this is one result. The other one is $-(\mathbf{i} - \mathbf{j})/\sqrt{2}$.

7. Almost any guess for $\mathbf{u}$, $\mathbf{v}$ and $\mathbf{w}$ will work. For example, $\mathbf{u} = \mathbf{i}$, $\mathbf{v} = \mathbf{i}$ and $\mathbf{w} = \mathbf{j}$. Then $(\mathbf{u} \times \mathbf{v}) \times \mathbf{w} = (\mathbf{i} \times \mathbf{i}) \times \mathbf{j} = \mathbf{0} \times \mathbf{w} = \mathbf{0}$, but $\mathbf{u} \times (\mathbf{v} \times \mathbf{w}) = \mathbf{i} \times (\mathbf{i} \times \mathbf{j}) = \mathbf{i} \times \mathbf{k} = -\mathbf{j}$.

9. The volume of the parallelepiped is equal to the absolute value of the scalar triple product of given vectors. From

$$(3\mathbf{i} + \mathbf{j} - \mathbf{k}) \cdot ((-\mathbf{i} - \mathbf{j}) \times (\mathbf{j} + \mathbf{k})) = \begin{vmatrix} 3 & 1 & -1 \\ -1 & -1 & 0 \\ 0 & 1 & 1 \end{vmatrix} = 3(-1) - 1(-1) - 1(-1) = -1,$$

it follows that the volume is 1.

11. Write $\mathbf{a} = (a_1, a_2, a_3)$, $\mathbf{b} = (b_1, b_2, b_3)$, $\mathbf{c} = (c_1, c_2, c_3)$. We will prove the identity first for $\mathbf{a} = \mathbf{i}$, $\mathbf{a} = \mathbf{j}$ and $\mathbf{a} = \mathbf{k}$, and then use properties of the dot and the cross products (Theorems 1.2 and 1.9) to show how it works in general case. So, let $\mathbf{a} = \mathbf{i}$. Then

$$\mathbf{a} \times (\mathbf{b} \times \mathbf{c}) = \mathbf{i} \times \begin{vmatrix} \mathbf{i} & \mathbf{j} & \mathbf{k} \\ b_1 & b_2 & b_3 \\ c_1 & c_2 & c_3 \end{vmatrix} = \mathbf{i} \times \left((b_2 c_3 - b_3 c_2)\mathbf{i} - (b_1 c_3 - b_3 c_1)\mathbf{j} + (b_1 c_2 - b_2 c_1)\mathbf{k} \right)$$

$$= -(b_1 c_3 - b_3 c_1)\mathbf{k} - (b_1 c_2 - b_2 c_1)\mathbf{j},$$

since $\mathbf{i} \times \mathbf{i} = \mathbf{0}$, $\mathbf{i} \times \mathbf{j} = \mathbf{k}$, and $\mathbf{i} \times \mathbf{k} = -\mathbf{j}$. The right side is computed to be

$$(\mathbf{a} \cdot \mathbf{c})\mathbf{b} - (\mathbf{a} \cdot \mathbf{b})\mathbf{c} = (\mathbf{i} \cdot \mathbf{c})\mathbf{b} - (\mathbf{i} \cdot \mathbf{b})\mathbf{c} = c_1(b_1\mathbf{i} + b_2\mathbf{j} + b_3\mathbf{k}) - b_1(c_1\mathbf{i} + c_2\mathbf{j} + c_3\mathbf{k})$$

$$= (b_2 c_1 - b_1 c_2)\mathbf{j} + (b_3 c_1 - b_1 c_3)\mathbf{k}.$$

Thus, $\mathbf{a} \times (\mathbf{b} \times \mathbf{c}) = (\mathbf{a} \cdot \mathbf{c})\mathbf{b} - (\mathbf{a} \cdot \mathbf{b})\mathbf{c}$ holds for $\mathbf{a} = \mathbf{i}$. In the same way, we prove that it holds for $\mathbf{a} = \mathbf{j}$ and $\mathbf{a} = \mathbf{k}$. In general,

$$\mathbf{a} \times (\mathbf{b} \times \mathbf{c}) = (a_1\mathbf{i} + a_2\mathbf{j} + a_3\mathbf{k}) \times (\mathbf{b} \times \mathbf{c})$$

$$= a_1(\mathbf{i} \times (\mathbf{b} \times \mathbf{c})) + a_2(\mathbf{j} \times (\mathbf{b} \times \mathbf{c})) + a_3(\mathbf{k} \times (\mathbf{b} \times \mathbf{c}))$$

$$= a_1((\mathbf{i} \cdot \mathbf{c})\mathbf{b} - (\mathbf{i} \cdot \mathbf{b}))\mathbf{c} + a_2((\mathbf{j} \cdot \mathbf{c})\mathbf{b} - (\mathbf{j} \cdot \mathbf{b}))\mathbf{c} + a_3((\mathbf{k} \cdot \mathbf{c})\mathbf{b} - (\mathbf{k} \cdot \mathbf{b}))\mathbf{c}$$

$$= ((a_1\mathbf{i}) \cdot \mathbf{c})\mathbf{b} - ((a_1\mathbf{i}) \cdot \mathbf{b})\mathbf{c} + ((a_2\mathbf{j}) \cdot \mathbf{c})\mathbf{b} - ((a_2\mathbf{j}) \cdot \mathbf{b})\mathbf{c}$$

$$+ ((a_3\mathbf{k}) \cdot \mathbf{c})\mathbf{b} - ((a_3\mathbf{k}) \cdot \mathbf{b})\mathbf{c}$$
$$= ((a_1\mathbf{i} + a_2\mathbf{j} + a_3\mathbf{k}) \cdot \mathbf{c})\mathbf{b} - ((a_1\mathbf{i} + a_2\mathbf{j} + a_3\mathbf{k}) \cdot \mathbf{b})\mathbf{c}$$
$$= (\mathbf{a} \cdot \mathbf{c})\mathbf{b} - (\mathbf{a} \cdot \mathbf{b})\mathbf{c}.$$

13. Let $\mathbf{v}$ be the vector from $(0, 2)$ to $(3, 2)$ (so $\mathbf{v} = 3\mathbf{i}$) and $\mathbf{w}$ the vector from $(0, 2)$ to $(1, 1)$ (so $\mathbf{w} = \mathbf{i} - \mathbf{j}$). The area of the parallelogram spanned by $\mathbf{v}$ and $\mathbf{w}$ is

$$\|\mathbf{v} \times \mathbf{w}\| = \left\| \begin{vmatrix} \mathbf{i} & \mathbf{j} & \mathbf{k} \\ 3 & 0 & 0 \\ 1 & -1 & 0 \end{vmatrix} \right\| = \| - 3\mathbf{k}\| = 3.$$

15. Using four given points, we will construct three vectors (all starting at the same point) and compute their scalar triple product. The three vectors (and hence all four points) belong to the same plane if and only if the scalar triple product is zero. Let $\mathbf{u}$ be the vector from $(1, 0, 0)$ to $(1, 2, -1)$ (so $\mathbf{u} = (0, 2, -1)$), $\mathbf{v}$ the vector from $(1, 0, 0)$ to $(0, 2, -4)$ (so $\mathbf{v} = (-1, 2, -4)$) and $\mathbf{w}$ the vector from $(1, 0, 0)$ to $(2, -1, 0)$ (so $\mathbf{w} = (1, -1, 0)$). Then

$$\mathbf{u} \cdot (\mathbf{v} \times \mathbf{w}) = \begin{vmatrix} 0 & 2 & -1 \\ -1 & 2 & -4 \\ 1 & -1 & 0 \end{vmatrix} = -7,$$

and therefore the four points do not belong to the same plane.

17. Construct the vectors $\mathbf{v}$ (from $(1, 0, 0)$ to $(-3, 2, 2)$, so $\mathbf{v} = (-4, 2, 2)$) and $\mathbf{w}$ (from $(1, 0, 0)$ to $(3, -1, -1)$, so $\mathbf{w} = (2, -1, -1)$). From $\mathbf{v} \times \mathbf{w} = \mathbf{0}$ it follows that $\mathbf{v}$ and $\mathbf{w}$ are parallel, and since they have the point $(1, 0, 0)$ in common, all three points must lie on the same line.

19. In order to find an equation of the plane, we need a point and a vector $\mathbf{n}$ normal (perpendicular) to it. We have infinitely many points to choose from, since both ℓ_1 and ℓ_2 lie in the plane. Take, for example, $\ell_2(1) = (3, 1, 0)$. A vector that is perpendicular to the plane must be perpendicular to both lines; i.e., must be perpendicular to their direction vectors $(-1, 0, 1)$ and $(3, 0, -1)$. Hence

$$\mathbf{n} = \begin{vmatrix} \mathbf{i} & \mathbf{j} & \mathbf{k} \\ -1 & 0 & 1 \\ 3 & 0 & -1 \end{vmatrix} = 2\mathbf{j},$$

and so $0(x - 3) + 2(y - 1) + 0(z - 0) = 0$; i.e., $y = 1$ represents the desired equation.

21. Let $\mathbf{v}$ be the vector from $(3, 0, 1)$ to $(0, 4, -2)$, i.e., $\mathbf{v} = (-3, 4, -3)$, and $\mathbf{w}$ the vector from $(3, 0, 1)$ to $(2, -3, 0)$, i.e., $\mathbf{v} = (-1, -3, -1)$. Their cross product

$$\mathbf{n} = \begin{vmatrix} \mathbf{i} & \mathbf{j} & \mathbf{k} \\ -3 & 4 & -3 \\ -1 & -3 & -1 \end{vmatrix} = -13\mathbf{i} + 13\mathbf{k}$$

is normal to the plane (to simplify calculations, we will use $-\mathbf{i} + \mathbf{k}$ instead). Using the point $(3, 0, 1)$, we compute the equation of the plane to be $-1(x - 3) + 0(y - 0) + 1(z - 1) = 0$, i.e., $-x + z + 2 = 0$.

23. A normal vector to the plane in question is

$$\mathbf{n} = \mathbf{v} \times \mathbf{w} = \begin{vmatrix} \mathbf{i} & \mathbf{j} & \mathbf{k} \\ 2 & 4 & 0 \\ 0 & 1 & -1 \end{vmatrix} = -4\mathbf{i} + 2\mathbf{j} + 2\mathbf{k}.$$

Since the plane contains the origin, its equation is $-4(x - 0) + 2(y - 0) + 2(z - 0) = 0$; i.e., $2x - y - z = 0$. We have to draw the line through $(3, 2, 0)$ that is perpendicular to the plane and then measure the distance from $(3, 2, 0)$ to the point of intersection of the line and the plane. The vector $\mathbf{n}$ (or any vector parallel to it, like $2\mathbf{i} - \mathbf{j} - \mathbf{k}$) defines the direction of the line, and hence $\ell(t) = (3, 2, 0) + t(2, -1, -1) = (3 + 2t, 2 - t, -t)$, $t \in \mathbb{R}$, is its parametric equation. Substituting $x = 3 + 2t$, $y = 2 - t$ and $z = -t$ into the equation of the plane, we get $2(3 + 2t) - (2 - t) + t = 0$ and $t = -2/3$. It follows that $\ell(-2/3) = (5/3, 8/3, 2/3)$ is the point of intersection of the plane and the line $\ell(t)$ normal to it. Finally, the distance $d((3, 2, 0), (5/3, 8/3, 2/3)) = \sqrt{8/3}$ between $(3, 2, 0)$ and $(5/3, 8/3, 2/3)$ is the distance between the point $(3, 2, 0)$ and the plane.

25. The torque of $\mathbf{F} = 2\mathbf{k}$ is given by $\mathbf{T} = \mathbf{r} \times \mathbf{F}$, where $\mathbf{r}$ is the position vector of the point where $\mathbf{F}$ is applied. Since $\|\mathbf{r}\| = 1$ and $\|\mathbf{F}\| = 2$ it follows that

$$\|\mathbf{T}\| = \|\mathbf{r}\| \|\mathbf{F}\| \sin \theta = 2 \sin \theta,$$

where θ is the angle between the position vector of a point and $\mathbf{F}$. The magnitude of $\mathbf{T}$ is the largest when $\theta = \pi/2$; i.e., along the equator (that is, at the points on the intersection of the sphere and the xy-plane). $\|\mathbf{T}\|$ is the smallest when $\theta = 0$ or $\theta = \pi$; i.e., on the North Pole and on the South Pole (i.e., at $(0, 0, \pm 1)$).

27. Since the angular speed of the disk is 10 rad/s, the angular velocity is parallel to the axis of rotation, and so $\mathbf{w} = \pm 10\mathbf{k}$. Due to the counterclockwise orientation, we must choose $\mathbf{w} = 10\mathbf{k}$ ("right-hand" rule). The tangential velocity vector is $\mathbf{v} = \mathbf{r} \times \mathbf{w}$ ($\mathbf{r}$ is the position vector of a point on the disk) and its magnitude is $\|\mathbf{v}\| = \|\mathbf{r}\| \|\mathbf{w}\| \sin \theta = 10 \|\mathbf{r}\| \sin \theta$, where θ is the angle between the position vector and the z-axis (that is the axis of rotation). Since the disk lies in the xy-plane, $\theta = \pi/2$ and hence $\|\mathbf{v}\| = 10 \|\mathbf{r}\|$. Consequently, $\|\mathbf{v}\|$ has the largest value when $\mathbf{r}$ has the largest value (i.e., on the edge of the disk), and the smallest value at the center of the disk.

Chapter Review

True/false Quiz

1. True. Follows from $\|\mathbf{v}\| = \sqrt{1/3 + 1/3 + 1/3} = 1$.

3. False. Eliminating t from $x = t - 1$, $y = -2t + 4$ we get $y = -2(x + 1) + 4$, i.e., $y = -2x + 2$.

5. False. If $\mathbf{w} = -\mathbf{v} \neq \mathbf{0}$, then

$$\mathbf{v} \cdot \mathbf{w} = \mathbf{v} \cdot (-\mathbf{v}) = -\mathbf{v} \cdot \mathbf{v} = -\|\mathbf{v}\|^2 = -\|\mathbf{v}\| \|\mathbf{v}\| = -\|\mathbf{v}\| \| -\mathbf{v}\| = -\|\mathbf{v}\| \|\mathbf{w}\|.$$

7. False. Form vectors $\mathbf{u} = (-1, -2, 0)$ (from $(1, 1, 1)$ to $(0, -1, 1)$), $\mathbf{v} = (-3, -2, 3)$ (from $(1, 1, 1)$ to $(-2, -1, 4)$), and $\mathbf{w} = (-2, 0, 3)$ (from $(0, -1, 1)$ to $(-2, -1, 4)$). Since all dot products $\mathbf{u} \cdot \mathbf{v} = 7$, $\mathbf{u} \cdot \mathbf{w} = 2$ and $\mathbf{v} \cdot \mathbf{w} = 15$ are nonzero, we conclude that the given triangle is not a right triangle.

9. False. For instance, take non-zero matrices $A = \begin{bmatrix} 0 & 1 \\ 0 & 0 \end{bmatrix}$ and $B = \begin{bmatrix} 2 & 0 \\ 0 & 3 \end{bmatrix}$. Then $AB = \begin{bmatrix} 0 & 3 \\ 0 & 0 \end{bmatrix}$, and $\det(AB) = 0$.

11. True. $A^2 = \begin{bmatrix} 1 & 0 \\ 0 & 2 \end{bmatrix} \begin{bmatrix} 1 & 0 \\ 0 & 2 \end{bmatrix} = \begin{bmatrix} 1 & 0 \\ 0 & 4 \end{bmatrix}$ and $A^3 = A^2 A = \begin{bmatrix} 1 & 0 \\ 0 & 4 \end{bmatrix} \begin{bmatrix} 1 & 0 \\ 0 & 2 \end{bmatrix} = \begin{bmatrix} 1 & 0 \\ 0 & 8 \end{bmatrix}$.

13. True. Follows from the fact that $\|\mathbf{v} \times \mathbf{w}\| = \|\mathbf{v}\| \, \|\mathbf{w}\| \sin(\pi/2) = \|\mathbf{v}\| \, \|\mathbf{w}\| > 0$.

Review Exercises and Problems

1. Using the fact that $\|\mathbf{a}\|^2 = \mathbf{a} \cdot \mathbf{a}$ for any vector $\mathbf{a}$, we expand the left side as

$$\|\mathbf{v} - \mathbf{w}\|^2 + \|\mathbf{v} + \mathbf{w}\|^2 = (\mathbf{v} - \mathbf{w}) \cdot (\mathbf{v} - \mathbf{w}) + (\mathbf{v} + \mathbf{w}) \cdot (\mathbf{v} + \mathbf{w})$$

$$= \mathbf{v} \cdot \mathbf{v} - \mathbf{w} \cdot \mathbf{v} - \mathbf{v} \cdot \mathbf{w} + \mathbf{w} \cdot \mathbf{w} + \mathbf{v} \cdot \mathbf{v} + \mathbf{w} \cdot \mathbf{v} + \mathbf{v} \cdot \mathbf{w} + \mathbf{w} \cdot \mathbf{w}.$$

The four terms containing the dot product of $\mathbf{v}$ and $\mathbf{w}$ cancel. What is left is $2\mathbf{v} \cdot \mathbf{v} + 2\mathbf{w} \cdot \mathbf{w} = 2(\|\mathbf{v}\|^2 + \|\mathbf{w}\|^2)$, and we are done.

Consider the parallelogram spanned by $\mathbf{v}$ and $\mathbf{w}$. The vectors $\mathbf{v} - \mathbf{w}$ and $\mathbf{v} + \mathbf{w}$ represent its diagonals; the identity we proved states that the sum of the squares of the lengths of the two diagonals in a parallelogram is equal to the sum of the squares of lengths of its four sides.

3. By the formula for the angle between two vectors, the angle α between $\mathbf{v} = v_1 \mathbf{i} + v_2 \mathbf{j} + v_3 \mathbf{k}$ and the positive x-axis satisfies $\cos\alpha = \mathbf{v} \cdot \mathbf{i}/(\|\mathbf{v}\| \, \|\mathbf{i}\|) = v_1/\|\mathbf{v}\|$. Similarly, the angle β between $\mathbf{v}$ and the positive y-axis satisfies $\cos\beta = \mathbf{v} \cdot \mathbf{j}/(\|\mathbf{v}\| \, \|\mathbf{j}\|) = v_2/\|\mathbf{v}\|$; the angle γ between $\mathbf{v}$ and the positive z-axis satisfies $\cos\gamma = \mathbf{v} \cdot \mathbf{k}/(\|\mathbf{v}\| \, \|\mathbf{k}\|) = v_3/\|\mathbf{v}\|$. Since $\|\mathbf{v}\|^2 = v_1^2 + v_2^2 + v_3^2$, we get

$$\cos^2\alpha + \cos^2\beta + \cos^2\gamma = \frac{v_1^2}{\|\mathbf{v}\|^2} + \frac{v_2^2}{\|\mathbf{v}\|^2} + \frac{v_3^2}{\|\mathbf{v}\|^2} = \frac{\|\mathbf{v}\|^2}{\|\mathbf{v}\|^2} = 1.$$

Take any vector $\mathbf{v} \in \mathbb{R}^3$; then

$$\mathbf{v} = v_1 \mathbf{i} + v_2 \mathbf{j} + v_3 \mathbf{k} = \|\mathbf{v}\| \left(\frac{v_1}{\|\mathbf{v}\|} \mathbf{i} + \frac{v_2}{\|\mathbf{v}\|} \mathbf{j} + \frac{v_3}{\|\mathbf{v}\|} \mathbf{k} \right)$$

$$= \|\mathbf{v}\|(\cos\alpha \, \mathbf{i} + \cos\beta \, \mathbf{j} + \cos\gamma \, \mathbf{k})$$

is a required representation of a vector in terms of its direction cosines.

5. We will construct the line $\mathbf{l}$ perpendicular to the given plane that goes through $(2, 3, -4)$, find where it intersects the plane and then compute the distance from that intersection point to the point $(2, 3, -4)$.

The vector normal to the plane, $(1, -2, -1)$, is the direction vector for the line $\mathbf{l}$. Thus, $\mathbf{l}(t) = (2, 3, -4) + t(1, -2, -1)$, $t \in \mathbb{R}$, parametrizes $\mathbf{l}$. Let (x, y, z) be the point of intersection of

$\mathbf{l}$ and the given plane. Then $x = 2+t$, $y = 3-2t$ and $z = -4-t$, and from $x-2y-z = 11$ we get $2+t-2(3-2t)-(-4-t) = 11$, i.e., $t = 11/6$. Thus, $\mathbf{l}(11/6) = (2,3,-4)+(11/6)(1,-2,-1) = (23/6, -4/6, -35/6)$ is the point of intersection of $\mathbf{l}$ and the plane.

The distance d between the given point and the plane is equal to the length of the vector $\mathbf{l}(11/6) - (2,3,-4) = (11/6)(1,-2,-1)$. Thus, $d^2 = (121/36)(6) = 121/6$, and so $d = 11/\sqrt{6} \approx 4.491$.

7. Let $A = \begin{bmatrix} 0 & 1 \\ 0 & 0 \end{bmatrix}$. Then $A^2 = AA = \begin{bmatrix} 0 & 1 \\ 0 & 0 \end{bmatrix}\begin{bmatrix} 0 & 1 \\ 0 & 0 \end{bmatrix} = \begin{bmatrix} 0 & 0 \\ 0 & 0 \end{bmatrix}$. I.e, the square of a non-zero matrix could be a zero matrix.

Let $A = \begin{bmatrix} a & b \\ c & d \end{bmatrix}$. then

$$AA^t = \begin{bmatrix} a & b \\ c & d \end{bmatrix}\begin{bmatrix} a & c \\ b & d \end{bmatrix} = \begin{bmatrix} a^2 + b^2 & ac + bd \\ ac + bd & c^2 + d^2 \end{bmatrix}.$$

So, if $AA^t = 0$, then $a^2 + b^2 = 0$ (and so $a = b = 0$), and $c^2 + d^2 = 0$ (and so $c = d = 0$); i.e., $A = 0$.

9. Most of the work in finding Π is done once we find a normal vector. Using given points, form the vectors $\mathbf{v}$ (from $(1,2,0)$ to $(2,0,4)$, hence $\mathbf{v} = (1,-2,4)$) and $\mathbf{w}$ (from $(1,2,0)$ to $(-1,-1,2)$, hence $\mathbf{w} = (-2,-3,2)$). Since $\mathbf{v}$ and $\mathbf{w}$ belong to Π, a normal to Π is given by their cross product

$$\mathbf{v} \times \mathbf{w} = \begin{vmatrix} \mathbf{i} & \mathbf{j} & \mathbf{k} \\ 1 & -2 & 4 \\ -2 & -3 & 2 \end{vmatrix} = 8\mathbf{i} - 10\mathbf{j} - 7\mathbf{k}.$$

Consequently, an equation of Π is $8(x-1) - 10(y-2) - 7(z-0) = 0$; i.e., $8x - 10y - 7z + 12 = 0$. A normal (call it $\mathbf{n}$) of a plane that is perpendicular to Π must belong to Π (i.e., be parallel to it). Since $\mathbf{v}$ and $\mathbf{w}$ span Π, $\mathbf{n} = t\mathbf{v} + s\mathbf{w}$ for some real numbers s and t. Hence $\mathbf{n} = t(1,-2,4) + s(-2,-3,2) = (t - 2s, -2t - 3s, 4t + 2s)$ and

$$(t - 2s)(x - 1) + (-2t - 3s)(y - 2) + (4t + 2s)(z - 0) = 0;$$

i.e.,

$$(t - 2s)x - (2t + 3s)y + (4t + 2s)z + 3t + 8s = 0,$$

where $t, s \in \mathbb{R}$ (and at least one of s or t is non-zero), represents all planes through $(1,2,0)$ that are perpendicular to Π.

11. The volume of the parallelepiped spanned by $\mathbf{a}$, $\mathbf{b}$ and $\mathbf{c}$ is the absolute value of the determinant

$$\begin{vmatrix} 1 & -1 & 0 \\ 2 & -3 & 0 \\ 0 & 0 & 1 \end{vmatrix} = -1,$$

i.e., equal to 1. The volume of the parallelepiped spanned by $\mathbf{a}$, $\mathbf{b}$ and $\mathbf{a} + \mathbf{c}$ is the absolute value of the determinant

$$\begin{vmatrix} 1 & -1 & 0 \\ 2 & -3 & 0 \\ 1 & -1 & 1 \end{vmatrix} = -1,$$

which is 1.

The volume of the parallelepiped spanned by vectors $\mathbf{a}$, $\mathbf{b}$ and $\mathbf{c}$ is equal to $|a \cdot (\mathbf{b} \times \mathbf{c})|$. Since

$$|a \cdot (\mathbf{b} \times (\mathbf{a} + \mathbf{c}))| = |a \cdot (\mathbf{b} \times \mathbf{a} + \mathbf{b} \times \mathbf{c})| = |a \cdot (\mathbf{b} \times \mathbf{a}) + a \cdot (\mathbf{b} \times \mathbf{c})| = |a \cdot (\mathbf{b} \times \mathbf{c})|,$$

the two parallelepipeds have the same volume (recall that $a \cdot (\mathbf{b} \times \mathbf{a}) = 0$).

The volume of a parallelepiped is equal to the product of its base area and height. Both parallelepipeds in question have the same base area (since in both cases it is the parallelogram spanned by $\mathbf{a}$ and $\mathbf{b}$). Because the vector $\mathbf{a}$ belongs to the plane spanned by $\mathbf{a}$ and $\mathbf{b}$, it does not contribute to the height. I.e., the tips of both $\mathbf{c}$ and $\mathbf{a} + \mathbf{c}$ are at the same distance from the plane spanned by $\mathbf{a}$ and $\mathbf{b}$. Thus, the two parallelepipeds have the same height, and thus the same volume.

13. Let $A = [a_{ij}]$, $i, j = 1, 2, 3$. Then

$$B = \begin{bmatrix} a_{11} & a_{12} & a_{13} \\ a_{21} & a_{22} & a_{23} \\ \alpha a_{11} + \beta a_{21} & \alpha a_{12} + \beta a_{22} & \alpha a_{13} + \beta a_{23} \end{bmatrix}$$

and

$$\det(B) = a_{11}(a_{22}(\alpha a_{13} + \beta a_{23}) - (\alpha a_{12} + \beta a_{22})a_{23})$$
$$- a_{12}(a_{21}(\alpha a_{13} + \beta a_{23}) - (\alpha a_{11} + \beta a_{21})a_{23})$$
$$+ a_{13}(a_{21}(\alpha a_{12} + \beta a_{22}) - (\alpha a_{11} + \beta a_{21})a_{22})$$
$$= \alpha(a_{11}a_{22}a_{13} - a_{11}a_{12}a_{23} - a_{12}a_{21}a_{13} + a_{12}a_{11}a_{23} + a_{13}a_{21}a_{12} - a_{13}a_{11}a_{22})$$
$$+ \beta(a_{11}a_{22}a_{23} - a_{11}a_{22}a_{23} - a_{12}a_{21}a_{23} + a_{12}a_{21}a_{23} + a_{13}a_{21}a_{22} - a_{13}a_{21}a_{22}) = 0.$$

The determinant of B can be interpreted as the scalar triple product of $\mathbf{u} = (a_{11}, a_{12}, a_{13})$, $\mathbf{v} = (a_{21}, a_{22}, a_{23})$, and $\mathbf{w} = (\alpha a_{11} + \beta a_{21}, \alpha a_{12} + \beta a_{22}, \alpha a_{13} + \beta a_{23})$. In that context, the absolute value of $\det(B)$ represents the volume of the parallelepiped spanned by $\mathbf{u}$, $\mathbf{v}$ and $\mathbf{w}$. Since $\mathbf{w} = \alpha \mathbf{u} + \beta \mathbf{v}$, all three vectors belong to the same plane and the volume is zero. Consequently, $\det(B) = 0$.

15. (a) For example,

$$A = \begin{bmatrix} 0 & 2 \\ 0 & 1 \end{bmatrix} \qquad \text{and} \qquad \mathbf{v} = \begin{bmatrix} 3 \\ 0 \end{bmatrix},$$

or

$$A = \begin{bmatrix} 2 & -1 \\ 4 & -2 \end{bmatrix} \qquad \text{and} \qquad \mathbf{v} = \begin{bmatrix} 3 \\ 6 \end{bmatrix}$$

satisfy $\mathbf{F}_A(\mathbf{v}) = A \cdot \mathbf{v} = \mathbf{0}$.

(b) Let

$$A = \begin{bmatrix} a_{11} & a_{12} \\ a_{21} & a_{22} \end{bmatrix} \qquad \text{and} \qquad \mathbf{v} = \begin{bmatrix} v_1 \\ v_2 \end{bmatrix}.$$

Since $\mathbf{F}_A(\mathbf{v}) = \mathbf{0}$ it follows that

$$\begin{bmatrix} a_{11} & a_{12} \\ a_{21} & a_{22} \end{bmatrix} \cdot \begin{bmatrix} v_1 \\ v_2 \end{bmatrix} = \begin{bmatrix} a_{11}v_1 + a_{12}v_2 \\ a_{21}v_1 + a_{22}v_2 \end{bmatrix} = \begin{bmatrix} 0 \\ 0 \end{bmatrix};$$

i.e., $a_{11}v_1 + a_{12}v_2 = 0$ and $a_{21}v_1 + a_{22}v_2 = 0$. We have obtained a system of two equations in two unknowns v_1 and v_2. The matrix of the system is A, and since $\det(A) \neq 0$, the only solution is the trivial solution $v_1 = 0$ and $v_2 = 0$. Hence $\mathbf{v} = \mathbf{0}$.

17. (a) The velocities of the two cars are $\mathbf{v}_A = -50\mathbf{i}$ and $\mathbf{v}_B = 80\mathbf{j}$. Their positions at time t are $\mathbf{p}_A(t) = (10 - 50t)\mathbf{i}$ and $\mathbf{p}_B(t) = 80t\mathbf{j}$.

(b) The relative position of the two cars at time t is $\mathbf{p}(t) = \mathbf{p}_A(t) - \mathbf{p}_B(t) = (10 - 50t)\mathbf{i} - 80t\mathbf{j}$, and their distance is

$$d(t) = \|\mathbf{p}(t)\| = \sqrt{(10 - 50t)^2 + (80t)^2} = \sqrt{8900t^2 - 10000t + 100} = 10\sqrt{89t^2 - 10t + 1}.$$

(c) See Figure 17.

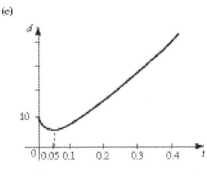

Figure 17

(d) Looking at the plot, we see that the distance is smallest when $t \approx 0.05$ hours. Using calculus, we will improve this approximation. From

$$d'(t) = 10\frac{178t - 10}{2\sqrt{89t^2 - 10t + 1}} = 0$$

it follows that $178t - 10 = 0$ and $t = 10/178 \approx 0.05618$. Since $d'(t) < 0$ when $t < 10/178$ and $d'(t) > 0$ when $t > 10/178$, we conclude that $t = 10/178$ is a point where $d(t)$ attains its minimum. The minimum distance is $d(10/178) \approx 8.47998$ km.

(e) We have to solve $d(t) = 200$ for t; we can try to read an approximation from the graph, see Figure 17. From $10\sqrt{89t^2 - 10t + 1} = 200$ we get $89t^2 - 10t - 399 = 0$, and (drop the negative solution) $t \approx 2.17427$ h.

2. CALCULUS OF FUNCTIONS OF SEVERAL VARIABLES

2.1. Real-Valued and Vector-Valued Functions of Several Variables

1. If $\mathbf{r} = (x, y, z)$, then

$$\mathbf{G}(\mathbf{r}) = 2\|\mathbf{r}\|\mathbf{r} = 2\sqrt{x^2 + y^2 + z^2}(x, y, z).$$

Thus, the components of $\mathbf{G}(\mathbf{r})$ are

$$2x\sqrt{x^2 + y^2 + z^2}, \quad 2y\sqrt{x^2 + y^2 + z^2} \quad \text{and} \quad 2z\sqrt{x^2 + y^2 + z^2}.$$

From (2.1) we get

$$\mathbf{F}(\mathbf{r}) = -\frac{GMm}{\|\mathbf{r}\|^2}\frac{\mathbf{r}}{\|\mathbf{r}\|} = -\frac{GMm}{x^2 + y^2 + z^2}\frac{(x, y, z)}{\sqrt{x^2 + y^2 + z^2}} = -\frac{GMm}{(x^2 + y^2 + z^2)^{3/2}}(x, y, z).$$

So, the component functions of $\mathbf{F}(\mathbf{r})$ are

$$-\frac{GMmx}{(x^2 + y^2 + z^2)^{3/2}}, \quad -\frac{GMmy}{(x^2 + y^2 + z^2)^{3/2}} \quad \text{and} \quad -\frac{GMmz}{(x^2 + y^2 + z^2)^{3/2}}.$$

3. The domain of f is the subset of $\mathbb{R}^2$ defined by $x^2 + y^2 - 1 \neq 0$; i.e., $x^2 + y^2 \neq 1$. Geometrically, it consists of all points in the plane $\mathbb{R}^2$ that do not belong to the circle of radius 1 centered at the origin.

5. The function $f(x, y) = \arctan(x/y)$ is defined whenever $y \neq 0$; that is, the domain of f is the plane $\mathbb{R}^2$ with the x-axis removed.

7. The range of $f(x, y) = 3x + y - 7$ is the set of all real numbers. To prove that, pick any $C \in \mathbb{R}$. Then $f(C/3, 7) = 3(C/3) + 7 - 7 = C$, which shows that C belongs to the range of f. (Note: there are other values for x and y that will give C. For instance, $f(0, C + 7) = C$.)

In general, we consider $f(x, y) = ax + by + c$. If $a = b = 0$, then $f(x, y) = c$ for all (x, y), and so the range of f consists of the single value c. Now assume that $a \neq 0$ and pick any real number C. In this case,

$$F(C/a - c/a, 0) = a\left(\tfrac{C}{a} - \tfrac{c}{a}\right) + b(0) + c = C - c + c = C.$$

(Again, there are other values for x and y that such that $f(x, y) = C$.) Thus, the range of f is $\mathbb{R}$. Similarly, $f(0, C/b - c/b) = C$ proves that the range is $\mathbb{R}$ in the case $b \neq 0$.

9. Since the exponential function e^x is defined for all real numbers, it follows that the domain of $f(x, y, z) = e^{-(x^2 + y^2 + z^2)}$ is $\mathbb{R}^3$.

Recall that $e^c > 0$ for all $c \in \mathbb{R}$. Since $-(x^2 + y^2 + z^2) \leq 0$, it follows that $e^{-(x^2 + y^2 + z^2)} \leq 1$ for all x, y and z. Thus, the range of $f(x, y, z)$ is the interval $(0, 1]$.

11. The denominator $x^2 + y^2$ is zero only when $x = 0$ and $y = 0$, hence the domain of f is the set $\{(x, y) \mid x \neq 0 \text{ and } y \neq 0\} = \mathbb{R}^2 - \{(0, 0)\}$. From $0 \leq x^2 \leq x^2 + y^2$ it follows that (divide all sides by $x^2 + y^2$) $0 \leq x^2/(x^2 + y^2) \leq 1$ and (multiply by 3) $0 \leq f(x, y) \leq 3$. Consequently, the range of f is $[0, 3]$.

13. Since $\ln a$ is defined for $a > 0$, the domain of $\mathbf{F}$ consists of all ordered pairs (x, y) such that $x > 0$ and $y > 0$. The range of $\ln a$ is $\mathbb{R}$, and so the range of $\mathbf{F}$ consists of all ordered pairs (c, d) such that $c \in \mathbb{R}$ and $d \geq 0$.

15. The absolute value function is defined for all real numbers, and therefore the domain of f is $\mathbb{R}^2$. Since $|a| \geq 0$ for any real number a, it follows that the range of f is $[0, \infty)$.

17. Since $|x| = 0$ only when $x = 0$, it follows that the domain of f is $\{(x, y) \mid x \neq 0, y \in \mathbb{R}\}$. The fact that $f(1, c) = c$ (for any $c \in \mathbb{R}$) implies that the range is $\mathbb{R}$.

19. Note: the units of $W(T, v)$ are not units of temperature. However, when we look at the wind chill information, we usually compare the wind chill number to the actual air temperature.

(a) Clearly, $W(0, 0) = 13.12$. Using a calculator, we compute $W(10, 0) = 19.335$, $W(-10, 0) = 6.905$, and $W(-20, 0) = 0.69$. In the absence of wind (i.e., when $v = 0$) the wind chill shows values that are considerably higher than the actual temperature. I.e., these values of the wind chill would imply that, when there is no wind, we would feel warmer than it actually is.

(b) Using the formula for $W(T, v)$, we compute $W(0, 5) = -1.589$, $W(-5, 5) = -7.262$ and $W(-10, 5) = -12.934$. In words, when the wind speed is 5 km/h, wind chill indicates that we will feel that it is a bit colder than what we would expect, given the temperature.

Thus, W makes sense for $v = 5$, and so we might require that $v \geq 5$ in defining the domain of W. For T, we consider some reasonable interval, say $-70 \leq T \leq 10$.

(c) We compute

$$W(-5, v) = 13.12 + 0.6215(-5) - 11.37v^{0.16} + 0.3965(-5)v^{0.16} = 10.0125 - 13.3525v^{0.16}.$$

Thus, as v increases, W will decrease. This fact makes sense, since for a fixed temperature, (in this case, $T = -5$) the stronger the wind, the lower the wind chill will get.

(d) We compute

$$W(T, 20) = 13.12 + 0.6215T - 11.37(20)^{0.16} + 0.3965T(20)^{0.16} = -5.2422 + 1.2618T.$$

So, W is an increasing function of T. Thus (with constant wind speed) higher temperature will give higher wind chill; i.e., if temperature decreases (and wind blows at constant speed), the wind chill will decrease as well.

21. Since $-1 \leq \sin \alpha \leq 1$ and $-1 \leq \cos \beta \leq 1$, it follows that the range of the function $\mathbf{F}(\alpha, \beta) = (\sin \alpha, \cos \beta)$ is the subset of the xy-plane defined by $-1 \leq x \leq 1$ and $-1 \leq y \leq 1$. Geometrically, it is the square in the xy-plane whose vertices are at $(1, 1)$, $(-1, 1)$, $(-1, -1)$ and $(1, -1)$, together with the boundary line segments.

23. The direction of the vector field in question (call it $\mathbf{F}$) is that of the vector $\mathbf{v} = (x - 1)\mathbf{i} + (y - 2)\mathbf{j} + (z + 2)\mathbf{k}$ going from $(1, 2, -2)$ to (x, y, z). Since $\mathbf{F}$ must be of unit length, it follows that

$$\mathbf{F} = \frac{\mathbf{v}}{\|\mathbf{v}\|} = \frac{(x-1)\mathbf{i} + (y-2)\mathbf{j} + (z+2)\mathbf{k}}{\sqrt{(x-1)^2 + (y-2)^2 + (z+2)^2}}.$$

25. $\mathbf{F}(\mathbf{x}) = \mathbf{x}/\|\mathbf{x}\|$ is the unit vector in the direction of the vector $\mathbf{x}$ (i.e., in the radial direction away from the origin); see Figure 25 (vectors not drawn to scale).

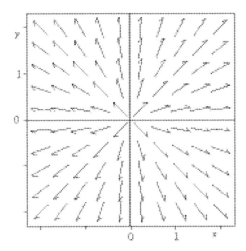

Figure 25

27. $\mathbf{F}(\mathbf{x})$ is a unit vector from $\mathbf{x}_0$ to $\mathbf{x}$, see Figure 27 (vectors not drawn to scale). It is the vector field of Exercise 25, translated by the vector $\mathbf{i} + 2\mathbf{j}$, so that the origin gets moved to $(1, 2)$.

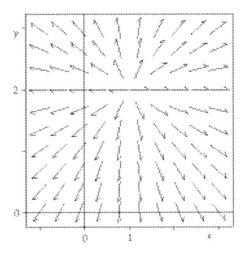

Figure 27

29. Since $\mathbf{F}(x, y)$ does not depend on y, its value at all points on a vertical line is the same. See Figure 29 (vectors not drawn to scale).

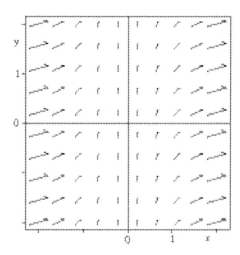

Figure 29

31. In words, the vector field $\mathbf{F}$ assigns to a vector $x\mathbf{i} + y\mathbf{j}$ the vector $x\mathbf{i} - y\mathbf{j}$, which is its mirror image with respect to the x-axis. See Figure 31 (vectors not drawn to scale).

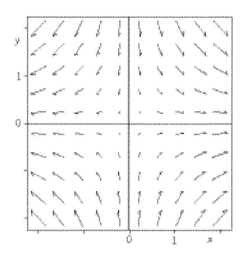

Figure 31

33. (a) Since
$$\mathbf{F}_1 \cdot \mathbf{r} = \frac{-y\mathbf{i} + x\mathbf{j}}{x^2 + y^2} \cdot (x\mathbf{i} + y\mathbf{j}) = 0,$$
it follows that $\mathbf{F}_1$ is perpendicular to a radial direction (i.e., to the radius of the circle, at the point (x, y)). Thus, $\mathbf{F}_1$ is tangent to the circle centred at the origin that goes through (x, y).

(b) We compute
$$\|\mathbf{F}_1(x, y)\| = \frac{1}{x^2 + y^2} \|-y\mathbf{i} + x\mathbf{j}\| = \frac{1}{\sqrt{x^2 + y^2}}.$$

Thus, the flow $\mathbf{F}_1$, viewed along a fixed circle centred at the origin, has constant speed, which is inversely proportional to the radius of the circle.

(c) See figure below (vectors not drawn to scale).

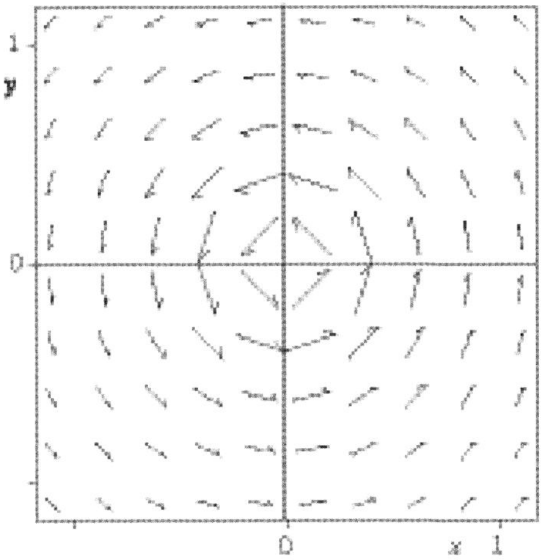

Figure 33

2.2. Graph of a Function of Several Variables

1. There are no points such that $f(x,y) = c \neq 1$, and $f(x,y) = c = 1$ holds for all (x,y). It follows that there are no level curves for $c \neq 1$ and the "level curve" (now the term *level set* comes in handy) of value $c = 1$ is the xy-plane.

3. From $f(x,y) = 3 - x^2 - y^2 = c$ it follows that $x^2 + y^2 = 3 - c$. Since $x^2 + y^2 \geq 0$, it follows that there are no level curves if $c > 3$. If $c = 3$, then $x^2 + y^2 = 0$ implies that $(x,y) = (0,0)$, i.e., the "level curve" of value $c = 3$ is the origin. If $c < 3$, then the level curve of value c is the circle, centered at the origin, of radius $\sqrt{3 - c}$.

5. Since $e^{xy} > 0$, it follows that $f(x,y) = e^{xy}$ has no level curves for $c \leq 0$. If $c > 0$ then $e^{xy} = c$ implies $xy = \ln c$ and hence $y = (\ln c)/x$. If $c > 1$ then $\ln c > 0$ and $y = (\ln c)/x$ is a hyperbola with branches in the first and the third quadrants; if $c < 1$ then $\ln c < 0$ and $y = (\ln c)/x$ is a hyperbola with branches in the second and the fourth quadrants; if $c = 1$ then $y = (\ln 1)/x = 0$ is the x-axis.

7. From $f(x,y) = x/y = c$ it follows that $y = x/c$ and $y \neq 0$. Therefore, the level curve of value c (if $c \neq 0$) is the line of slope $1/c$ going through the origin (the origin is not included since $y \neq 0$). If $c = 0$ then $x/y = 0$ implies that $x = 0$. Consequently, the level curve of value c is the y-axis (again, without the origin).

9. The level curve of value c is defined by $f(x,y) = y - \sin x = c$; i.e., $y = \sin x + c$. It is obtained by moving the graph of $y = \sin x$ up c units if $c \geq 0$ or down $|c|$ units if $c < 0$.

11. The equation $x^2 - y^2 = c$, $c \neq 0$, represents a family of hyperbolas with asymptotes $y = \pm x$.

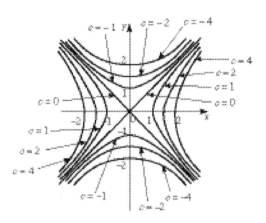

Figure 11

If $c > 0$, the x-intercepts are at $(\pm\sqrt{c}, 0)$ — therefore all hyperbolas with $c > 0$ are located in the regions to the left and to the right of the origin; see Figure 11. If $c < 0$, then there are no x-intercepts; the y-intercepts are at $(0, \pm\sqrt{-c})$. Consequently, the hyperbolas are located in the regions above and below the origin, as shown in Figure 11. If $c = 0$, then $x^2 - y^2 = 0$ implies that $y = \pm x$. The corresponding "level curve" consists of two lines, $y = x$ and $y = -x$, that intersect at the origin (they are also asymptotes of all hyperbolas).

13. (a) Let $f(x, y) = ax + by + c$. From the contour diagram, we see that $f(0, 0) = 8$, $f(1, 0) = f(0, 1) = 6$, etc. Based on these values, we compute the coefficients a, b and c. From $f(0, 0) = 8$ it follows that $c = 8$. From $f(1, 0) = f(0, 1) = 6$ we get that $a + c = b + c = 6$, i.e., $a = b$. Since $c = 8$, we get $a = b = -2$. Thus, $f(x, y) = -2x - 2y + 8$.

Alternatively, we could argue in the following way: the level curve of value 6 has the equation $x + y = 1$. Comparing with $f(x, y) = ax + by + c = 6$, we conclude that $a = b$ (note that we cannot conclude that $a = b = 1$!). Now we proceed as before: from $f(0, 0) = 8$ we get that $c = 8$, and then calculate $a = b = -2$.

(b) Let $f(x, y) = ax + by + c$. From $f(0, 0) = 2$ it follows that $c = 2$. From $f(1, 0) = 3$ we get $a + c = 3$, and so $a = 1$. Finally, from $f(0, 2) = 1$ we get $2b + c = 1$, i.e., $b = -1/2$. Thus, $f(x, y) = x - y/2 + 2$.

Here is an alternative: if we take $y = 0$, we obtain the linear function $g(x) = f(x, 0) = ax + c$. Looking at the x-axis (and lines that cross it), we see that $g(0) = 2$, $g(1) = 3$, $g(2) = 4$, etc., from which we conclude that $g(x) = x + 2$. Now take $x = 0$, and consider $h(y) = f(0, y) = by + c$. From $h(0) = 2$, $h(2) = 1$, $h(4) = 0$, we get that $h(y) = -y/2 + 2$. Putting $g(x)$ and $h(y)$ together, we get $f(x, y) = x - y/2 + 2$.

15. From the definition of f it follows that $0 \le f(x,y) \le 1$. The level curve of value $c = 0$ is the set $\{(x,y) \mid x^2 + y^2 \ge 1\}$; i.e., it is the xy-plane with the region inside the circle of radius 1 centered at the origin removed from it. For $0 < c < 1$, the level curve is given by $1 - x^2 - y^2 = c$; it is the circle of radius $\sqrt{1-c}$ centered at the origin. When $c = 1$ the level curve consists of the single point $(0,0)$.

17. From $\sqrt{1 - 9x^2 - 4y^2} = c$ it follows that $9x^2 + 4y^2 = 1 - c^2$. There are no level curves if $c < 0$ or if $c > 1$ (since, in that case, $1 - c^2 < 0$). If $c = 1$, then the corresponding level curve consists of the point $(0,0)$. For $0 \le c < 1$ we get

$$\frac{x^2}{(1 - c^2)/9} + \frac{y^2}{(1 - c^2)/4} = 1;$$

this equation represents an ellipse with semi-axes $\sqrt{1 - c^2}/3$ and $\sqrt{1 - c^2}/2$. Several level curves are shown in Figure 17.

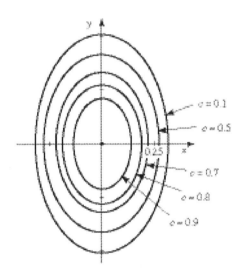

Figure 17

19. From $f(x,y,z) = y - x^2 = c$ it follows that $y = x^2 + c$. In the xy-plane, this is the parabola $y = x^2$ moved up c units if $c \ge 0$ and down $|c|$ units if $c < 0$. Since there are no conditions on z, the surface $y = x^2 + c$ is a parabolic sheet obtained by glueing copies of $y = x^2 + c$ on top of each other (and below each other); see Figure 19. In other words, $y = x^2 + c$ is a surface all of whose horizontal cross-sections are the same curve, namely the parabola $y = x^2 + c$.

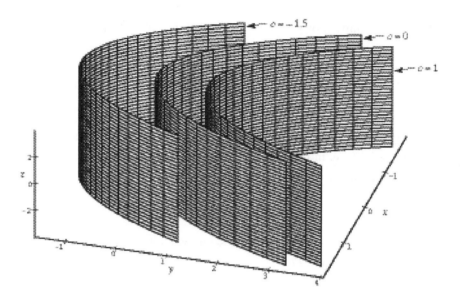

Figure 19

21. From $f(x, y, z) = c$ it follows that $4x^2 + 4y^2 + z^2 = c - 1$. There are no level surfaces for $c < 1$, since $4x^2 + 4y^2 + z^2 \geq 0$. The level surface of value $c = 1$ degenerates to a point (the origin). If $c > 1$, rewrite the equation as (divide by $c - 1$)

$$\frac{x^2}{(\sqrt{c-1}/2)^2} + \frac{y^2}{(\sqrt{c-1}/2)^2} + \frac{z^2}{(\sqrt{c-1})^2} = 1.$$

This equation represents the ellipsoid with semiaxes $\sqrt{c-1}/2$ (in the x-direction), $\sqrt{c-1}/2$ (in the y-direction) and $\sqrt{c-1}$ (in the z-direction), symmetric with respect to the origin (i.e., its center is at the origin).

23. From $3x + 4 = c$ it follows that $x = (c - 4)/3$. So the level surface of value c is the plane parallel to the yz-plane that crosses the x-axis at $((c - 4)/3, 0, 0)$.

25. (a) In two dimensions, the equation $x = a$ represents the line, parallel to the y-axis, going through $(a, 0)$. In three dimensions, it describes the plane, parallel to the yz-plane, going through $(a, 0, 0)$.

(b) In two dimensions, $y = b$ represents the line parallel to the x-axis, going through $(0, b)$. In $\mathbb{R}^3$, it is the plane, parallel to the xz-plane, going through $(0, b, 0)$.

27. (a) Since $x^2 + y^2 + z^2$ is the distance squared from the origin, the equation $x^2 + y^2 + z^2 = a^2$ represents all points whose distance from the origin is a. It is a sphere, centered at the origin, of radius a.

(b) This equation states that (compute the square root of both sides) the distance from a point (x, y, z) to the fixed point (m, n, p) is a. It represents a sphere, centered at (m, n, p), of radius a.

29. Consider the level curves $z = \sqrt{2 - x^2 - y^2} = c$ first. If $c < 0$, there are no level curves. Since $2 - x^2 - y^2 = 2 - (x^2 + y^2) \leq 2$, it follows that $z = \sqrt{2 - x^2 - y^2} \leq \sqrt{2}$, and so there are no level curves for $c > \sqrt{2}$ either. If $c = \sqrt{2}$, the "level curve" is the origin. If $0 \leq c < \sqrt{2}$, then $\sqrt{2 - x^2 - y^2} = c$ implies that $x^2 + y^2 = 2 - c^2$ — consequently, the level curve of value c is a circle of radius $\sqrt{2 - c^2}$.

The trace in the xz-plane is (take $y = 0$) $z = \sqrt{2 - x^2}$; i.e., $x^2 + z^2 = 2$ and $z \geq 0$. It is the upper semi-circle of radius $\sqrt{2}$ in the xz-plane. Similarly, the trace in the yz-plane is (let $x = 0$) $z = \sqrt{2 - y^2}$; it is the upper semi-circle of radius $\sqrt{2}$ in the yz-plane. In words, the surface under investigation is made of circles, smaller ones placed on top of larger circles, from $z = 0$ to $z = \sqrt{2}$. The traces in the xz-plane and in the yz-plane are semi-circles — therefore, the surface in question is the upper hemisphere; (i.e., the part of the sphere centered at the origin of radius $\sqrt{2}$ that lies above and in the xy-plane).

31. The level curve of value c is the circle $x^2 + y^2 = 9$, no matter what c is. The trace in the xz-plane is (substitute $y = 0$) $x^2 = 9$; i.e., $x = \pm 3$ (two vertical lines in the xz-plane). The trace in the yz-plane is $y^2 = 9$; i.e., $y = \pm 3$ (two vertical lines in the yz-plane). The surface is built of circles (of radius 3) stacked vertically (since the traces are vertical lines) on top of each other. It is a cylinder of radius 3, whose axis of (rotational) symmetry is the z-axis.

33. The equation $z = 4$ represents all points whose z-coordinates are 4. It is the plane, parallel to the xy-plane, 4 units above it. The trace in the xz-plane is the horizontal line $z = 4$ and the trace in the yz-plane is the horizontal line $z = 4$.

35. There are no level curves for $c > 2$, since $z = 2 - x^2 - y^2 \leq 2$. The level curve of value $c = 2$ is the point $(0, 0)$. For $c < 2$ the level curve is $2 - x^2 - y^2 = c$, or $x^2 + y^2 = 2 - c$ — it is a circle of radius $\sqrt{2 - c}$. The trace in the xz-plane is the parabola $z = 2 - x^2$, and the trace in the yz-plane is the parabola $z = 2 - y^2$.

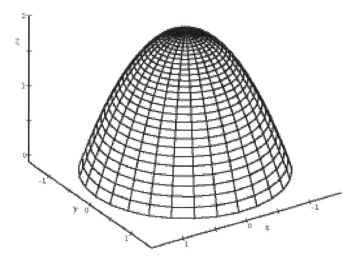

Figure 35

It follows that the surface is built of circles that decrease in size as the height increases. The circles "decrease" in such a way that the vertical cross-sections (i.e., the traces) are parabolas. The surface is a paraboloid, symmetric with respect to the z-axis, turned upside-down, with the vertex at $(0, 0, 2)$; see Figure 35.

37. Since $z = \sqrt{x^2 + y^2}$, it follows that the surface in question lies above the xy-plane; it touches it at $(0, 0, 0)$ (since the "level curve" of value $c = 0$ satisfies $\sqrt{x^2 + y^2} = 0$; i.e., $x = 0$ and $y = 0$). If $c > 0$ then $\sqrt{x^2 + y^2} = c$ implies $x^2 + y^2 = c^2$; so the level curve of value c is a circle of radius c. The trace in the xz-plane (substitute $y = 0$) is $z = \sqrt{x^2} = |x|$; i.e., consists of those parts of the lines $z = \pm x$ that lie above the xy-plane. Similarly, the trace in the yz-plane (substitute $x = 0$) is $z = |y|$.

It follows that the surface in question is built of circles that increase in size as the height increases. Since the traces are lines, it is a cone; see Figure 37.

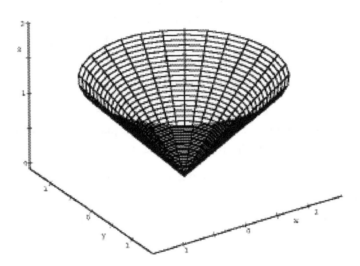

Figure 37

39. Setting $p(T, V) = RnT/V = c$, and assuming that $c \neq 0$, we get that $V = RnT/c = (Rn/c)T$ (keep in mind that $V > 0$). Thus, if $c \neq 0$, level curve is a line of slope Rn/c through the origin in TV-coordinate system. Because $V > 0$, the level curves are: part of the line of slope Rn/c that goes through the origin in the third quadrant, if $c < 0$; part of the line of slope Rn/c that goes through the origin in the first quadrant, if $c > 0$. If $c = 0$, then $T = 0$, and so the level curve is the (positive) V-axis.

41. (a) Setting $f(x, y) = x/(x^2 + y^2 + 4) = 0$, we get $x = 0$. Thus, the level curve of value 0 is the y-axis.

(b) From $f(x, y) = x/(x^2 + y^2 + 4) = c$ we conclude that $cx^2 + cy^2 + 4c = x$, ie, $x^2 - x/c + y^2 = -4$. Completing the square, we get

$$\left(x - \frac{1}{2c}\right)^2 + y^2 = \frac{1}{4c^2} - 4.$$

Thus, whenever $\frac{1}{4c^2} - 4 > 0$ (that is, when $1/4c^2 > 4$, i.e., $|c| < 1/4$) the level curve is a circle centered at $(1/2c, 0)$ of radius $\sqrt{1/4c^2 - 4}$. When $c = \pm 1/4$, the level curve collapses to $(1/2c, 0) = (\pm 2, 0)$. When $|c| > 1/4$, the level set $f(x, y) = x/(x^2 + y^2 + 4) = c$ is empty.

(c) When $c = -1/4$, the level curve is the point $(-2, 0)$. For $-1/4 < c < 0$, level curves are circles that increase in size as their centers (all lie on the x-axis) approach $-\infty$. As $c \to 0$, the circles get larger and larger, and in the limit we get the y-axis. When $0 < c < 1/4$, circles desrease in size, as their centers move (along the x-axis) from $+\infty$ toward $(2, 0)$. Finally, when c reaches $1/4$, the circles collapse to the point $(2, 0)$.

43. Let $f(x, y) = ax + by + d$. The fact that the contour curve $f(x, y) = 0$ goes through the origin implies that $f(0, 0) = 0$, and so $d = 0$. From $f(0, 1) = 3$ (i.e., we use $c = 3$) we get $b(1) = 3$, i.e., $b = 3$. Similarly, from $f(1, 1) = 3$ (again, use $c = 3$) we get $a + b = 3$, i.e., $a = 0$. It follows that $f(x, y) = 3y$.

45. A plane with normal vector $\mathbf{n} = (2, 3, -4)$ has equation $2x + 3y - 4z + d = 0$, where $d \in \mathbb{R}$. So, we define $f(x, y, z) = 2x + 3y - 4z + d$, where d can be any real number. Clearly, $f(x, y, z) = c$ is a plane whose normal vector is $\mathbf{n} = (2, 3, -4)$.

2.3. Limits and Continuity

1. The function in (a) is not defined at a. The function in (b) is defined at a, and $f(a) \neq L$. The function in (c) is defined at a, and $f(a) = L$.

3. From $|f(x, y) - 1| < 0.01$ it follows that

$$|e^{-(x^2+y^2)} - 1| < 0.01.$$

Now $-(x^2 + y^2) \leq 0$ implies that $e^{-(x^2+y^2)} \leq 1$; i.e., the quantity inside the absolute value is negative. Therefore, from $|e^{-(x^2+y^2)} - 1| < 0.01$ we get $-(e^{-(x^2+y^2)} - 1) < 0.01$ or $e^{-(x^2+y^2)} > 0.99$. So $-(x^2 + y^2) > \ln 0.99$ and $x^2 + y^2 < -\ln 0.99 \approx 0.01005$. Finally,

$$\sqrt{x^2 + y^2} < \sqrt{-\ln 0.99} \approx 0.10025.$$

It follows that (x, y) must belong to a ball $B((0, 0), r)$, with $r < \sqrt{-\ln 0.99} \approx 0.10025$, in order to satisfy $|f(x, y) - 1| < 0.01$.

5. See Figure 5.

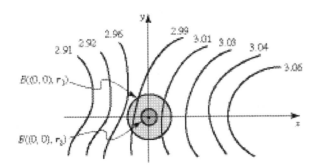

Figure 5

7. Choosing the approach $x = 0$ (i.e., along the y-axis), we compute

$$\lim_{(x,y)\to(0,0)} \arctan\left(0.2x/y\right) = \lim_{y\to 0} \arctan 0 = 0.$$

Now, let $y = 0.2x$ (clearly, $x \to 0$ implies $y \to 0$). Then

$$\lim_{(x,y)\to(0,0)} \arctan\left(0.2x/y\right) = \lim_{x\to 0} \arctan 1 = \pi/4.$$

Thus, $\lim_{(x,y)\to(0,0)} \arctan\left(0.2x/y\right)$ does not exist.

9. $\lim_{(x,y)\to(0,0)} \left(1 - y - e^{-x^2-y^2}\cos x\right) = 1 - 0 - e^0 \cos 0 = 0.$

11. Using the substitution $u = x + y$, we get

$$\lim_{(x,y)\to(0,0)} (x + y)e^{-1/(x+y)} = \lim_{u\to 0} ue^{-1/u}.$$

We have to examine one-sided limits. If $u \to 0^+$, then $-1/u \to -\infty$ and $e^{-1/u} \to 0$. Therefore, $\lim_{u\to 0^+} ue^{-1/u} = 0$. To compute the left limit, use L'Hôpital's Rule:

$$\lim_{u\to 0^-} ue^{-1/u} = \lim_{u\to 0^-} \frac{e^{-1/u}}{1/u} = \lim_{u\to 0^-} \frac{e^{-1/u}(1/u^2)}{-1/u^2} = \lim_{u\to 0^-} - e^{-1/u} = -e^{+\infty} = -\infty.$$

Consequently, $\lim_{u\to 0} ue^{-1/u}$ does not exist.

13. Using polar coordinates, we get

$$\lim_{(x,y)\to(0,0)} \frac{x^2 y}{x^2 + y^2} = \lim_{r\to 0} \frac{r^2 \cos^2\theta \, r\sin\theta}{r^2} = \lim_{r\to 0} r\cos^2\theta \sin\theta = 0,$$

no matter what θ is.

Alternatively,

$$0 \le \left| \frac{x^2 y}{x^2 + y^2} - 0 \right| = \frac{x^2|y|}{x^2 + y^2} \le \frac{(x^2 + y^2)\,|y|}{x^2 + y^2} = |y|$$

(the following identities/inequalities were used: $|x^2 y| = |x^2| \cdot |y| = x^2|y|$ and $x^2 \le x^2 + y^2$). As $(x, y) \to (0, 0)$, $|y| \to 0$, and therefore

$$\lim_{(x,y)\to(0,0)} \frac{x^2 y}{x^2 + y^2} = 0.$$

15. Check the approach along the lines $y = mx$:

$$\lim_{(x,y)\to(0,0)} \frac{2x^2 - y^2}{x^2 + 2y^2} = \lim_{x\to 0} \frac{2x^2 - m^2 x^2}{x^2 + 2m^2 x^2} = \lim_{x\to 0} \frac{2 - m^2}{1 + 2m^2} = \frac{2 - m^2}{1 + 2m^2}.$$

Since the value depends on m (and m is the slope of a line along which we approach the origin) it follows that different approaches give different results. So the limit does not exist.

17. Check the approach along the lines $y = mx$:

$$\lim_{(x,y)\to(0,0)} \frac{2xy}{2x^2 + y^2} = \lim_{x\to 0} \frac{2x^2 m}{2x^2 + m^2 x^2} = \lim_{x\to 0} \frac{2m}{2 + m^2} = \frac{2m}{2 + m^2}.$$

Since the value depends on m (m is the slope of a line along which we approach the origin), the limit does not exist.

19. If $y = mx$ (i.e., we approach the origin along straight lines other than the y-axis), then

$$\lim_{(x,y)\to(0,0)} \frac{x^3 y}{x^6 + y^2} = \lim_{x\to 0} \frac{x^3 mx}{x^6 + m^2 x^2} = \lim_{x\to 0} \frac{mx^2}{x^4 + m^2} = 0,$$

assuming that $m \neq 0$. If $m = 0$, then $y = 0$, and

$$\lim_{(x,y)\to(0,0)} \frac{x^3 y}{x^6 + y^2} = \lim_{x\to 0} \frac{0}{x^6} = 0.$$

When $x = 0$,

$$\lim_{(x,y)\to(0,0)} \frac{x^3 y}{x^6 + y^2} = \lim_{y\to 0} \frac{0}{y^2} = 0.$$

This still does not prove anything. As a matter of fact, this limit does not exist, since the approach $y = x^3$ gives

$$\lim_{(x,y)\to(0,0)} \frac{x^3 y}{x^6 + y^2} = \lim_{(x,y)\to(0,0)} \frac{x^6}{x^6 + x^6} = \tfrac{1}{2}.$$

21. There is no difference between analyzing limits of functions of two, three, or more variables. If we use the approach $y = 0$, $z = 0$, we get

$$\lim_{(x,y,z)\to(0,0,0)} \frac{xyz}{x^3 + y^3 + z^3} = \lim_{x\to 0} \frac{0}{x^3} = 0.$$

However, along $x = y = z$,

$$\lim_{(x,y,z)\to(0,0,0)} \frac{xyz}{x^3 + y^3 + z^3} = \lim_{x\to 0} \frac{x^3}{3x^3} = \tfrac{1}{3}.$$

Therefore, the limit in question does not exist.

23. $\mathbf{F}$ is a rational function (i.e., a fraction), and is therefore continuous at all points where the denominator is not equal to zero. Since $x^2 + y^2 + z^2 = 0$ implies that $x = 0$, $y = 0$ and $z = 0$, it follows that $\mathbf{F}$ is not continuous only at $(0,0,0)$.

25. Both the numerator and the denominator are continuous functions for all x and all y. Since $\sin x \le 1$ and $\cos x \le 1$ it follows that $\sin x \cos x \le 1$ and hence the denominator $3 - \sin x \cos x$ is never equal to zero. So the function $f(x, y)$ is continuous for all $x, y \in \mathbb{R}$.

27. Both the numerator and the denominator are continuous functions for all $\mathbf{x} \in \mathbb{R}^m$. If $\|\mathbf{x} - \mathbf{x}_0\| = 0$, then $\mathbf{x} = \mathbf{x}_0$; i.e., f is not continuous at $\mathbf{x} = \mathbf{x}_0$ (and is continuous if $\mathbf{x} \ne \mathbf{x}_0$).

29. The limit of f does not exist, since along $y = 0$

$$\lim_{(x,y)\to(0,0)} \frac{xy^3}{x^2 + y^6} = \lim_{x\to0} \frac{0}{x^2} = 0,$$

but along $x = y^3$

$$\lim_{(x,y)\to(0,0)} \frac{xy^3}{x^2 + y^6} = \lim_{y\to0} \frac{y^6}{2y^6} = \tfrac{1}{2}.$$

Therefore, it is impossible to define $f(0,0)$ so as to make f continuous at $(0,0)$.

31. The condition $xy \ne 0$ implies that $x \ne 0$ and $y \ne 0$. Hence U is obtained from $\mathbb{R}^2$ by removing the x-axis and the y-axis. All points in U are interior points. The boundary of U consists of the x-axis and the y-axis.

33. Let $\mathbf{x} = (x_1, \ldots, x_m)$ and $\mathbf{a} = (a_1, \ldots, a_m)$. The dot product function $f(\mathbf{x}) = \mathbf{x} \cdot \mathbf{a} = x_1 a_1 + \ldots + x_m a_m$ is a first-degree (i.e., linear) polynomial in $x_1, \ldots, x_m$ (with coefficients $a_1, \ldots, a_m$), and therefore continuous.

35. Since $\lim_{x\to0} (\sin x / x) = 1$, it follows that

$$\lim_{(x,y)\to(0,2)} \mathbf{F}(x, y) = \lim_{(x,y)\to(0,2)} \left(\frac{y \sin x}{x}, y e^x \right) = (2, 2).$$

So, if we define $\mathbf{F}(0, 2) = (2, 2)$, we will make $\mathbf{F}$ continuous at $(0, 2)$.

37. From $\lim_{y\to0} (\cos y - 1)/y = 0$, it follows that

$$\lim_{(x,y)\to(1,0)} \mathbf{F}(x, y) = \lim_{(x,y)\to(1,0)} \left(\sin(x + y), \frac{\cos y - 1}{xy}, e^{xy} \right) = (\sin 1, 0, 1).$$

Therefore, defining $\mathbf{F}(1, 0) = (\sin 1, 0, 1)$ will make $\mathbf{F}$ continuous at $(1, 0)$.

39. We compute

$$\lim_{(x,y)\to(-1,-2)} \frac{\sin^2(xy - 2)}{xy - 2} = \lim_{(x,y)\to(-1,-2)} \frac{\sin(xy - 2)}{xy - 2} \sin(xy - 2) = (1) \sin 0 = 0.$$

Thus, by defining $f(-1, -2) = 0$, we make f continuous at $(-1, -2)$.

2.4. Derivatives

1. The set U is the region in $\mathbb{R}^2$ between the circles of radii $\sqrt{2}$ and $\sqrt{3}$ centered at the origin. Since U does not contain the circles, all points in U are interior points. Consequently, U is open in $\mathbb{R}^2$.

3. U is the set of all points that belong to the line $x + y = 2$. No ball centered at a point on that line is completely contained in the line, and therefore U is not open in $\mathbb{R}^2$.

5. The set U consists of the four octants where either all of the coordinates x, y and z are positive, or two are negative and the third one is positive. Since the possibility that the product xyz is zero is not allowed, the three coordinate planes do not belong to U. In other words, all points in U are interior points, and therefore, U is open in $\mathbb{R}^3$.

7. (a) From the diagram we see that $f(5,3) = 11$. As we move away from $(5,3)$ in the horizontal direction (i.e, keeping the y coordinate at 3), we see that the value of f decreases, and reaches 10 at $(10, 3)$. Thus, $(\partial f / \partial x)(5, 3) < 0$.

(b) We know that $f(10, 3) = 10$. Moving horizontally in the direction of the x-axis, we see that f decreases to 8 as we move about 3 units from $(10, 3)$. Thus,

$$\frac{\partial f}{\partial x}(10, 3) \approx \frac{8 - 10}{13 - 10} = -\frac{2}{3}.$$

Similarly, from $f(10, 5) = 12$ and $f(15, 5) = 11$ we estimate

$$\frac{\partial f}{\partial x}(10, 5) \approx \frac{11 - 12}{15 - 10} = -\frac{1}{5}.$$

Thus, $(\partial f / \partial x)(10, 5) > (\partial f / \partial x)(10, 3)$.

9. Keeping y fixed, we get $f_x = yx^{y-1} + y/x$; keeping x fixed, $f_y = x^y \ln x + \ln x$.

11. By the chain rule,

$$f_x = \frac{1}{x + y + z^2} \cdot 1 = \frac{1}{x + y + z^2}.$$

Similarly,

$$f_z = \frac{1}{x + y + z^2} \cdot 2z = \frac{2z}{x + y + z^2}.$$

13. Using the product rule and the quotient rule, we get

$$f_x = e^{xy} y \cos x \sin y + e^{xy}(-\sin x) \sin y = e^{xy} \sin y (y \cos x - \sin x)$$

and

$$f_y = e^{xy} x \cos x \sin y + e^{xy} \cos x \cos y = e^{xy} \cos x (x \sin y + \cos y).$$

15. Since $f(x_1, \ldots, x_m) = (x_1^2 + \cdots + x_m^2)^{1/2}$, it follows that

$$\frac{\partial f}{\partial x_i} = \frac{1}{2}(x_1^2 + \cdots + x_m^2)^{-1/2} 2x_i = \frac{x_i}{\sqrt{x_1^2 + \cdots + x_m^2}},$$

for $i = 1, \ldots, m$.

17. By the Fundamental Theorem of Calculus,

$$f_x = \frac{\partial}{\partial x}\left(\int_0^x te^{-t^2}\,dt\right) = xe^{-x^2}.$$

Since $\int_0^x te^{-t^2}\,dt$ is a function of x only, it follows that $f_y = 0$.

19. Using the "prime" notation ($'$) for the derivatives of f and g, we get $z_x = f'(x) + 0 = f'(x)$, since $g(y)$ is constant when x is viewed as a variable. Similarly, $z_y = 0 + g'(y) = g'(y)$.

21. Keeping y constant (so that $g(y)$ is constant as well), we get $z_x = f'(x)/g(y)$. Keeping x fixed and thinking of z as $z = f(x)g(y)^{-1}$, we get

$$z_y = f(x)(-1)g(y)^{-2}g'(y) = -\frac{f(x)g'(y)}{g(y)^2}.$$

23. We have to compute the rates of change of z with respect to x and y at the given point. From

$$z_x(2, 1, 11) = -2(x - 3)\Big|_{(2,1,11)} = 2$$

and

$$z_y(2, 1, 11) = -8(y - 2)^3\Big|_{(2,1,11)} = 8$$

it follows that the hill is steeper in the y (i.e., northern) direction.

25. The function in question is $f(x, y) = -xe^{-x^2 - 2y^2}$.

(a) By the chain rule,

$$f_y(2, 3) = -xe^{-x^2 - 2y^2}(-4y)\Big|_{(2,3)} = 4xye^{-x^2 - 2y^2}\Big|_{(2,3)} = 24e^{-22}.$$

(b) Substituting $x = 2$ into the formula for $z = f(x, y)$, we get the equation $z = -2e^{-4 - 2y^2}$ of the curve that is the intersection of the graph of f and $x = 2$. Its slope at $y = 3$ is

$$z'(3) = 8ye^{-4 - 2y^2}\Big|_{y=3} = 24e^{-22}.$$

(c) The partial derivative of $f(x, y)$ with respect to y at $(2, 3)$ is the slope of the tangent line (at the point where $y = 3$) to the curve $z = f(2, y)$ that is the intersection of the graph of $z = f(x, y)$ and the vertical plane $x = 2$.

27. The derivative of $\mathbf{F}$ at $(0, 0)$ is the 3×2 matrix

$$D\mathbf{F}(0, 0) = \begin{bmatrix} 0 & 1 \\ 1 & 0 \\ 0 & 0 \end{bmatrix}_{at\ (0,0)} = \begin{bmatrix} 0 & 1 \\ 1 & 0 \\ 0 & 0 \end{bmatrix}.$$

29. The derivative of $\mathbf{F}$ at $(1, 1, 0)$ is the 2×3 matrix

$$D\mathbf{F}(1, 1, 0) = \begin{bmatrix} 2x/(x^2 + y^2 + z^2) & 2y/(x^2 + y^2 + z^2) & 2z/(x^2 + y^2 + z^2) \\ 2y & 2x & 1 \end{bmatrix}_{at\ (1,1,0)}$$

$$= \begin{bmatrix} 1 & 1 & 0 \\ 2 & 2 & 1 \end{bmatrix}.$$

31. Since $f(x, y, z) = x^2 + y^2 + z^2$, it follows that

$$Df = \begin{bmatrix} 2x & 2y & 2z \end{bmatrix}_{(a_1, a_2, a_3)} = \begin{bmatrix} 2a_1 & 2a_2 & 2a_3 \end{bmatrix}.$$

33. Recall that $\mathbf{F}(\mathbf{r}) = \frac{1}{4\pi\epsilon_0} \frac{Qq}{\|\mathbf{r}\|^2} \frac{\mathbf{r}}{\|\mathbf{r}\|}$, and $V(\mathbf{r}) = \frac{1}{4\pi\epsilon_0} \frac{Qq}{\|\mathbf{r}\|}$. Since $\frac{1}{4\pi\epsilon_0}$, Q and q are constant, it suffices to show that $-\nabla(1/\|\mathbf{r}\|) = \mathbf{r}/\|\mathbf{r}\|^3$.

From $1/\|\mathbf{r}\| = (x^2 + y^2 + z^2)^{-1/2}$ we get that

$$-\frac{\partial}{\partial x}\left(\frac{1}{\|\mathbf{r}\|}\right) = \frac{1}{2}(x^2 + y^2 + z^2)^{-3/2}\, 2x == \frac{x}{(x^2 + y^2 + z^2)^{3/2}}.$$

Thus, the $\mathbf{i}$-component of $-\nabla(1/\|\mathbf{r}\|)$ is equal to $x/\|\mathbf{r}\|^3$, which is the $\mathbf{i}$-component of $\mathbf{r}/\|\mathbf{r}\|^3$. In the same way, we prove the equality of $\mathbf{j}$ and $\mathbf{k}$ components.

35. Let $\mathbf{x} = (x, y, z)$. Then $f(\mathbf{x}) = \sqrt{x^2 + y^2 + z^2}$ and (by the chain rule)

$$\nabla f(\mathbf{x}) = Df(\mathbf{x}) = \begin{bmatrix} \frac{x}{\sqrt{x^2+y^2+z^2}} & \frac{y}{\sqrt{x^2+y^2+z^2}} & \frac{z}{\sqrt{x^2+y^2+z^2}} \end{bmatrix}.$$

Hence (interpreting $\nabla f(\mathbf{x})$ as a vector, rather than a matrix),

$$\nabla f(\mathbf{x}) = \frac{1}{\sqrt{x^2 + y^2 + z^2}} \begin{bmatrix} x & y & z \end{bmatrix} = \mathbf{x}/\|\mathbf{x}\|.$$

The domain of ∇f is $\mathbb{R}^3 - \{(0,0,0)\}$.

37. From $f(\mathbf{a}) = f(2, -1) = \ln 4$,

$$f_x(2, -1) = \frac{3}{3x + 2y}\bigg|_{(2,-1)} = \frac{3}{4} \quad \text{and} \quad f_y(2, -1) = \frac{2}{3x + 2y}\bigg|_{(2,-1)} = \frac{1}{2}$$

we get

$$L_{(2,-1)}(x, y) = \ln 4 + \tfrac{3}{4}(x - 2) + \tfrac{1}{2}(y + 1).$$

39. From $f(3, 2) = 8$,

$$f_x(3, 2) = 2x - y\big|_{(3,2)} = 4 \quad \text{and} \quad f_y(3, 2) = -x + y\big|_{(3,2)} = -1,$$

we get

$$L_{(3,2)}(x, y) = 8 + 4(x - 3) - (y - 2) = 4x - y - 2.$$

41. From $f(\mathbf{a}) = f(0, 1, 1) = \sqrt{2}$,

$$f_x(0, 1, 1) = \frac{x}{\sqrt{x^2 + y^2 + z^2}}\bigg|_{(0,1,1)} = 0,$$

$$f_y(0, 1, 1) = \frac{y}{\sqrt{x^2 + y^2 + z^2}}\bigg|_{(0,1,1)} = \frac{1}{\sqrt{2}}$$

and

$$f_z(0, 1, 1) = \frac{z}{\sqrt{x^2 + y^2 + z^2}}\bigg|_{(0,1,1)} = \frac{1}{\sqrt{2}}$$

we get

$$L_{(0,1,1)}(x, y, z) = \sqrt{2} + \frac{1}{\sqrt{2}}(y - 1) + \frac{1}{\sqrt{2}}(z - 1) = \frac{y + z}{\sqrt{2}}.$$

43. Let $f(x, y) = xy(x + y)^{-1}$. In order to verify the given approximation we use the fact that $f(x, y) \approx L_{(2,3)}(x, y)$ for (x, y) close to $(2, 3)$. From $f(2, 3) = 6/5$,

$$f_x(2, 3) = \left. \frac{y(x + y) - xy}{(x + y)^2} \right|_{(2,3)} = \left. \frac{y^2}{(x + y)^2} \right|_{(2,3)} = \frac{9}{25}$$

and

$$f_y(2, 3) = \left. \frac{x(x + y) - xy}{(x + y)^2} \right|_{(2,3)} = \left. \frac{x^2}{(x + y)^2} \right|_{(2,3)} = \frac{4}{25}$$

it follows that

$$\frac{xy}{x + y} = f(x, y) \approx L_{(2,3)}(x, y) = \frac{6}{5} + \frac{9}{25}(x - 2) + \frac{4}{25}(y - 3).$$

45. (a) Since the given limit is 0, all approaches to (a, b) must yield 0. In particular, we use approach $y = b$, $x \to a$:

$$\lim_{(x,y) \to (a,b)} \frac{\left| f(x, y) - \overline{L}(x, y) \right|}{\sqrt{(x - a)^2 + (y - b)^2}} = \lim_{x \to a} \frac{\left| f(x, b) - f(a, b) - m(x - a) \right|}{\sqrt{(x - a)^2}}$$

$$= \lim_{x \to a} \left| \frac{f(x, b) - f(a, b)}{x - a} - m \right| = 0.$$

It follows that

$$\lim_{x \to a} \frac{f(x, b) - f(a, b)}{x - a} = m,$$

i.e., $m = (\partial f / \partial x)(a, b)$.

(b) Taking the approach $x = a$, $y \to b$, and proceeding as in (a), we get that $n = (\partial f / \partial y)(a, b)$. Thus,

$$\overline{L}(x, y) = f(a, b) + m(x - a) + n(y - b) = f(a, b) + \frac{\partial f}{\partial x}(a, b)(x - a) + \frac{\partial f}{\partial y}(a, b)(y - b),$$

i.e., $\overline{L}(x, y)$ is equal to the linear approximation $L_{(a,b)}(x, y)$ of $f(x, y)$ at (a, b).

47. Let $f(x, y) = \sqrt{x^3 + y^3}$; we are asked to approximate $f(0.99, 2.02)$. Consider the linear approximation of f at $(1, 2)$ (notice that $(1, 2)$ is near $(0.99, 2.02)$). From $f(1, 2) = \sqrt{9} = 3$,

$$f_x(1, 2) = \left. \frac{3x^2}{2\sqrt{x^3 + y^3}} \right|_{(1,2)} = \frac{1}{2}$$

and

$$f_y(1, 2) = \left. \frac{3y^2}{2\sqrt{x^3 + y^3}} \right|_{(1,2)} = 2,$$

it follows that

$$L_{(1,2)}(x, y) = 3 + \frac{1}{2}(x - 1) + 2(y - 2).$$

Recall that "linear approximation" means that $f(x, y) \approx L_{(1,2)}(x, y)$ for points *near* $(1, 2)$. In particular,

$$f(0.99, 2.02) \approx L_{(1,2)}(0.99, 2.02) = 3 + \tfrac{1}{2}(-0.01) + 2(0.02) = 3.035.$$

The calculator value is 3.0352441 (last digit(s) could differ, depending on the precision of the calculator).

49. Let $f(x, y) = x \ln y$. We will approximate $f(7.95, 1.02)$ using the linear approximation of f at $(8, 1)$. From $f(8, 1) = 8 \ln 1 = 0$, $f_x(8, 1) = \ln y\big|_{(8,1)} = 0$ and $f_y(8, 1) = (x/y)\big|_{(8,1)} = 8$ it follows that $L_{(8,1)}(x, y) = 8(y - 1)$.

Since $f(x, y) \approx L_{(8,1)}(x, y)$ for (x, y) near $(8, 1)$, it follows that

$$f(7.95, 1.02) = 7.95 \ln 1.02 \approx L_{(8,1)}(7.95, 1.02) = 8(1.02 - 1) = 0.16.$$

Therefore, $7.95 \ln 1.02 \approx 0.16$. The calculator value is 0.1574309.

51. Let $f(x, y) = \int_x^y e^{-t^2} dt$; then $\int_{0.995}^{1.02} e^{-t^2} dt = f(0.995, 1.02)$. We will use the fact that $f(0.995, 1.02) \approx L_{(1,1)}(0.995, 1.02)$, where $L_{(1,1)}(x, y)$ is the linear approximation of $f(x, y)$ at $(1, 1)$. So we have to compute $L_{(1,1)}(x, y)$ first.

The value of f at $(1, 1)$ is $f(1, 1) = \int_1^1 e^{-t^2} dt = 0$. From the Fundamental Theorem of Calculus we get

$$f_x = \frac{\partial}{\partial x}\left(\int_x^y e^{-t^2} dt\right) = \frac{\partial}{\partial x}\left(-\int_y^x e^{-t^2} dt\right) = -e^{-x^2}, \qquad \text{hence} \qquad f_x(1, 1) = -e^{-1}.$$

Similarly,

$$f_y = \frac{\partial}{\partial y}\left(\int_x^y e^{-t^2} dt\right) = e^{-y^2}, \qquad \text{and hence} \qquad f_y(1, 1) = e^{-1}.$$

It follows that

$$L_{(1,1)}(x, y) = -e^{-1}(x - 1) + e^{-1}(y - 1) = e^{-1}(y - x).$$

Therefore, $f(0.995, 1.02) \approx L_{(1,1)}(0.995, 1.02) = 0.025e^{-1}$. In order to check this approximation, one has to use a numerical integration method.

53. From $\Delta f = f(a + \Delta x, b + \Delta y) - f(a, b)$, with $a = 0$, $b = 1$, $\Delta x = 0.3$ and $\Delta y = 0.01$ we get

$$\Delta f = f(0.3, 1.01) - f(0, 1) = e^{0.3} - 1.01e^{1.01} - (1 - e) = 0.2950836.$$

On the other hand, from $df = f_x(a, b)\Delta x + f_y(a, b)\Delta y$ we get

$$df = e^x\big|_{(0,1)}(0.3) - (e^y + ye^y)\big|_{(0,1)}(0.01) = 0.3 - 0.02e \approx 0.2456344.$$

Comparing the two, we see that $\Delta f \approx df$.

55. From

$$\Delta f = f(a + \Delta x, b + \Delta y, c + \Delta z) - f(a, b, c) = f(0.98, 2.01, -0.98) - f(1, 2, -1)$$

we get $\Delta f \approx 2.919616 - 3 = -0.080384$. On the other hand, from $df = f_x(a, b, c)\Delta x + f_y(a, b, c)\Delta y + f_z(a, b, c)\Delta z$ we get

$$df = (2xy - yz)\Big|_{(1,2,-1)}(-0.02) + (x^2 - xz)\Big|_{(1,2,-1)}(0.01) - (-xy + 3z^2)\Big|_{(1,2,-1)}(0.02)$$
$$= 6(-0.02) + 2(0.01) + 1(0.02) \approx -0.08.$$

Comparing the results, we see that $\Delta f \approx df$.

57. From $P(T, V) = RnT/V$ we compute $dP = (Rn/V)dT - (RnT/V^2)dV$. The coefficient Rn/V of dT is positive, since the increase in temperature (keeping the volume V fixed) will increase the pressure. The coefficient $-RnT/V^2$ of dV is negative, since the increase in volume (keeping the temperature T fixed) will decrease the pressure.

59. The volume of a cylinder is given by $V(r, h) = \pi r^2 h$, where r is the radius and h the height. The error ΔV is equal to the difference of the measured volume and the exact volume. Therefore, $\Delta V = V(r + \Delta r, h + \Delta h) - V(r, h)$. Since ΔV is approximately equal to the differential dV, it follows that

$$\Delta V \approx dV = 2\pi rh\Delta r + \pi r^2\Delta h.$$

The error in h is 1.5%, hence $\Delta h = 0.015h$. Similarly, $\Delta r = 0.015r$ and so

$$\Delta V \approx 2\pi rh(0.015r) + \pi r^2(0.015h) = 0.045\pi r^2 h = 0.045V.$$

Consequently, the error in computing the volume is approximately 4.5%.

61. The area of a rectangle is $A = \ell w$, where ℓ is its length and w is its width. From $\Delta A \approx dA$ it follows that

$$\Delta A \approx A_\ell\Delta\ell + A_w\Delta w = w\Delta\ell + \ell\Delta w.$$

The error in ℓ is 2%, hence $\Delta\ell = 0.02\ell$; the error in w is 3%, hence $\Delta w = 0.03w$. Therefore

$$\Delta A \approx w(0.02\ell) + \ell(0.03w) = 0.05w\ell = 0.05A.$$

It follows that the error in computing the area is approximately 5%.

63. We are actually asked to compute the linear approximation of $z = 6 - x^2 - y^2$ at $(1, 2)$. From $z(1, 2) = 1$, $z_x(1, 2) = -2x\Big|_{(1,2)} = -2$, and $z_y(1, 2) = -2y\Big|_{(1,2)} = -4$, we get

$$L_{(1,2)}(x, y) = 1 - 2(x - 1) - 4(y - 2) = -2x - 4y + 11;$$

so the equation of the tangent plane is $z = -2x - 4y + 11$.

65. The partial derivative f_x is given by $f_x(x, y) = 2xy/(x^2 + y^2)$ if $(x, y) \neq (0, 0)$. By definition,

$$f_x(0, 0) = \lim_{h \to 0} \frac{f(h, 0) - f(0, 0)}{h} = \lim_{h \to 0} \frac{0}{h} = 0,$$

and hence f_x is defined for all (x, y). We will show that $\lim_{(x,y)\to(0,0)} f_x(x, y)$ does not exist, and therefore f_x is not continuous at $(0, 0)$ (by definition, it should be equal to $f_x(0, 0) = 0$ if f_x were to be continuous).

If $y = mx$ then

$$\lim_{(x,y)\to(0,0)} \frac{2xy}{x^2 + y^2} = \lim_{x\to 0} \frac{2mx^2}{x^2 + m^2 x^2} = \frac{2m}{1 + m^2}.$$

Since the result depends on m, it depends on the way we approach $(0,0)$. Consequently, the limit does not exist.

67. The components of $\mathbf{F}(x,y) = (x + y^2, 2xy)$ are $F_1 = x + y^2$ and $F_2 = 2xy$. All partial derivatives of the components of $\mathbf{F}$ exist at $(0,0)$:

$$\frac{\partial F_1}{\partial x}(0,0) = 1, \quad \frac{\partial F_1}{\partial y}(0,0) = 2y\Big|_{(0,0)} = 0,$$

$$\frac{\partial F_2}{\partial x}(0,0) = 2y\Big|_{(0,0)} = 0 \quad \text{and} \quad \frac{\partial F_2}{\partial y}(0,0) = 2x\Big|_{(0,0)} = 0.$$

It follows that

$$D\mathbf{F}(0,0) = \begin{bmatrix} 1 & 0 \\ 0 & 0 \end{bmatrix}$$

and

$$D\mathbf{F}(0,0) \cdot (x,y) = \begin{bmatrix} 1 & 0 \\ 0 & 0 \end{bmatrix} \begin{bmatrix} x \\ y \end{bmatrix} = \begin{bmatrix} x \\ 0 \end{bmatrix}.$$

Now

$$\lim_{(x,y)\to(0,0)} \frac{\|\mathbf{F}(x,y) - \mathbf{F}(0,0) - D\mathbf{F}(0,0) \cdot (x,y)\|}{\|(x,y)\|} = \lim_{(x,y)\to(0,0)} \frac{\|(x + y^2, 2xy) - (x,0)\|}{\|(x,y)\|}$$

$$= \lim_{(x,y)\to(0,0)} \frac{\|(y^2, 2xy)\|}{\|(x,y)\|} = \lim_{(x,y)\to(0,0)} \frac{\sqrt{y^4 + 4x^2 y^2}}{\sqrt{x^2 + y^2}}.$$

Passing to polar coordinates, we get

$$\lim_{(x,y)\to(0,0)} \frac{\sqrt{y^4 + 4x^2 y^2}}{\sqrt{x^2 + y^2}} = \lim_{r\to 0} \frac{\sqrt{r^4 \sin^4 \theta + 4r^4 \sin^2 \theta \cos^2 \theta}}{r}$$

$$= \lim_{r\to 0} r\sqrt{\sin^4 \theta + 4 \sin^2 \theta \cos^2 \theta} = 0.$$

By definition, it follows that $\mathbf{F}$ is differentiable at $(0,0)$.

69. (a) By definition (using L'Hôpital's rule),

$$f_x(0,0) = \lim_{h\to 0} \frac{f(h,0) - f(0,0)}{h} = \lim_{h\to 0} \frac{\ln h^2 - 0}{h} = \lim_{h\to 0} \frac{\ln h^2}{h} = \pm\infty,$$

depending on the approach (left limit is $+\infty$, whereas the right limit is $-\infty$). Thus, f_x does not exist at $(0,0)$ and so f cannot be differentiable at $(0,0)$. (For the record, $f_y(0,0) = \pm\infty$.)

(b) No conclusion about the continuity of f at $(0,0)$ can be made from (a).

(c) Since

$$\lim_{(x,y)\to(0,0)} f(x,y) = \lim_{(x,y)\to(0,0)} \ln(x^2 + y^2) = \ln 0 = -\infty,$$

it follows that f is not continuous at $(0,0)$.

2.5. Paths and Curves in $\mathbb{R}^2$ and $\mathbb{R}^3$

1. Let $\mathbf{a} = (3, 1, -2)$ and let $\mathbf{v}$ be the vector from $(3, 1, -2)$ to $(0, 5, 0)$. It follows that $\mathbf{v} = (-3, 4, 2)$ and the required equation is

$$\mathbf{c}(t) = \mathbf{a} + t\mathbf{v} = (3, 1, -2) + t(-3, 4, 2) = (3 - 3t, 1 + 4t, -2 + 2t),$$

where $t \in \mathbb{R}$.

3. There are infinitely many answers. For example, $\mathbf{c}(t) = (\sqrt{5}\cos t, \sqrt{5}\sin t)$, $t \in [0, 2\pi]$; or, $\mathbf{c}(t) = (\sqrt{5}\sin(2t), \sqrt{5}\cos(2t))$, $t \in [0, \pi]$.

5. The parametrization $\overline{\mathbf{c}} = (3\cos t, \sin t)$, $t \in [0, 2\pi]$ gives the right semi-axes, but the ellipse is centered at $(0, 0)$. So we move it to $(-2, -1)$:

$$\mathbf{c}(t) = (-2, -1) + (3\cos t, \sin t) = (-2 + 3\cos t, -1 + \sin t), \qquad t \in [0, 2\pi].$$

7. Let $y = t$; then $x = 1 + t^2$ and a parametrization is $\mathbf{c}(t) = (1 + t^2, t)$, $0 \le t \le 1$.

9. Let $x = t$; then $y^{2/3} = 1 - t^{2/3}$ and $y = \pm(1 - t^{2/3})^{3/2}$. Since $x^{2/3} + y^{2/3} = 1$ implies that $-1 \le x \le 1$ (and $-1 \le y \le 1$, but we do not need it here), it follows that $-1 \le t \le 1$. So

$$\mathbf{c}(t) = (t, \pm(1 - t^{2/3})^{3/2}), \qquad -1 \le t \le 1,$$

is one possible parametrization.

There is a more elegant way to get a parametrization: let $x = \cos^3 t$ and $y = \sin^3 t$; then $x^{2/3} + y^{2/3} = (\cos^3 t)^{2/3} + (\sin^3 t)^{2/3} = 1$! It follows that we can parametrize the given curve as $\mathbf{c}(t) = (\cos^3 t, \sin^3 t)$, $t \in [0, 2\pi]$.

11. Eliminating t from $x = t - 3$ and $y = t^2$ we get $y = (x + 3)^2$. Since $0 \le t \le 2$, it follows that $-3 \le x \le -1$, and therefore $\mathbf{c}(t)$ represents the part of the parabola $y = (x + 3)^2$ (that is, the graph of $y = x^2$ moved 3 units to the left) from $(-3, 0)$ to $(-1, 4)$; see Figure 11.

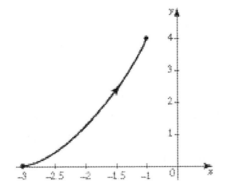

Figure 11

13. From $x = 2\cosh t$ and $y = 2\sinh t$ it follows that $x^2 - y^2 = 4$. Hence $\mathbf{c}$ is a part (or all, we have not figured it out yet) of the hyperbola with asymptotes $y = \pm x$ and x-intercepts at $(\pm 2, 0)$. Now $x = 2\cosh t \ge 0$ for all t, but $y = 2\sinh t$ can be any real number. It follows that **c** is the right branch of the hyperbola $x^2 - y^2 = 4$. It has no endpoints, and is oriented so that, as $t \to \infty$, both components of $\mathbf{c}(t)$ approach $+\infty$ (i.e., as t increases, $\mathbf{c}(t)$ gets closer and closer to its asymptote $y = x$ in the first quadrant); see Figure 13. The orientation can also be determined from the fact that for negative values of t, $\mathbf{c}(t)$ lies below the x-axis, and for $t > 0$, $\mathbf{c}(t)$ is above the x-axis.

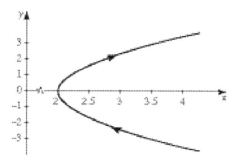

Figure 13

15. Combining $x = t$ and $y = e^{3t}$ we get $y = e^{3x}$. It follows that $\mathbf{c}(t)$ represents the part of the graph of the exponential function $y = e^{3x}$ between $(0, 1)$ and $(\ln 2, 8)$. It is oriented from $(0, 1)$ to $(\ln 2, 8)$; see Figure 15.

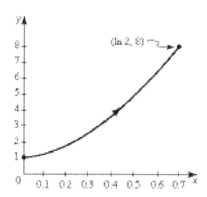

Figure 15

17. From $x = \cos t$, $y = \sin t$ it follows that $x^2 + y^2 = 1$. Since $z = 3$, it follows that $\mathbf{c}(t)$ represents a circle of radius 1 in the plane $z = 3$ with the center at $(0, 0, 3)$ on the z-axis. Its initial and terminal points are at $(1, 0, 3)$, and the orientation is counterclockwise, as seen by somebody standing on the plane $z = 3$.

19. From $x = \cos t$, $y = \sin t$, it follows that $x^2 + y^2 = 1$; i.e., $\mathbf{c}(t)$ lies on the cylinder of radius 1 whose axis of rotation is the z-axis. As t changes from 0 to π, the height (the z-coordinate) changes from 0 to π^3. So a point on $\mathbf{c}(t)$ has to "rotate" counterclockwise, and, at the same time, increase its height according to $z = t^3$. It follows that $\mathbf{c}(t)$ represents a circular helix (see Figure 19) with initial point $\mathbf{c}(0) = (1, 0, 0)$ and terminal point $\mathbf{c}(\pi) = (-1, 0, \pi^3)$.

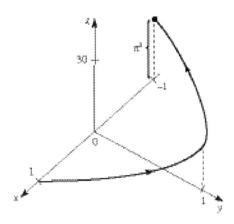

Figure 19

21. Eliminating t, we get $y = \arctan x$. So $\mathbf{c}(t)$ represents the part of the graph of the function $y = \arctan x$ between $(-1, \arctan(-1)) = (-1, -\pi/4)$ and $(1, \arctan(1)) = (1, \pi/4)$; image of $\mathbf{c}(t)$ is shown in Figure 21.

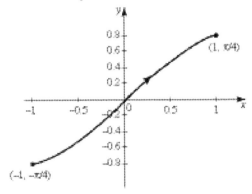

Figure 21

23. We have to eliminate t to get an equation in x and y. The trick is to square and subtract: from $x^2 = \left(t + t^{-1}\right)^2 = t^2 + 2 + t^{-2}$ and $y^2 = \left(t - t^{-1}\right)^2 = t^2 - 2 + t^{-2}$ it follows that $x^2 - y^2 = 4$. This equation represents the hyperbola with asymptotes $y = \pm x$ and x-intercepts at $(\pm 2, 0)$. When $t = 1$, $\mathbf{c}(1) = (2, 0)$, and when $t = 2$, $\mathbf{c}(2) = (5/2, 3/2)$. It follows that $\mathbf{c}(t)$ is the part of the right branch of the hyperbola $x^2 - y^2 = 4$, with endpoints $(2, 0)$ and $(5/2, 3/2)$, oriented from $(2, 0)$ to $(5/2, 3/2)$.

25. A parametrization of the form $\mathbf{c}(t) = (a \sin t, a \cos t)$, $a > 0$, $t \in [0, 2\pi]$ represents the circle in the xy-plane of radius a, oriented clockwise, with initial and terminal points at $(0, a)$. The parameter t represents the angle measured clockwise from the positive y-axis. As t changes from 0 to 2π, the point $(0, a)$ is rotated, thus describing a circle.

This is not what is needed, but will help. Replace a by $e^{t/4}$ and think of t as described above. This time, as t changes, the "radius" keeps increasing from e^0 (when $t = 0$) to $e^{\pi/2}$

(when $t = 2\pi$). It follows that $\mathbf{c}(t)$ describes a clockwise spiral, starting at $(0,1)$ and ending at $(0, e^{\pi/2} \approx 4.81048)$; see Figure 25.

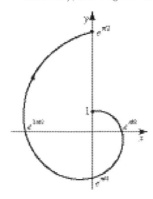

Figure 25

27. In all four cases, $x^2 + y^2 = 4$.

(a) The initial point of $\mathbf{c}_1$ is $\mathbf{c}_1(0) = (0,2)$; as t starts increasing from 0, both components of $\mathbf{c}_1$ become positive; i.e., $\mathbf{c}_1$ goes from $(0,2)$ into the first quadrant. In other words, $\mathbf{c}_1$ is oriented clockwise. Since sin and cos have period 2π, $\mathbf{c}_1$ winds around the circle once.

(b) The initial point of $\mathbf{c}_2$ is $\mathbf{c}_2(0) = (2,0)$ on the x-axis. Since the y-coordinate of $\mathbf{c}_2$ is positive as t starts increasing from 0, it follows that $\mathbf{c}_2$ is oriented counterclockwise; it winds around the circle once.

(c) The parametrization $\mathbf{c}_3$ gives a clockwise orientation to the circle $x^2 + y^2 = 4$. The initial and terminal points are at $(0,2)$. Since the period of $\sin 3t$ and $\cos 3t$ is $2\pi/3$, $\mathbf{c}_3$ winds around the circle three times.

(d) The initial point of $\mathbf{c}_4$ is the point $\mathbf{c}_4(0) = (-2,0)$ on the negative x-axis. As t starts increasing from 0, $y = 2\sin(t/2) > 0$; so $\mathbf{c}_4$ is oriented clockwise. The period of $\cos(t/2)$ and $\sin(t/2)$ is 4π, hence $\mathbf{c}_4$ winds around the circle once.

There are infinitely many possible parametrizations. For example,

$$\mathbf{c}_5(t) = (2\cos 5t, 2\sin 5t), \qquad t \in [0, 4\pi]$$

(counterclockwise, winds around the circle ten times), or, more "exotic",

$$\mathbf{c}_6(t) = (\sqrt{2}(\sin t + \cos t), \sqrt{2}(\sin t - \cos t)), \qquad t \in [0, 2\pi]$$

(counterclockwise, winds around the circle once).

29. Using the fact that $f(t) = t^{1/3}$ is not differentiable at $t = 0$ (recall that $f'(t) = t^{-2/3}/3$) we get a parametrization $\mathbf{c}(t) = (t^{1/3}, 2t^{1/3})$, $t \in \mathbb{R}$, of the line $y = 2x$ that is not differentiable at $t = 0$.

Remark: we could have used $\mathbf{c}_1(t) = (t^{3/5}, 2t^{3/5})$ (not differentiable at $t = 0$); or, $\mathbf{c}_1(t) = ((t+1)^{2/3}, 2(t+1)^{2/3})$ (not differentiable at $t = -1$), etc.

31. The parametrization $c(t)$ is continuous, but since $c'(t) = (t^{-2/3}/3, 1)$ it is not differentiable at $t = 0$. For the same reason it is not C^1 at $t = 0$.

33. The parametrization $c(t)$ is continuous, since both of its components are continuous. If $t \geq 0$, then $c(t) = (t^2, t^6)$ and $c'(t) = (2t, 6t^5)$; if $t < 0$, then $c(t) = (-t^2, -t^6)$ and $c'(t) = (-2t, -6t^5)$. It follows that $c(t)$ is differentiable and C^1 if $t \neq 0$. Since $c'(0) = (0, 0)$ (i.e., $c'(0)$ exists) it follows that c is differentiable at $t = 0$ as well. Now $\lim_{t \to 0} c'(t) = (0, 0) = c'(0)$ implies that c is C^1 at $t = 0$.

35. The velocity is

$$Dc(t) = c'(t) = \begin{bmatrix} e^t + te^t \\ -e^t + (1-t)e^t \\ e^t \end{bmatrix} = \begin{bmatrix} (t+1)e^t \\ -te^t \\ e^t \end{bmatrix}$$

and the acceleration is

$$c''(t) = \begin{bmatrix} e^t + (t+1)e^t \\ -e^t - te^t \\ e^t \end{bmatrix} = \begin{bmatrix} (t+2)e^t \\ -(t+1)e^t \\ e^t \end{bmatrix}.$$

37. The velocity is $v(t) = c'(t) = (3t^2, -t^{-2}, 0)$, and the speed is $\|c'(t)\| = (9t^4 + t^{-4})^{1/2}$. The acceleration is $a(t) = c''(t) = (6t, 2t^{-3}, 0)$.

39. The velocity is computed by the product rule:

$$v(t) = c'(t) = (2e^{2t}\sin(2t) + 2e^{2t}\cos(2t))i + (2e^{2t}\cos(2t) - 2e^{2t}\sin(2t))j$$
$$= 2e^{2t}[(\sin(2t) + \cos(2t))i + (\cos(2t) - \sin(2t))j].$$

It follows that the speed is

$$\|c'(t)\| = 2e^{2t}(\sin^2(2t) + \cos^2(2t) + 2\sin(2t)\cos(2t) + \sin^2(2t)$$
$$+ \cos^2(2t) - 2\sin(2t)\cos(2t))^{1/2} = 2\sqrt{2}e^{2t}.$$

Using the product rule again, we get

$$c''(t) = 4e^{2t}[(\sin(2t) + \cos(2t))i + (\cos(2t) - \sin(2t))j]$$
$$+ 2e^{2t}[(2\cos(2t) - 2\sin(2t))i + (-2\sin(2t) - 2\cos(2t))j]$$
$$= 4e^{2t}(2\cos(2t)i - 2\sin(2t)j)$$
$$= 8e^{2t}(\cos(2t)i - \sin(2t)j).$$

41. The velocity is $v(t) = c'(t) = (t^{-1/2}/2, 1, 3t^{1/2}/2)$; the speed is

$$\|c'(t)\| = \sqrt{t^{-1}/4 + 1 + 9t/4} = \tfrac{1}{2}\sqrt{t^{-1} + 4 + 9t};$$

the acceleration is $a(t) = (-t^{-3/2}/4, 0, 3t^{-1/2}/4)$.

2.6. Properties of Derivatives

1. Using the chain rule, we get

$$\frac{\partial f}{\partial y} = D_1 g(-2y) + D_2 g(2y) \qquad \text{and} \qquad \frac{\partial f}{\partial x} = D_1 g(2x) + D_2 g(-2x)$$

(note: we used shorter notation $D_1 g$ and $D_2 g$ instead of $D_1 g(x^2 - y^2, y^2 - x^2)$ and $D_2 g(x^2 - y^2, y^2 - x^2)$). It follows that

$$x\frac{\partial f}{\partial y} + y\frac{\partial f}{\partial x} = -2xyD_1 g + 2xyD_2 g + 2xyD_1 g - 2xyD_2 g = 0.$$

3. By the chain rule,

$$g'(t) = (D_1 f)(t\sin t)' + (D_2 f)(t\cos t)' + (D_3 f)(t)',$$

where we use shorter notation $D_i f$ to denote $D_i f(t\sin t, t\cos t, t)$, $i = 1, 2, 3$. Thus,

$$g'(t) = (\sin t + t\cos t)D_1 f + (\cos t - t\sin t)D_2 f + D_3 f.$$

Alternatively, think of $g(t)$ as the composition $g = f \circ \mathbf{c}$, where $\mathbf{c}(t) = (t\sin t, t\cos t, t)$. Using (2.25), we get

$$\begin{aligned}
g'(t) &= D(f \circ \mathbf{c})(t) = \nabla f(\mathbf{c}(t)) \cdot \mathbf{c}'(t) \\
&= \Big(D_1 f, D_2 f, D_3 f\Big) \cdot (\sin t + t\cos t, \cos t - t\sin t, 1) \\
&= (\sin t + t\cos t)D_1 f + (\cos t - t\sin t)D_2 f + D_3 f.
\end{aligned}$$

5. From $(f \circ \mathbf{c})(t) = f(\mathbf{c}(t)) = f(\sin t, \cos t) = \sin^2 t \cos t$ we compute

$$(f \circ \mathbf{c})'(t) = (2\sin t\cos t)\cos t + \sin^2 t(-\sin t) = 2\sin t\cos^2 t - \sin^3 t.$$

Using (2.25), we get

$$\begin{aligned}
g'(t) &= D(f \circ \mathbf{c})(t) = \nabla f(\mathbf{c}(t)) \cdot \mathbf{c}'(t) \\
&= (2xy, x^2)\big|_{\mathbf{c}(t)} \cdot (\cos t, -\sin t) \\
&= (2\sin t\cos t, \sin^2 t) \cdot (\cos t, -\sin t) = 2\sin t\cos^2 t - \sin^3 t.
\end{aligned}$$

7. From $(f \circ \mathbf{c})(t) = f(\mathbf{c}(t)) = t^2\sin t + \cos t^2$, we compute

$$(f \circ \mathbf{c})'(t) = 2t\sin t + t^2\cos t - 2t\sin t^2.$$

Using (2.25), we get

$$\begin{aligned}
g'(t) &= D(f \circ \mathbf{c})(t) = \nabla f(\mathbf{c}(t)) \cdot \mathbf{c}'(t) \\
&= (y - 2x\sin(x^2 + z^2), x, -2z\sin(x^2 + z^2))\big|_{(t\sin t, t, t\cos t)} \cdot (\sin t + t\cos t, 1, \cos t - t\sin t) \\
&= (t - 2t\sin t\sin t^2, t\sin t, -2t\cos t\sin t^2) \cdot (\sin t + t\cos t, 1, \cos t - t\sin t) \\
&= t\sin t - 2t\sin^2 t\sin t^2 + t^2\cos t - 2t^2\sin t\sin t^2\cos t + t\sin t \\
&\quad - 2t\cos^2 t\sin t^2 + 2t^2\sin t\cos t\sin t^2
\end{aligned}$$

$$= 2t \sin t + t^2 \cos t - 2t \sin t^2.$$

9. Write $f = g(u, v, w, z)$, where $u = x^2 y$, $v = 2x + 5y$, $w = x$ and $z = y$. By the chain rule, $f_x = (\partial g / \partial u) 2xy + (\partial g / \partial v) 2 + (\partial g / \partial w) 1 + (\partial g / \partial z) 0$, where $\partial g / \partial u$, $\partial g / \partial v$, $\partial g / \partial w$ and $\partial g / \partial z$ are evaluated at $(x^2 y, 2x + 5y, x, y)$. Using the notation $D_i g$ for the derivative of g with respect to its i-th variable, we write

$$f_x = D_1 g \cdot 2xy + D_2 g \cdot 2 + D_3 g \cdot 1 + D_4 g \cdot 0 = 2xy D_1 g + 2D_2 g + D_3 g,$$

where $D_1 g$, $D_2 g$, etc. are evaluated at $(x^2 y, 2x + 5y, x, y)$. Similarly,

$$f_y = D_1 g \cdot x^2 + D_2 g \cdot 5 + D_3 g \cdot 0 + D_4 g \cdot 1 = x^2 D_1 g + 5D_2 g + D_4 g.$$

11. We use $D_i f$ for the derivative of f with respect to its i-th variable, $i = 1, 2, 3$. By the chain rule,

$$\begin{aligned}
F_x &= D_1 f(h(x), g(y), k(x, y)) \, h'(x) + D_2 f(h(x), g(y), k(x, y)) \cdot 0 \\
&\quad + D_3 f(h(x), g(y), k(x, y)) \frac{\partial k(x, y)}{\partial x} \\
&= D_1 f(h(x), g(y), k(x, y)) \, h'(x) + D_3 f(h(x), g(y), k(x, y)) \frac{\partial k(x, y)}{\partial x}
\end{aligned}$$

and

$$\begin{aligned}
F_y &= D_1 f(h(x), g(y), k(x, y)) \cdot 0 + D_2 f(h(x), g(y), k(x, y)) \, g'(y) \\
&\quad + D_3 f(h(x), g(y), k(x, y)) \frac{\partial k(x, y)}{\partial y} \\
&= D_2 f(h(x), g(y), k(x, y)) \, g'(y) + D_3 f(h(x), g(y), k(x, y)) \frac{\partial k(x, y)}{\partial y}.
\end{aligned}$$

Here is an alternative computation: $F(x, y) : \mathbb{R}^2 \to \mathbb{R}$ can be thought of as a composition $F(x, y) = f(G(x, y))$ of the function $G : \mathbb{R}^2 \to \mathbb{R}^3$ given by $G(x, y) = (h(x), g(y), k(x, y))$ and the function $f : \mathbb{R}^3 \to \mathbb{R}$. By the chain rule, $DF(x, y) = Df(G(x, y)) \cdot DG(x, y)$, where

$$DF(x, y) = \begin{bmatrix} F_x(x, y) & F_y(x, y) \end{bmatrix}$$

and

$$\begin{aligned}
&Df(G(x, y)) \cdot DG(x, y) \\
&= \begin{bmatrix} D_1 f(G(x, y)) & D_2 f(G(x, y)) & D_3 f(G(x, y)) \end{bmatrix} \cdot \begin{bmatrix} h'(x) & 0 \\ 0 & g'(y) \\ \partial k(x, y) / \partial x & \partial k(x, y) / \partial y \end{bmatrix} \\
&= \Big[D_1 f(G(x, y)) \, h'(x) + D_3 f(G(x, y)) \frac{\partial k(x, y)}{\partial x} \\
&\qquad D_2 f(G(x, y)) \, g'(y) + D_3 f(G(x, y)) \frac{\partial k(x, y)}{\partial y} \Big].
\end{aligned}$$

Comparing the corresponding entries in $DF(x, y)$ and $Df(G(x, y)) \cdot DG(x, y)$ and replacing $G(x, y)$ by $(h(x), g(y), k(x, y))$, we obtain the expressions for F_x and F_y.

13. By the product rule,

$$\nabla(fg)(x,y) = g(x,y)\nabla f(x,y) + f(x,y)\nabla g(x,y)$$
$$= \ln(xy)\,[\,2x+y \quad x\,] + (x^2+xy)\,[\,1/x \quad 1/y\,]$$
$$= [\,(2x+y)\ln(xy) + x + y \quad x\ln(xy) + (x^2+xy)/y\,].$$

By the quotient rule,

$$D\left(\frac{f}{g}\right)(2,2) = \frac{g(2,2)\nabla f(2,2) - f(2,2)\nabla g(2,2)}{g(2,2)^2}$$
$$= \frac{\ln 4\,[\,2x+y \quad x\,]_{at\ (2,2)} - 8\,[\,1/x \quad 1/y\,]_{at\ (2,2)}}{(\ln 4)^2}$$
$$= \frac{\ln 4\,[\,6 \quad 2\,] - 8\,[\,1/2 \quad 1/2\,]}{(\ln 4)^2}$$
$$= [\,(6\ln 4 - 4)/(\ln 4)^2 \quad (2\ln 4 - 4)/(\ln 4)^2\,] \approx [\,2.24672 \quad -0.63867\,].$$

15. Let $g(x,y,z) = x$. Then

$$D\,(f/x)\,(x,y,z) = D\,(f/g)\,(x,y,z) = \frac{g(x,y,z)\,Df(x,y,z) - f(x,y,z)\,Dg(x,y,z)}{g(x,y,z)^2},$$

and so

$$D\,(f/g)\,(1,\pi,-1) = \frac{1\,Df(1,\pi,-1) - (-2)\,Dg(1,\pi,-1)}{1}$$
$$= Df(1,\pi,-1) + 2Dg(1,\pi,-1).$$

From

$$Df(1,\pi,-1) = [\,2x \quad z\cos(yz) \quad y\cos(yz)\,]_{at\ (1,\pi,-1)} = [\,2 \quad 1 \quad -\pi\,]$$

and

$$Dg(1,\pi,-1) = [\,1 \quad 0 \quad 0\,]_{at\ (1,\pi,-1)} = [\,1 \quad 0 \quad 0\,]$$

we get

$$D\,(f/x)\,(1,\pi,-1) = D\,(f/g)\,(1,\pi,-1)$$
$$= [\,2 \quad 1 \quad -\pi\,] + 2\,[\,1 \quad 0 \quad 0\,] = [\,4 \quad 1 \quad -\pi\,].$$

By the product rule,

$$D(x^2yf)(2,0,1) = f(2,0,1)\,D(x^2y)(2,0,1) + x^2y\Big|_{(2,0,1)}\,Df(2,0,1)$$

(if this notation seems inconvenient, introduce a function $g(x,y,z) = x^2y$ and differentiate gf instead of x^2yf; that's how we did the first part: we used f/g with $g(x,y,z) = x$ instead of f/x). Since

$$D(x^2y)(2,0,1) = [\,2xy \quad x^2 \quad 0\,]_{at\ (2,0,1)} = [\,0 \quad 4 \quad 0\,]$$

and

$$Df(2,0,1) = [\,2x \quad z\cos(yz) \quad y\cos(yz)\,]_{at\ (2,0,1)} = [\,4 \quad 1 \quad 0\,],$$

it follows that

$$D(x^2 yf)(2,0,1) = 1 \cdot [\,0 \quad 4 \quad 0\,] + 0 \cdot [\,4 \quad 1 \quad 0\,] = [\,0 \quad 4 \quad 0\,].$$

17. From

$$\frac{\partial w}{\partial \rho} = \frac{\partial f}{\partial x}\frac{\partial x}{\partial \rho} + \frac{\partial f}{\partial y}\frac{\partial y}{\partial \rho} + \frac{\partial f}{\partial z}\frac{\partial z}{\partial \rho}$$

we get

$$\frac{\partial w}{\partial \rho} = \frac{\partial f}{\partial x} \sin\phi\cos\theta + \frac{\partial f}{\partial y} \sin\phi\sin\theta + \frac{\partial f}{\partial z} \cos\phi.$$

Similarly,

$$\frac{\partial w}{\partial \theta} = \frac{\partial f}{\partial x}\frac{\partial x}{\partial \theta} + \frac{\partial f}{\partial y}\frac{\partial y}{\partial \theta} + \frac{\partial f}{\partial z}\frac{\partial z}{\partial \theta}$$

$$= \frac{\partial f}{\partial x}(-\rho\sin\phi\sin\theta) + \frac{\partial f}{\partial y}\,\rho\sin\phi\cos\theta + \frac{\partial f}{\partial z}\cdot 0$$

$$= -\rho\sin\phi\left(\frac{\partial f}{\partial x}\sin\theta - \frac{\partial f}{\partial y}\cos\theta\right)$$

and

$$\frac{\partial w}{\partial \phi} = \frac{\partial f}{\partial x}\frac{\partial x}{\partial \phi} + \frac{\partial f}{\partial y}\frac{\partial y}{\partial \phi} + \frac{\partial f}{\partial z}\frac{\partial z}{\partial \phi}$$

$$= \frac{\partial f}{\partial x}(\rho\cos\phi\cos\theta) + \frac{\partial f}{\partial y}\,\rho\cos\phi\sin\theta - \frac{\partial f}{\partial z}\rho\sin\phi.$$

19. From

$$(\mathbf{v} \times \mathbf{w})(t) = \begin{vmatrix} \mathbf{i} & \mathbf{j} & \mathbf{k} \\ t^3 & 0 & te^t \\ 0 & -2t & 0 \end{vmatrix} = 2t^2 e^t \mathbf{i} - 2t^4 \mathbf{k}$$

we get

$$(\mathbf{v} \times \mathbf{w})'(t) = (4te^t + 2t^2 e^t)\mathbf{i} - 8t^3\mathbf{k} = 2te^t(2+t)\mathbf{i} - 8t^3\mathbf{k}.$$

Using the product rule,

$$(\mathbf{v} \times \mathbf{w})'(t) = \mathbf{v}'(t) \times \mathbf{w}(t) + \mathbf{v}(t) \times \mathbf{w}'(t)$$

$$= \begin{vmatrix} \mathbf{i} & \mathbf{j} & \mathbf{k} \\ 3t^2 & 0 & e^t + te^t \\ 0 & -2t & 0 \end{vmatrix} + \begin{vmatrix} \mathbf{i} & \mathbf{j} & \mathbf{k} \\ t^3 & 0 & te^t \\ 0 & -2 & 0 \end{vmatrix}$$

$$= 2t(e^t + te^t)\mathbf{i} - 6t^3\mathbf{k} + 2te^t\mathbf{i} - 2t^3\mathbf{k} = 2te^t(2+t)\mathbf{i} - 8t^3\mathbf{k}.$$

21. By the Chain Rule, $D(g \circ \mathbf{F})(0,0) = Dg(\mathbf{F}(0,0)) \cdot D\mathbf{F}(0,0)$. Since

$$Dg(\mathbf{F}(0,0)) = Dg(1,0,1) = [\,w \quad 2v \quad u\,]_{at\ (1,0,1)} = [\,1 \quad 0 \quad 1\,]$$

and

$$D\mathbf{F}(0,0) = \begin{bmatrix} e^x & 0 \\ y & x \\ 0 & e^y \end{bmatrix}_{at\ (0,0)} = \begin{bmatrix} 1 & 0 \\ 0 & 0 \\ 0 & 1 \end{bmatrix},$$

it follows that

$$D(g \circ \mathbf{F})(0,0) = [\,1 \quad 0 \quad 1\,] \cdot \begin{bmatrix} 1 & 0 \\ 0 & 0 \\ 0 & 1 \end{bmatrix} = [\,1 \quad 1\,].$$

Alternatively, $(g \circ \mathbf{F})(x, y) = g(\mathbf{F}(x, y)) = g(e^x, xy, e^y) = e^x e^y + x^2 y^2$, and

$$D(g \circ \mathbf{F})(0, 0) = [e^x e^y + 2xy^2 \quad e^x e^y + 2x^2 y]_{at \ (0,0)} = [1 \quad 1].$$

23. Since $w = f(x, g(x, z), z)$, it follows that (using $D_i f$ for the partial derivative of f with respect to its i-th variable)

$$\frac{\partial w}{\partial x} = D_1 f \frac{\partial x}{\partial x} + D_2 f \frac{\partial g(x, z)}{\partial x} + D_3 f \frac{\partial z}{\partial x} = D_1 f + D_2 f \frac{\partial g(x, z)}{\partial x}.$$

Similarly,

$$\frac{\partial w}{\partial z} = D_2 f \frac{\partial g(x, z)}{\partial z} + D_3 f,$$

where all partial derivatives $D_i f$, $i = 1, 2, 3$, are computed at $(x, g(x, z), z)$.

25. Let $a = [a_{ij}]$, $i, j = 1, 2$, and $\mathbf{x} = (x, y)$. Then

$$\mathbf{F}(\mathbf{x}) = A \cdot \mathbf{x} = \begin{bmatrix} a_{11} & a_{12} \\ a_{21} & a_{22} \end{bmatrix} \cdot \begin{bmatrix} x \\ y \end{bmatrix} = \begin{bmatrix} a_{11}x + a_{12}y \\ a_{21}x + a_{22}y \end{bmatrix},$$

and therefore

$$D\mathbf{F}(\mathbf{x}) = \begin{bmatrix} a_{11} & a_{12} \\ a_{21} & a_{22} \end{bmatrix} = A.$$

The components $a_{11}x + a_{12}y$ and $a_{21}x + a_{22}y$ of $\mathbf{F}$ are polynomials, and so have partial derivatives at all points. It remains to show that (let $\mathbf{x}_0 = (x_0, y_0)$)

$$\lim_{\mathbf{x} \to \mathbf{x}_0} \frac{\|\mathbf{F}(\mathbf{x}) - \mathbf{F}(\mathbf{x}_0) - D\mathbf{F}(\mathbf{x})(\mathbf{x} - \mathbf{x}_0)\|}{\|\mathbf{x} - \mathbf{x}_0\|} = 0.$$

The numerator of this expression is equal to

$$\|A \cdot \mathbf{x} - A \cdot \mathbf{x}_0 - A \cdot (\mathbf{x} - \mathbf{x}_0)\| = \|\mathbf{0}\| = 0,$$

by the distributivity property of matrix multiplication. Therefore,

$$\lim_{\mathbf{x} \to \mathbf{x}_0} \frac{\|\mathbf{F}(\mathbf{x}) - \mathbf{F}(\mathbf{x}_0) - D\mathbf{F}(\mathbf{x})(\mathbf{x} - \mathbf{x}_0)\|}{\|\mathbf{x} - \mathbf{x}_0\|} = \lim_{\mathbf{x} \to \mathbf{x}_0} \frac{0}{\|\mathbf{x} - \mathbf{x}_0\|} = 0,$$

and so $\mathbf{F}$ is differentiable at $\mathbf{x}_0$.

Comment: this solution is provided as an exercise in using the definition of differentiability (Definition 2.13). There is an easier way: since the components of $\mathbf{F}$ are polynomials, their partial derivatives are continuous (i.e., $\mathbf{F}$ is C^1). By Theorem 2.5 in Section 2.4, it follows that $\mathbf{F}$ is differentiable.

27. By the chain rule,

$$\frac{df}{dt} = \frac{\partial f}{\partial x} \frac{dx}{dt} + \frac{\partial f}{\partial y} \frac{dy}{dt}.$$

In order to compute dx/dt and dy/dt we have to use implicit differentiation. From $x^3 + tx = 8$ it follows that

$$3x^2 \frac{dx}{dt} + x + t \frac{dx}{dt} = 0 \qquad \text{and hence} \qquad \frac{dx}{dt} = \frac{-x}{3x^2 + t}.$$

Similarly, $ye^y = t$ implies

$$e^y \frac{dy}{dt} + ye^y \frac{dy}{dt} = 1 \qquad \text{and hence} \qquad \frac{dy}{dt} = \frac{1}{e^y(1+y)}.$$

It follows that

$$\frac{df}{dt} = 3x^2 y \frac{-x}{3x^2+t} + x^3 \frac{1}{e^y(1+y)}.$$

If $t = 0$, then $x^3 = 8$ implies $x = 2$; the equation $ye^y = 0$ implies $y = 0$. Therefore, $(df/dt)(0) = 0 + 8/1 = 8$.

29. By the chain rule,

$$Df(\mathbf{x}) = -\sin\left(\|\mathbf{F}(\mathbf{x})\|^2\right) \, D\left(\|\mathbf{F}(\mathbf{x})\|^2\right).$$

Let $g(x, y) = \|\mathbf{F}(\mathbf{x})\|^2 = (F_1(x, y))^2 + (F_2(x, y))^2$, where $\mathbf{F}(x, y) = (F_1(x, y), F_2(x, y))$. The derivative of g is

$$\begin{aligned}
Dg(\mathbf{x}) &= \left[\frac{\partial g}{\partial x}(\mathbf{x}) \quad \frac{\partial g}{\partial y}(\mathbf{x}) \right] \\
&= \left[2F_1(\mathbf{x}) \frac{\partial F_1}{\partial x}(\mathbf{x}) + 2F_2(\mathbf{x}) \frac{\partial F_2}{\partial x}(\mathbf{x}) \quad 2F_1(\mathbf{x}) \frac{\partial F_1}{\partial y}(\mathbf{x}) + 2F_2(\mathbf{x}) \frac{\partial F_2}{\partial y}(\mathbf{x}) \right] \\
&= 2\left[F_1(\mathbf{x}) \quad F_2(\mathbf{x}) \right] \cdot \begin{bmatrix} (\partial F_1/\partial x)(\mathbf{x}) & (\partial F_1/\partial y)(\mathbf{x}) \\ (\partial F_2/\partial x)(\mathbf{x}) & (\partial F_2/\partial y)(\mathbf{x}) \end{bmatrix} \\
&= 2\mathbf{F}(\mathbf{x}) D\mathbf{F}(\mathbf{x}).
\end{aligned}$$

Thus,

$$Df(\mathbf{x}) = -2\sin\left(\|\mathbf{F}(\mathbf{x})\|^2\right) \mathbf{F}(\mathbf{x}) \cdot D\mathbf{F}(\mathbf{x}).$$

2.7. Gradient and Directional Derivative

1. From the diagram we read $f(2, 1) = 16$. Moving in the $\mathbf{j}$ direction we arrive at $(2, 1.5)$, where $f(2, 1.5) = 18$. Thus, $D_{\mathbf{j}}f(2, 1) \approx (18 - 16)/(1.5 - 1) = 4$.

3. From the diagram we read $f(3, 2) = 18$. Moving in the direction of the vector $-2\mathbf{i} + \mathbf{j}$ we arrive at $(2, 2.5)$, where $f(2, 2.5) = 20$. The distance between $(3, 2)$ and $(2, 2.5)$ is $\sqrt{1 + 1/4} = \sqrt{5}/2$. Thus, $D_{-2\mathbf{i}+\mathbf{j}}f(3, 2) \approx (20 - 18)/(\sqrt{5}/2) = 4/\sqrt{5}$.

To make notation simpler, we write $D_{-2\mathbf{i}+\mathbf{j}}$ instead of using the unit vector $D_{(-2\mathbf{i}+\mathbf{j})/\sqrt{5}}$, as it will not be a cause of confusion.

5. From the diagram we read $f(4, 1) = 12$. Moving in the direction of the vector $-\mathbf{i}$ we arrive at $(3, 1)$, where $f(3, 1) = 14$. Thus, $D_{-\mathbf{i}}f(4, 1) \approx (14 - 12)/1 = 2$.

7. The unit vector in the direction of $\mathbf{v}$ is $\mathbf{u} = \mathbf{v}/\|\mathbf{v}\| = 2\mathbf{i}/\sqrt{5} - \mathbf{j}/\sqrt{5}$. The gradient of f is computed by the product rule:

$$\nabla f = (ye^{xy}(\cos x + \sin y) + e^{xy}(-\sin x))\mathbf{i} + (xe^{xy}(\cos x + \sin y) + e^{xy}\cos y)\mathbf{j},$$

and therefore $\nabla f(\pi/2, 0) = -\mathbf{i} + \mathbf{j}$. It follows that

$$D_\mathbf{u} f(\pi/2, 0) = \nabla f(\pi/2, 0) \cdot \mathbf{u} = -3/\sqrt{5}.$$

9. The gradient of f at $(0, -1, 2)$ is

$$\nabla f(0, -1, 2) = (-2xe^{-x^2-y^2-z^2}\mathbf{i} - 2ye^{-x^2-y^2-z^2}\mathbf{j} - 2ze^{-x^2-y^2-z^2}\mathbf{k})\Big|_{(0,-1,2)}$$

$$= 2e^{-5}\mathbf{j} - 4e^{-5}\mathbf{k}.$$

The rate of change of f in the direction of $\mathbf{v} = \mathbf{i} + \mathbf{j} + \mathbf{k}$ is (we need a unit vector, so let $\mathbf{u} = \mathbf{v}/\|\mathbf{v}\| = (\mathbf{i} + \mathbf{j} + \mathbf{k})/\sqrt{3}$)

$$D_\mathbf{u} f(0, -1, 2) = \nabla f(0, -1, 2) \cdot \mathbf{u} = (2e^{-5}\mathbf{j} - 4e^{-5}\mathbf{k}) \cdot (\mathbf{i} + \mathbf{j} + \mathbf{k})/\sqrt{3} = -2e^{-5}/\sqrt{3}.$$

11. The unit vector in the direction of $\mathbf{v}$ is $\mathbf{u} = \mathbf{v}/\|\mathbf{v}\| = 3\mathbf{i}/5 + 4\mathbf{j}/5$. The gradient of f is

$$\nabla f = \ln y^2 \mathbf{i} + (2 + 2x/y)\mathbf{j},$$

and therefore $\nabla f(1, 2) = \ln 4\,\mathbf{i} + 3\mathbf{j}$. It follows that

$$D_\mathbf{u} f(1, 2) = \nabla f(1, 2) \cdot \mathbf{u} = (\ln 4\,\mathbf{i} + 3\mathbf{j}) \cdot (\tfrac{3}{5}\mathbf{i} + \tfrac{4}{5}\mathbf{j}) = \tfrac{1}{5}(3\ln 4 + 12).$$

13. The polar form of a vector is $\mathbf{u} = \|\mathbf{u}\|(\cos\theta\,\mathbf{i} + \sin\theta\,\mathbf{j})$. In our case, $\|\mathbf{u}\| = 1$ and $\theta = \pi/2$, so $\mathbf{u} = \mathbf{j}$. From $\nabla f(\mathbf{p}) = ye^{xy}\mathbf{i} + xe^{xy}\mathbf{j}\big|_{(0,1)} = \mathbf{i}$ it follows that

$$D_\mathbf{u} f(0, 1) = \nabla f(0, 1) \cdot \mathbf{u} = \mathbf{i} \cdot \mathbf{j} = 0.$$

15. By definition,

$$D_\mathbf{u} f(0, 0) = \lim_{t \to 0} \frac{f((0, 0) + t(u, v)) - f(0, 0)}{t}$$

$$= \lim_{t \to 0} \frac{f(tu, tv) - f(0, 0)}{t} = \lim_{t \to 0} \frac{2tutv}{t^2 u^2 + t^2 v^2} \frac{1}{t} = \frac{2uv}{u^2 + v^2} \lim_{t \to 0} \frac{1}{t}.$$

Since $\lim_{t \to 0} 1/t$ does not exist, it follows that the directional derivative $D_\mathbf{u} f(0, 0)$ does not exist in any direction.

17. The maximum rate of change of f at $(\pi/4, \pi/4)$ occurs in the direction of

$$\nabla f(\pi/4, \pi/4) = (\sec x \tan x \tan y, \sec x \sec^2 y)\Big|_{(\pi/4, \pi/4)} = (\sqrt{2}, 2\sqrt{2}).$$

The maximum rate of change is $\|\nabla f(\pi/4, \pi/4)\| = \sqrt{10}$.

19. The maximum rate of change of f at $(1, 2, -1)$ occurs in the direction of

$$\nabla f(1, 2, -1) = (y^{-1} - zx^{-2}, -xy^{-2} + z^{-1}, -yz^{-2} + x^{-1})\Big|_{(1,2,-1)} = (3/2, -5/4, -1).$$

The maximum rate of change is $\|\nabla f(1, 2, -1)\| = \sqrt{77}/4$.

21. Near $\mathbf{p} = (3, -2)$, $f(x, y) = |xy| = -xy$, since $x > 0$ and $y < 0$. Therefore

$$\nabla f(3, -2) = (-y, -x)\Big|_{(3,-2)} = (2, -3).$$

Maximum rate of change equals $\|\nabla f(3, -2)\| = \sqrt{13}$, and it occurs in the direction of the vector $2\mathbf{i} - 3\mathbf{j}$.

23. (a) Since $\mathbf{u} = (\mathbf{i} + \mathbf{j})/\|\mathbf{i} + \mathbf{j}\| = (\mathbf{i} + \mathbf{j})/\sqrt{2}$, it follows that the rate of change of P at $(0, 1)$ in the direction of the vector $\mathbf{i} + \mathbf{j}$ is

$$D_{\mathbf{u}} P(0, 1) = \nabla P(0, 1) \cdot \mathbf{u}$$
$$= \left. (-200xe^{-x^2 - 2y^2}, -400ye^{-x^2 - 2y^2}) \right|_{(0,1)} \cdot (1/\sqrt{2}, 1/\sqrt{2})$$
$$= (0, -400e^{-2}) \cdot (1/\sqrt{2}, 1/\sqrt{2}) = -400e^{-2}/\sqrt{2}.$$

(b) The most rapid increase of P at $\mathbf{p}$ occurs in the direction $\nabla P(0, 1) = (0, -400e^{-2})$. The most rapid decrease of P at $\mathbf{p}$ occurs in the direction $-\nabla P(0, 1) = (0, 400e^{-2})$.

(c) Maximum rate of increase at $\mathbf{p}$ is $\|\nabla P(0, 1)\| = 400e^{-2}$.

(d) At $\mathbf{p}$, the pressure does not change in the directions perpendicular to the gradient $\nabla P(0, 1) = (0, -400e^{-2})$. These directions are $\mathbf{u} = \pm\mathbf{i}$.

25. Let $\mathbf{u} = (a, b)$ be a unit vector (so that $a^2 + b^2 = 1$). From

$$\nabla f(1, 2) = (2y, 2x) \Big|_{(1,2)} = (4, 2)$$

it follows that $D_{\mathbf{u}} f(1, 2) = \nabla f(1, 2) \cdot \mathbf{u} = 4a + 2b$. The requirement that $D_{\mathbf{u}} f(1, 2) = 4$ implies that $4a + 2b = 4$, i.e., $b = 2 - 2a$. Substitute the expression for b into $a^2 + b^2 = 1$, thus getting $a^2 + (2 - 2a)^2 = 1$; i.e., $5a^2 - 8a + 3 = 0$. Solving for a, we get $a = 3/5$ (and then $b = 4/5$) and $a = 1$ (and $b = 0$). Therefore, the directional derivative of f at $(1, 2)$ in the direction of the vector $(3/5, 4/5)$ or in the direction of the vector $(1, 0)$ is equal to 4.

27. The requirement is that $D_{\mathbf{u}} f(x, y) \geq 0.8 \|\nabla f(x, y)\|$, where $\mathbf{u}$ is a unit vector. Since $D_{\mathbf{u}} f(x, y) = \nabla f(x, y) \cdot \mathbf{u} = \|\nabla f(x, y)\| \cos\theta$ (where θ is the angle between $\nabla f(x, y)$ and $\mathbf{u}$), we conclude that $\cos\theta \geq 0.8$. Thus, $-\arccos 0.8 \leq \theta \leq \arccos 0.8$. With the definition of θ as the smaller of the two angles between $\mathbf{u}$ and $\nabla f(x, y)$ (that we adopted and usually use), the answer is $0 \leq \theta \leq \arccos 0.8$, i.e., $0 \leq \theta \leq 0.6435$ radians.

29. Let $\mathbf{u}$ be the unit vector pointing toward the west. By definition,

$$D_{\mathbf{u}} f(x, y) = \nabla f(x, y) \cdot \mathbf{u} = \|\nabla f(x, y)\| \cos\theta,$$

where $\theta = \pi/4$ is the angle between $\nabla f(x, y)$ and $\mathbf{u}$. Since $\|\nabla f(x, y)\| = 15/100 = 3/20$, it follows that $D_{\mathbf{u}} f(x, y) = (3/20)(\sqrt{2}/2) = 3\sqrt{2}/40 \approx 0.1061$. So, in the direction towards the west, the height of the hill increases by approximately 10.61 meters per 100 meters.

31. It is given that $D_{\mathbf{u}_1} f(0, 2) = 4$, where $\mathbf{u}_1 = (\mathbf{i} + 2\mathbf{j})/\sqrt{5}$ and $D_{\mathbf{u}_2} f(0, 2) = 12$, where $\mathbf{u}_2 = (2\mathbf{i} - \mathbf{j})/\sqrt{5}$. Let $\nabla f(0, 2) = (a, b)$. Then $D_{\mathbf{u}_1} f(0, 2) = 4$ implies that $\nabla f(0, 2) \cdot \mathbf{u}_1 = 4$; i.e., $a/\sqrt{5} + 2b/\sqrt{5} = 4$ or $a + 2b = 4\sqrt{5}$. Similarly, from $D_{\mathbf{u}_2} f(0, 2) = 12$ it follows that $2a - b = 12\sqrt{5}$. Solving for a and b, we get $a = 28/\sqrt{5}$ and $b = -4/\sqrt{5}$; i.e., $\nabla f(0, 2) =$

$(28/\sqrt{5}, -4/\sqrt{5})$. The directional derivative at $(0,2)$ in the direction of $3\mathbf{i} + 3\mathbf{j}$ is (we need a unit vector, hence $\mathbf{u} = (\mathbf{i} + \mathbf{j})/\sqrt{2}$)

$$D_{\mathbf{u}}f(0,2) = \nabla f(0,2) \cdot \mathbf{u} = (28/\sqrt{5}, -4/\sqrt{5}) \cdot (1/\sqrt{2}, 1/\sqrt{2}) = 24/\sqrt{10}.$$

33. Assume (for convenience) that $f, g: \mathbb{R}^2 \to \mathbb{R}$ are differentiable functions of two variables (the proof works for any number of variables). Then

$$\nabla(af \pm bg) = ((af \pm bg)_x, (af \pm bg)_y)$$
$$= (af_x \pm bg_x, af_y \pm bg_y) = (af_x, af_y) \pm (bg_x, bg_y) = a\nabla f \pm b\nabla g.$$

35. Assume, for simplicity, that $f, g: \mathbb{R}^2 \to \mathbb{R}$ (a proof in $\mathbb{R}^n$ is analogous). Using the definition of the gradient and the quotient rule we get

$$\nabla\left(\frac{f}{g}\right) = \left(\left(\frac{f}{g}\right)_x, \left(\frac{f}{g}\right)_y\right)$$
$$= \left(\frac{f_x g - f g_x}{g^2}, \frac{f_y g - f g_y}{g^2}\right) = \frac{g(f_x, f_y)}{g^2} - \frac{f(g_x, g_y)}{g^2} = \frac{g\nabla f - f\nabla g}{g^2}.$$

37. The angle between two surfaces at their common point is the angle between their tangent planes at that point — and that angle is equal to the angle between the normal vectors to the two surfaces (at that point).

The normal vector to $x^3 y^3 - 3yz = 8$ at $(1,2,0)$ is given by (think of the surface as a level surface of $f_1(x,y,z) = x^3 y^3 - 3yz$ of value 8)

$$\nabla f_1(1,2,0) = (3x^2 y^3 \mathbf{i} + (3x^3 y^2 - 3z)\mathbf{j} - 3y\mathbf{k})\Big|_{(1,2,0)} = 24\mathbf{i} + 12\mathbf{j} - 6\mathbf{k}.$$

Since we need the direction only, we will take $\mathbf{n}_1 = 4\mathbf{i} + 2\mathbf{j} - \mathbf{k}$ as a normal, to simplify computations. Similarly, the surface $x + 3z^2 = y^2 - 3$ can be viewed as a level surface of $f_2 = x + 3z^2 - y^2$ of value -3. It follows that

$$\nabla f_2(1,2,0) = (\mathbf{i} - 2y\mathbf{j} + 6z\mathbf{k})\Big|_{(1,2,0)} = \mathbf{i} - 4\mathbf{j};$$

i.e., $\mathbf{n}_2 = \mathbf{i} - 4\mathbf{j}$ is a normal vector to the surface $x + 3z^2 = y^2 - 3$ at the point $(1,2,0)$. Now

$$\cos\theta = \frac{\mathbf{n}_1 \cdot \mathbf{n}_2}{\|\mathbf{n}_1\|\,\|\mathbf{n}_2\|} = \frac{(4\mathbf{i} + 2\mathbf{j} - \mathbf{k}) \cdot (\mathbf{i} - 4\mathbf{j})}{\sqrt{21}\,\sqrt{17}} = \frac{-4}{\sqrt{357}}.$$

Hence $\theta = \arccos(-4/\sqrt{357}) \approx 1.78411 > \pi/2$ rad. So the acute angle is $\pi - \theta \approx 1.35748$ rad.

39. Let $f(x,y,z) = \sin(xy) - 2\cos(yz)$, and think of the given surface as a level surface $f(x,y,z) = 0$. A normal vector to that surface at $(\pi/2, 1, \pi/3)$ is

$$\nabla f(\pi/2, 1, \pi/3) = (y\cos(xy), x\cos(xy) + 2z\sin(yz), 2y\sin(yz))\Big|_{(\pi/2,1,\pi/3)} = (0, \pi\sqrt{3}/3, \sqrt{3}).$$

The tangent plane has the equation

$$0\left(x - \tfrac{\pi}{2}\right) + \tfrac{\pi\sqrt{3}}{3}\left(y - 1\right) + \sqrt{3}\left(z - \tfrac{\pi}{3}\right) = 0;$$

i.e., $\pi y + 3z - 2\pi = 0$.

41. Rewrite the given equation as $f(x, y, z) = z^2 - (4x - y)/(x + y + 1) = 0$ and think of it as a level surface of f of value 0. The vector

$$\nabla f(1, 0, \sqrt{2}) = \left(-\frac{5y + 4}{(x + y + 1)^2}, \frac{5x + 1}{(x + y + 1)^2}, 2z \right) \bigg|_{(1,0,\sqrt{2})} = (-1, 3/2, 2\sqrt{2}).$$

is perpendicular to the level surface $f(x, y, z) = 0$, and is therefore normal to the tangent plane. So

$$-1(x - 1) + \tfrac{3}{2}(y - 0) + 2\sqrt{2}(z - \sqrt{2}) = 0,$$

or $2x - 3y - 4\sqrt{2}z + 6 = 0$ is the equation of the tangent plane at $(1, 0, \sqrt{2})$.

43. A normal direction to the given surface at the point $(2, 4, \sqrt{3})$ is $\nabla f(2, 4, \sqrt{3})$, where $f(x, y, z) = (x - 2)^2 + (y - 3)^2 + z^2 - 4$ (and the surface is viewed as a level surface $f(x, y, z) = 0$). From

$$\nabla f(2, 4, \sqrt{3}) = (2(x - 2), 2(y - 3), 2z) \big|_{(2,4,\sqrt{3})} = (0, 2, 2\sqrt{3})$$

we get a parametric equation of the line normal to the surface at $(2, 4, \sqrt{3})$:

$$\ell_N(t) = (2, 4, \sqrt{3}) + t(0, 2, 2\sqrt{3}) = (2, 4 + 2t, \sqrt{3} + 2t\sqrt{3}),$$

where $t \in \mathbb{R}$. An equation of the tangent plane at $(2, 4, \sqrt{3})$ is

$$0(x - 2) + 2(y - 4) + 2\sqrt{3}(z - \sqrt{3}) = 0;$$

i.e., $y + \sqrt{3}z - 7 = 0$.

Here is another way of finding a normal line: the given surface is the sphere centered at $(2, 3, 0)$ of radius 2. The normal to the sphere is the line going through its center; so, to find its equation all we have to do is find a parametric equation of the line going through the point $(2, 3, 0)$ and the (given) point $(2, 4, \sqrt{3})$.

45. A normal direction to the surface at $(0, e, 1)$ is $\nabla f(0, e, 1)$, where $f(x, y, z) = \ln(x^2 + y^2) - 2$, and the surface is viewed as a level surface $f(x, y, z) = 0$. From

$$\nabla f(0, e, 1) = \left(\frac{2x}{x^2 + y^2}, \frac{2y}{x^2 + y^2}, 0 \right) \bigg|_{(0,e,1)} = (0, 2/e, 0)$$

we get a parametric equation of the line normal to the surface at $(0, e, 1)$ (to make things simpler, take $(0, 1, 0)$ as a normal)

$$\ell_N(t) = (0, e, 1) + t(0, 1, 0) = (0, e + t, 1),$$

where $t \in \mathbb{R}$. An equation of the tangent plane at $(0, e, 1)$ is

$$0(x - 0) + 1(y - 1) + 0(z - 1) = 0;$$

i.e., $y = 1$.

47. View the given curve as the level curve of the function $f(x,y) = 3x - 2y - 4$ of value 0. A normal direction at $(0, -2)$ is given by $\nabla f(0, -2) = (3, -2)\big|_{(0,-2)} = (3, -2)$. Consequently, the normal line has the equation

$$\ell_N(t) = (0, -2) + t(3, -2) = (3t, -2 - 2t),$$

$t \in \mathbb{R}$. To get a tangent direction, take any vector perpendicular to $(3, -2)$; i.e., take any vector whose dot product with $(3, -2)$ is zero — for example, $(2, 3)$. It follows that the tangent line at $(0, -2)$ has the equation

$$\ell_T(t) = (0, -2) + t(2, 3) = (2t, -2 + 3t),$$

where $t \in \mathbb{R}$.

49. View the given curve as the level curve of the function $f(x,y) = e^x \sin y$ of value 2. A normal direction at $(\ln 2, \pi/2)$ is given by

$$\nabla f(\ln 2, \pi/2) = (e^x \sin y, e^x \cos y)\big|_{(\ln 2, \pi/2)} = (2, 0).$$

Consequently, the normal line has the equation

$$\ell_N(t) = (\ln 2, \pi/2) + t(2, 0) = (2t + \ln 2, \pi/2),$$

$t \in \mathbb{R}$. To get a tangent direction, take any vector perpendicular to $(2, 0)$; for example, $(0, 1)$. So the tangent line at $(\ln 2, \pi/2)$ has the equation

$$\ell_T(t) = (\ln 2, \pi/2) + t(0, 1) = (\ln 2, t + \pi/2),$$

where $t \in \mathbb{R}$.

51. Parallel planes have parallel normal vectors. The paraboloid $z = x^2 + y^2 - 5$ can be thought of as the level surface of $f(x, y, z) = x^2 + y^2 - z$ of value 5. Its normal (and hence a normal to its tangent plane) is given by $\nabla f(x, y, z) = (2x, 2y, -1)$. We are asked to find all points on the paraboloid where $\nabla f(x, y, z) = (2x, 2y, -1)$ is parallel to $(1, 3, -1)$. From $(2x, 2y, -1) = a(1, 3, -1)$ it follows that $2x = a$, $2y = 3a$ and $-1 = -a$. Therefore, $a = 1$, $x = 1/2$ and $y = 3/2$. So $z = 1/4 + 9/4 - 5 = -10/4$ and it follows that $(1/2, 3/2, -5/2)$ is the only point with the required property.

53. Let $f(x, y, z) = \cos(xy) - e^z$; the given surface is the level surface $f(x, y, z) = -1$. The gradient of f is $\nabla f(x, y, z) = (-y \sin(xy), -x \sin(xy), -e^z)$. At $(1, \pi/2, 0)$, the vector $\nabla f(1, \pi/2, 0) = (-\pi/2, -1, -1)$ defines a direction normal to the surface. The unit normals are

$$\pm \frac{\nabla f(1, \pi/2, 0)}{\|\nabla f(1, \pi/2, 0)\|} = \pm \frac{(-\pi/2, -1, -1)}{\sqrt{(\pi^2 + 8)/4}} = \pm \frac{(\pi, 2, 2)}{\sqrt{\pi^2 + 8}}.$$

55. The equation $(x - m)^2 + (y - n)^2 + (z - p)^2 = r^2$ represents the sphere of radius r centered at (m, n, p). Let $f(x, y, z) = (x - m)^2 + (y - n)^2 + (z - p)^2$ and (x_0, y_0, z_0) be a point on the sphere. Then $\nabla f(x_0, y_0, z_0) = (2(x_0 - m), 2(y_0 - n), 2(z_0 - p))$ is normal to the sphere and

$$\ell(t) = (x_0, y_0, z_0) + 2t(x_0 - m, y_0 - n, z_0 - p),$$

$t \in \mathbb{R}$, is the equation of the normal line through (x_0, y_0, z_0). Since $\ell(-1/2) = (m, n, p)$, the normal line goes through the center of the sphere.

57. Let $f(x, y) = xy - k$; then $\nabla f(x, y) = (y, x)$ is a vector field perpendicular to the family of curves $xy = k$. Similarly, defining $g(x, y) = x^2 - y^2 - m$, we obtain that $\nabla g(x, y) = (2x, -2y)$ is a vector field perpendicular to the family of curves $x^2 - y^2 = m$. Now let (x, y) be a point of intersection of two curves, one from each family. Since $\nabla f(x, y) \cdot \nabla g(x, y) = 0$, it follows that $\nabla f(x, y)$ is perpendicular to $\nabla g(x, y)$; therefore, the two curves are orthogonal to each other.

59. We use the fact that the gradient is perpendicular to a level surface. The normal to $x^2 + y^2 + z^2 = 16$ is $\mathbf{n}_1 = (2x, 2y, 2z)$ and the normal to $(x - 5)^2 + y^2 + z^2 = 9$ is $\mathbf{n}_2 = (2(x - 5), 2y, 2z)$. Hence $\mathbf{n}_1 \cdot \mathbf{n}_2 = 4x^2 - 20x + 4y^2 + 4z^2$. The two spheres intersect at points (x, y, z) whose coordinates satisfy $x^2 + y^2 + z^2 = 16$ and $(x-5)^2 + y^2 + z^2 = 9$. Subtracting the second equation from the first equation, we get $10x - 25 = 7$; i.e., $x = 32/10 = 3.2$. Then $y^2 + z^2 = 16 - (3.2)^2 = 5.76$ and

$$\mathbf{n}_1 \cdot \mathbf{n}_2 = 4x^2 - 20x + 4y^2 + 4z^2 = 4(3.2)^2 - 20(3.2) + 4(5.76) = 0.$$

61. From $\nabla V = -\mathbf{F} = -e^{xy}(1 + xy)\mathbf{i} - x^2 e^{xy}\mathbf{j}$ it follows that $V_x = -e^{xy}(1 + xy)$ and $V_y = -x^2 e^{xy}$. Integrating the second equation with respect to y, we get $V = -xe^{xy} + C(x)$, where $C(x)$ is a real number or a function of x only. Substitute V into the equation for V_x to get

$$-e^{xy} - xe^{xy}y + C'(x) = -e^{xy}(1 + xy)$$

and $C'(x) = 0$. Therefore, $C(x) = C$ (C is a constant) and $V(x, y) = -xe^{xy} + C$.

63. From $\nabla V = -\mathbf{F} = (2xy^2 + 3x^2 z, 2x^2 y - z^3, x^3 - 3yz^2)$ it follows that $\partial V/\partial x = 2xy^2 + 3x^2 z$, $\partial V/\partial y = 2x^2 y - z^3$ and $\partial V/\partial z = x^3 - 3yz^2$. Integrating the first equation with respect to x we get

$$V(x, y, z) = x^2 y^2 + x^3 z + C(y, z),$$

where the "constant" $C(y, z)$ of integration might depend on y and z, since these two variables were viewed as constants. We managed to partially recover V. To compute $C(y, z)$ substitute the expression for $V(x, y, z)$ into $\partial V/\partial y$, thus getting

$$2x^2 y + \frac{\partial C(y, z)}{\partial y} = 2x^2 y - z^3,$$

which implies that $\partial C(y, z)/\partial y = -z^3$ and $C(y, z) = -yz^3 + C(z)$, by integration with respect to y (the variable z was kept fixed, so the integration "constant" might still depend on z). Hence

$$V(x, y, z) = x^2 y^2 + x^3 z - yz^3 + C(z).$$

Finally, substituting this expression into the equation for $\partial V/\partial z$, we get

$$x^3 - 3yz^2 + C'(z) = x^3 - 3yz^2,$$

so that $C(z) = C$ after integrating with respect to z (C is a real number, not a function any longer). It follows that any function of the form

$$V(x, y, z) = x^2 y^2 + x^3 z - yz^3 + C$$

(where C is a real number) is a potential function for the given vector field.

65. From $\nabla V = -\mathbf{F} = -(xy^2 + 3x^2 y)\mathbf{i} - (x^3 + yx^2)\mathbf{j}$ it follows that

$$V_x = -xy^2 - 3x^2 y \qquad \text{and} \qquad V_y = -x^3 - yx^2.$$

Integrating the first equation with respect to x, we get $V = -x^2 y^2/2 - x^3 y + C(y)$, where $C(y)$ is a real number or a function of y only. Substitute V into the equation for V_y, thus getting $-x^2 y - x^3 + C'(y) = -x^3 - x^2 y$ and $C'(y) = 0$. Therefore, $C(y) = C$ (C is a constant) and so $V(x, y) = -x^2 y^2/2 - x^3 y + C$.

2.8. Cylindrical and Spherical Coordinate Systems

1. The point $(-4, 0)$ in the xy-plane has polar coordinates $(4, \pi)$, and hence the cylindrical coordinates of $(-4, 0, 0)$ are $(4, \pi, 0)$. The cylindrical coordinates of $(0, 0, 3)$ are $(0, \theta, 3)$ for any θ, $0 \leq \theta < 2\pi$, since $(0, 0)$ is represented in polar coordinates as $(0, \theta)$ for any θ. The polar coordinates of $(0, 2)$ are $(2, \pi/2)$; so the cylindrical coordinates of $(0, 2, 4)$ are $(2, \pi/2, 4)$. Finally, the polar coordinates of $(2, -3)$ are $r = \sqrt{13} \approx 3.60555$ and $\theta = \arctan(-3/2) + 2\pi \approx 5.30039$; therefore, the point $(2, -3, -1)$ is represented as $(3.60555, 5.30039, -1)$ in cylindrical coordinates.

3. View T as the composition $T_2 \circ T_1$ of the maps $T_1(r, \theta, z) = (2r, \theta, z)$ and $T_2(r, \theta, z) = (r, \theta + \pi, z)$. The map T_1 does not change the z-coordinate, and so the height of the image of the cube by T_1 will remain the same. Since T_1 doubles the distance to the z-axis, it maps the given cube of side a onto a rectangular box whose sides are $2a$ and the height is a. The map T_2 now rotates a point in the counterclockwise sense for π rad about the z-axis. It follows that the map T maps the given cube onto a rectangular box whose sides are $2a$ and the height is a and then rotates it for π rad about the z-axis.

5. In cylindrical coordinates, $x^2 + y^2 = r^2$, and so the paraboloid $z = 4 - x^2 - y^2$ is represented as $z = 4 - r^2$. The equation of the plane $x + 2y - z = 0$ is transformed to $r\cos\theta + 2r\sin\theta - z = 0$ in cylindrical coordinates. In spherical coordinates, $x^2 + y^2 = \rho^2 \sin^2\phi$, and so $\rho\cos\phi = 4 - \rho^2 \sin^2\phi$ represents the paraboloid $z = 4 - x^2 - y^2$. The plane $x + 2y - z = 0$ is represented as $\rho\sin\phi\cos\theta + 2\rho\sin\phi\sin\theta - \rho\cos\phi = 0$, or as $\sin\phi\cos\theta + 2\sin\phi\sin\theta - \cos\phi = 0$.

7. The coordinate surface $r = C$ (C is a constant) is the cylinder $x^2 + y^2 = C$ of radius C (if $C > 0$) whose axis of rotation is the z-axis. The "coordinate surface" $r = 0$ is the z-axis. The coordinate surface $\theta = C$ is the plane perpendicular to the xy-plane that contains the

z-axis and whose intersection with the xy-plane makes the angle θ with respect to the positive x-axis. The coordinate surface $z = C$ is the plane parallel to the xy-plane, C units above it if $C \geq 0$ and $|C|$ units below it if $C < 0$.

The coordinate curve $r = C_1$, $\theta = C_2$ (C_1 and C_2 are constants) is the line perpendicular to the xy-plane, crossing it at the point with polar coordinates (C_1, C_2). The coordinate curve $r = C_1$, $z = C_2$ is the circle of radius C_1 in the plane $z = C_2$ with the center at $(0, 0, C_2)$. The coordinate curve $\theta = C_1$, $z = C_2$ is the line in the plane $z = C_2$ crossing the z-axis at $(0, 0, C_2)$. When that line is projected onto the xy-plane, it makes the angle θ with respect to the positive x-axis.

9. We must solve equations (2.38) for $\mathbf{i}$, $\mathbf{j}$ and $\mathbf{k}$:

$$\mathbf{e}_\rho = \sin\phi\cos\theta\mathbf{i} + \sin\phi\sin\theta\mathbf{j} + \cos\phi\mathbf{k} \qquad (2.38a)$$

$$\mathbf{e}_\theta = -\sin\theta\mathbf{i} + \cos\theta\mathbf{j} \qquad (2.38b)$$

$$\mathbf{e}_\phi = \cos\phi\cos\theta\mathbf{i} + \cos\phi\sin\theta\mathbf{j} - \sin\phi\mathbf{k}. \qquad (2.38c)$$

Multiply (2.38a) by $\sin\phi$, (2.38c) by $\cos\phi$ and add them up, to get

$$\sin\phi\mathbf{e}_\rho + \cos\phi\mathbf{e}_\phi = \cos\theta\mathbf{i} + \sin\theta\mathbf{j}.$$

Now multiply this identity by $\cos\theta$:

$$\cos\theta\sin\phi\mathbf{e}_\rho + \cos\theta\cos\phi\mathbf{e}_\phi = \cos^2\theta\mathbf{i} + \sin\theta\cos\theta\mathbf{j};$$

multiply (2.38b) by $-\sin\theta$:

$$-\sin\theta\mathbf{e}_\theta = \sin^2\theta\mathbf{i} - \sin\theta\cos\theta\mathbf{j},$$

and add up the two, thus obtaining

$$\mathbf{i} = \cos\theta\sin\phi\mathbf{e}_\rho + \cos\theta\cos\phi\mathbf{e}_\phi - \sin\theta\mathbf{e}_\theta.$$

From (2.38b) it follows that

$$\mathbf{j} = \frac{1}{\cos\theta}\mathbf{e}_\theta + \frac{\sin\theta}{\cos\theta}\mathbf{i} = \sin\phi\sin\theta\mathbf{e}_\rho + \cos\phi\sin\theta\mathbf{e}_\phi + \cos\theta\mathbf{e}_\theta.$$

Similarly, from either (2.38a) or (2.38c) we get that $\mathbf{k} = \cos\phi\mathbf{e}_\rho - \sin\phi\mathbf{e}_\phi$.

11. In cylindrical coordinates, using (2.34), we get

$$\mathbf{F}(r, \theta, z) = \cos\theta\mathbf{e}_r - \sin\theta\mathbf{e}_\theta + \sin\theta\mathbf{e}_r + \cos\theta\mathbf{e}_\theta = (\cos\theta + \sin\theta)\mathbf{e}_r + (\cos\theta - \sin\theta)\mathbf{e}_\theta.$$

To express F in spherical coordinates, we use formulas derived in Exercise 9, and get

$$\mathbf{F}(\rho, \theta, \phi) = \sin\phi\cos\theta\mathbf{e}_\rho - \sin\theta\mathbf{e}_\theta + \cos\phi\cos\theta\mathbf{e}_\phi + \sin\phi\sin\theta\mathbf{e}_\rho + \cos\phi\sin\theta\mathbf{e}_\phi + \cos\theta\mathbf{e}_\theta$$

$$= \sin\phi(\cos\theta + \sin\theta)\mathbf{e}_\rho + (\cos\theta - \sin\theta)\mathbf{e}_\theta + \cos\phi(\cos\theta + \sin\theta)\mathbf{e}_\phi.$$

13. In cylindrical coordinates, using (2.34), we get

$$\mathbf{F}(r, \theta, z) = r\sin\theta(\cos\theta\mathbf{e}_r - \sin\theta\mathbf{e}_\theta) - r\cos\theta(\sin\theta\mathbf{e}_r + \cos\theta\mathbf{e}_\theta) = -r\mathbf{e}_\theta.$$

To express $\mathbf{F}$ in spherical coordinates, we use formulas derived in Exercise 9, and get

$$\mathbf{F}(\rho,\theta,\phi) = \rho\sin\phi\sin\theta(\sin\phi\cos\theta\mathbf{e}_\rho - \sin\theta\mathbf{e}_\theta + \cos\phi\cos\theta\mathbf{e}_\phi)$$
$$- \rho\sin\phi\cos\theta(\sin\phi\sin\theta\mathbf{e}_\rho + \cos\theta\mathbf{e}_\theta + \cos\phi\sin\theta\mathbf{e}_\phi)$$
$$= -\rho\sin\phi\mathbf{e}_\theta.$$

15. From (2.38), using the product and the chain rules, we get

$$\frac{d\mathbf{e}_\theta}{dt} = -\cos\theta\frac{d\theta}{dt}\mathbf{i} - \sin\theta\frac{d\theta}{dt}\mathbf{j} = -(\cos\theta\mathbf{i} + \sin\theta\mathbf{j})\frac{d\theta}{dt}.$$

To express $d\mathbf{e}_\theta/dt$ in terms of unit vectors $\mathbf{e}_r$, $\mathbf{e}_\theta$ and $\mathbf{e}_\phi$ we use the formulas of Exercise 9; thus

$$\frac{d\mathbf{e}_\theta}{dt} = -\Big[\cos\theta(\sin\phi\cos\theta\mathbf{e}_\rho - \sin\theta\mathbf{e}_\theta + \cos\phi\cos\theta\mathbf{e}_\phi)$$
$$+ \sin\theta(\sin\phi\sin\theta\mathbf{e}_\rho + \cos\theta\mathbf{e}_\theta + \cos\phi\sin\theta\mathbf{e}_\phi)\Big]\frac{d\theta}{dt}$$
$$= -(\sin\phi\mathbf{e}_\rho + \cos\phi\mathbf{e}_\phi)\frac{d\theta}{dt}.$$

Similarly,

$$\frac{d\mathbf{e}_\phi}{dt} = \frac{d}{dt}(\cos\phi\cos\theta\mathbf{i} + \cos\phi\sin\theta\mathbf{j} - \sin\phi\mathbf{k})$$
$$= -\sin\phi\frac{d\phi}{dt}\cos\theta\mathbf{i} - \cos\phi\sin\theta\frac{d\theta}{dt}\mathbf{i} - \sin\phi\frac{d\phi}{dt}\sin\theta\mathbf{j} + \cos\phi\cos\theta\frac{d\theta}{dt}\mathbf{j} - \cos\phi\frac{d\phi}{dt}\mathbf{k}$$
$$= -(\sin\phi\cos\theta\mathbf{i} + \sin\phi\sin\theta\mathbf{j} + \cos\phi\mathbf{k})\frac{d\phi}{dt} + \cos\phi(-\sin\theta\mathbf{i} + \cos\theta\mathbf{j})\frac{d\theta}{dt}$$
$$= -\frac{d\phi}{dt}\mathbf{e}_\rho + \cos\phi\frac{d\theta}{dt}\mathbf{e}_\theta.$$

17. The differential of $x = r\cos\theta$ is $dx = \cos\theta dr - r\sin\theta d\theta$, and the differential of $y = r\sin\theta$ is $dy = \sin\theta dr + r\cos\theta d\theta$. Hence $ds^2 = dx^2 + dy^2 + dz^2 = (\cos\theta dr - r\sin\theta d\theta)^2 + (\sin\theta dr + r\cos\theta d\theta)^2 + dz^2 = dr^2 + r^2 d\theta^2 + dz^2$.

Another way of computing ds^2 is the following: in cylindrical coordinates, $\mathbf{r} = x\mathbf{i} + y\mathbf{j} + z\mathbf{k} = r\cos\theta\mathbf{i} + r\sin\theta\mathbf{j} + z\mathbf{k}$. The differential $d\mathbf{r}$ of $\mathbf{r}$ is

$$d\mathbf{r} = \left(\frac{\partial\mathbf{r}}{\partial r}\right)dr + \left(\frac{\partial\mathbf{r}}{\partial\theta}\right)d\theta + \left(\frac{\partial\mathbf{r}}{\partial z}\right)dz$$
$$= (\cos\theta\mathbf{i} + \sin\theta\mathbf{j})dr + (-r\sin\theta\mathbf{i} + r\cos\theta\mathbf{j})d\theta + \mathbf{k}dz$$
$$= (\cos\theta dr - r\sin\theta d\theta)\mathbf{i} + (\sin\theta dr + r\cos\theta d\theta)\mathbf{j} + dz\mathbf{k},$$

and so $d\mathbf{r}\cdot d\mathbf{r} = dr^2 + r^2 d\theta^2 + dz^2$.

The differentials in the case of spherical coordinates are computed as follows: from $x = \rho\sin\phi\cos\theta$ we get

$$dx = \sin\phi\cos\theta d\rho + \rho\cos\phi\cos\theta d\phi - \rho\sin\phi\sin\theta d\theta;$$

$y = \rho\sin\phi\sin\theta$ implies

$$dy = \sin\phi\sin\theta d\rho + \rho\cos\phi\sin\theta d\phi + \rho\sin\phi\cos\theta d\theta,$$

and $z = \rho \cos \phi$ implies

$$dz = \cos \phi d\rho - \rho \sin \phi d\phi.$$

Thus

$$ds^2 = dx^2 + dy^2 + dz^2$$

$$= (\sin \phi \cos \theta d\rho + \rho \cos \phi \cos \theta d\phi - \rho \sin \phi \sin \theta d\theta)^2$$

$$+ (\sin \phi \sin \theta d\rho + \rho \cos \phi \sin \theta d\phi + \rho \sin \phi \cos \theta d\theta)^2$$

$$+ (\cos \phi d\rho - \rho \sin \phi d\phi)^2.$$

After simplifying, we get $= d\rho^2 + \rho^2 \sin^2 \phi d\theta^2 + \rho^2 d\phi^2$.

Chapter Review

True/false Quiz

1. False. Domain and range of a vector field must be of the same dimension.

3. True. From $ax + by + cz + d = h$, $h \in \mathbb{R}$, it follows that the level surfaces are planes that all share the same normal vector (a, b, c); therefore, they are all parallel to each other.

5. False. The given function is not linear.

7. True. Since all partial derivatives of f are zero, it follows that $df = f_x dx + f_y dy + f_z dz = 0$.

9. True. Since $\mathbf{j}$ is perpendicular to the direction of the gradient, the rate of change of f in $\mathbf{j}$ direction must be zero.

11. True. Initial point is $\mathbf{c}(0) = (1, 0)$. As t starts increasing from 0, the x-coordinate of $\mathbf{c}(t)$ is positive, whereas its y coordinate is negative. Thus, from $(1, 0)$, the path moves into the fourth quadrant, so it is oriented clockwise.

13. False. For instance, take $\mathbf{c}(t) = (\cos t, \sin t)$. Its velocity $\mathbf{c}'(t) = (-\sin t, \cos t)$ is clearly not constant, but the speed is: $\|\mathbf{c}'(t)\| = \|(-\sin t, \cos t)\| = 1$.

Review Exercises and Problems

1. For instance,

$$\mathbf{G}(\mathbf{r}) = \left(\frac{1}{\|\mathbf{r} - (2, 3)\|}, \frac{1}{\|\mathbf{r} - (2, 3)\|} \right),$$

where $\mathbf{r} = (x, y)$. Since $\|\mathbf{G}(\mathbf{r})\| = \sqrt{2}/\|\mathbf{r} - (2, 3)\|$, we conclude that $\|\mathbf{G}(\mathbf{r})\|$ decreases as $\mathbf{r}$ moves away from $(2, 3)$.

3. The level curves are shown in figure below. Keeping L fixed at 1.3, to increase the production from 1.2 to 1.4, to 1.6, to 1.8 (i.e., at a constant rate) we need to increase K at an increasing rate. The same behaviour is observed when K is kept fixed.

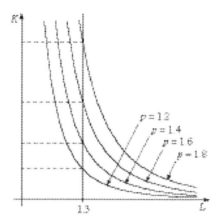

Figure 3

5. Let $f(x,y) = 1 - (x^2 + y^2 - 9)^2 = c$; i.e., $(x^2 + y^2 - 9)^2 = 1 - c$. It follows that there are no level curves if $c > 1$. If $c = 1$, then $x^2 + y^2 - 9 = 0$ and the level curve is the circle centered at the origin of radius 3. If $c < 1$, then $x^2 + y^2 = 9 \pm \sqrt{1-c}$. As long as $c > -80$, $1 - c < 81$ and $\sqrt{1-c} < 9$. Therefore, for $-80 < c < 1$, the level curve consists of two circles $x^2 + y^2 = 9 + \sqrt{1-c}$ and $x^2 + y^2 = 9 - \sqrt{1-c}$. When $c = -80$, we get $x^2 + y^2 = 9 \pm 9$; i.e., the level curve consists of a circle of radius $\sqrt{18}$ and the point $(0,0)$. Finally, when $c < -80$, the level curve is the circle $x^2 + y^2 = 9 + \sqrt{1-c}$.

7. Using $x = r\cos\theta$ and $y = r\sin\theta$ we get

$$\lim_{(x,y)\to(0,0)} f(x,y) = \lim_{r\to 0} \frac{r^5 \cos\theta \sin^4\theta}{r^2 \cos^2\theta + r^6 \sin^6\theta} = \lim_{r\to 0} \frac{r^3 \cos\theta \sin^4\theta}{\cos^2\theta + r^4 \sin^6\theta} = 0.$$

So, by defining $f(0,0) = 0$, we will make f continuous at $(0,0)$.

9. (a) The rate of change is given by

$$\frac{\partial T}{\partial x}(3/2, 2) = 30e^{-2t}\frac{\pi}{2}\cos(\pi x/2)\bigg|_{(3/2,2)} = 15\pi e^{-4}\cos(3\pi/4) \approx -0.6103.$$

When $t = 2$, we get the curve $y(x) = T(x, 2) = 30e^{-4}\sin(\pi x/2)$. It is a sine curve of period 4 and amplitude $30e^{-4} \approx 0.54947$; see Figure 9. The graph shows the temperature at points on the rod. For example, at the points $x = 0$, $x = 2$ and $x = 4$ the temperature is zero. The rod is warmest at $x = 1$ and coldest at $x = 3$. The fact that $(\partial T/\partial x)(3/2, 2) \approx -0.6103$ means that, as we move along the rod from $x = 3/2$ forward, we will experience a decrease in temperature of approximately 0.6103 degrees per unit distance.

(b) The required rate of change is

$$\frac{\partial T}{\partial t}(3/2, 2) = -60e^{-2t}\sin(\pi x/2)\bigg|_{(3/2,2)} = -60e^{-4}\sin(3\pi/4) \approx -0.77707.$$

When $x = 3/2$, we get the curve $y(t) = T(3/2, t) = 30e^{-2t}\sin(3\pi/4) = 15\sqrt{2}e^{-2t}$; see Figure 9. It is an exponentially decreasing function and represents the way the temperature changes at the location $x = 3/2$ on the rod. As $t \to \infty$, the temperature approaches 0. The fact that

$(\partial T/\partial t)(3/2, 2) \approx -0.77707$ means that, at $t = 2$, the temperature decreases at a rate of approximately 0.77707 degrees per unit time.

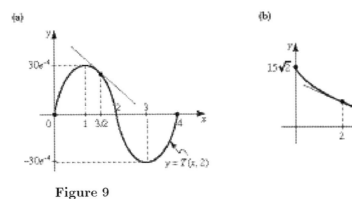

Figure 9

11. Substitute $x = y = 0$ into the assumption for f to get $|f(0,0)| \leq 0$; so $f(0,0) = 0$. Consider the partial derivative

$$\frac{\partial f}{\partial x}(0,0) = \lim_{h \to 0} \frac{f(h,0) - f(0,0)}{h} = \lim_{h \to 0} \frac{f(h,0)}{h}.$$

By assumption, $|f(h,0)| \leq h^2 + 0^2 = h^2$, and hence

$$0 \leq \frac{|f(h,0)|}{h} \leq \frac{h^2}{h} = h,$$

and therefore $|f(h,0)|/h \to 0$ as $h \to 0$. So $\lim_{h \to 0} f(h,0)/h = 0$ and therefore $(\partial f/\partial x)(0,0) = 0$. Similarly, $(\partial f/\partial y)(0,0) = 0$. To complete the proof, we must show that

$$\lim_{(x,y) \to (0,0)} \frac{|f(x,y) - f(0,0) - Df(0,0) \cdot (x,y)|}{\sqrt{x^2 + y^2}} = 0.$$

But $Df(0,0) = [\partial f/\partial x \quad \partial f/\partial y] = [0 \quad 0]$, so that the numerator is equal to $|f(x,y)|$. Using the assumption $|f(x,y)| \leq x^2 + y^2$ again, we get

$$0 \leq \frac{|f(x,y)|}{\sqrt{x^2 + y^2}} \leq \frac{x^2 + y^2}{\sqrt{x^2 + y^2}} = \sqrt{x^2 + y^2},$$

and, as $x \to 0$ and $y \to 0$, $|f(x,y)|/\sqrt{x^2 + y^2} \to 0$. It follows that f is differentiable at $(0,0)$.

13. By the chain rule,

$$\frac{\partial F}{\partial t} = \frac{\partial F}{\partial m}\frac{\partial m}{\partial t} + \frac{\partial F}{\partial r}\frac{\partial r}{\partial t} = gR^2(R+r)^{-2}\frac{\partial m}{\partial t} + mgR^2(-2)(R+r)^{-3}\frac{\partial r}{\partial t}$$

$$= \frac{gR^2}{(R+r)^2}\frac{\partial m}{\partial t} - \frac{2mgR^2}{(R+r)^3}\frac{\partial r}{\partial t}.$$

15. Since $f(c) \neq 0$, we can select an interval $(f(c) - \epsilon, f(c) + \epsilon)$, where $\epsilon > 0$, that does not contain 0. By definition of continuity of f at c, there must be a number $\delta > 0$ such that, whenever $|x - c| < \delta$, the value $f(x)$ lies within $(f(c) - \epsilon, f(c) + \epsilon)$; i.e., $f(x) \neq 0$. In other words, for any x in the interval $(c - \delta, c + \delta)$ the value $f(x) \neq 0$.

Now assume that $f: U \subseteq \mathbb{R}^m \to \mathbb{R}$ is continuous at $\mathbf{a} \in U$ and that $f(\mathbf{a}) \neq 0$. An argument very similar to the one above shows that there is an open ball $B(\mathbf{a}, r)$ centered at $\mathbf{a}$ such that $f(\mathbf{x}) \neq 0$ for all $\mathbf{x}$ in $B(\mathbf{a}, r)$.

17. Differentiating $f(x, y) = 0$ with respect to x, we get $D_1 f(x, y) + D_2 f(x, y)(dy/dx) = 0$, where $D_1 f$ and $D_2 f$ denote the derivatives of f with respect to its first and second variables respectively. Thus $dy/dx = -D_1 f(x, y)/D_2 f(x, y)$. On the other hand, differentiating $f(x, y) = 0$ with respect to t, we get

$$D_1 f(x, y) \frac{dx}{dt} + D_2 f(x, y) \frac{dy}{dt} = 0,$$

and thus

$$\frac{D_1 f(x, y)}{D_2 f(x, y)} = -\frac{dy/dt}{dx/dt}.$$

Combining the two identities, we get

$$\frac{dy}{dx} = \frac{dy/dt}{dx/dt},$$

provided, of course, that $dx/dt \neq 0$. Using the chain rule, we get

$$\frac{d^2 y}{dx^2} = \frac{d}{dx}\left(\frac{dy}{dx}\right) = \frac{d}{dt}\left(\frac{dy}{dx}\right)\frac{dt}{dx}.$$

By the quotient rule,

$$\frac{d}{dt}\left(\frac{dy}{dx}\right) = \frac{d}{dt}\left(\frac{dy/dt}{dx/dt}\right) = \frac{(d^2 y/dt^2)(dx/dt) - (dy/dt)(d^2 x/dt^2)}{(dx/dt)^2}.$$

Since $dx/dt = 1/(dt/dx)$ (by the Inverse Function Theorem), we get

$$\frac{d^2 y}{dx^2} = \frac{(d^2 y/dt^2)(dx/dt) - (dy/dt)(d^2 x/dt^2)}{(dx/dt)^3}.$$

19. Let us compute the derivative of $f(t\mathbf{x}) = t^p f(\mathbf{x})$ with respect to t. The left side is (let $\mathbf{x} = (x, y, z)$) $f(t\mathbf{x}) = f(tx, ty, tz)$, and, by the chain rule,

$$\frac{\partial}{\partial t} f(tx, ty, tz) = D_1 f(tx, ty, tz)\, x + D_2 f(tx, ty, tz)\, y + D_3 f(tx, ty, tz)\, z$$
$$= (D_1 f(tx, ty, tz), D_2 f(tx, ty, tz), D_3 f(tx, ty, tz)) \cdot (x, y, z)$$
$$= \nabla f(tx, ty, tz) \cdot \mathbf{x}.$$

The derivative on the right side is $p\, t^{p-1} f(x, y, z)$, and hence

$$\nabla f(tx, ty, tz) \cdot \mathbf{x} = p\, t^{p-1} f(x, y, z).$$

Now substitute $t = 1$ to get $\nabla f(x, y, z) \cdot \mathbf{x} = pf(x, y, z)$; i.e., $\mathbf{x} \cdot \nabla f(\mathbf{x}) = pf(\mathbf{x})$.

21. From $D_{(\mathbf{i}+\mathbf{j})/\sqrt{2}} f(\mathbf{a}) = 2$ it follows that

$$D_{(\mathbf{i}+\mathbf{j})/\sqrt{2}} f(\mathbf{a}) = \nabla f \cdot \left(\frac{\mathbf{i}+\mathbf{j}}{\sqrt{2}}\right) = \frac{1}{\sqrt{2}}\left(\nabla f \cdot \mathbf{i} + \nabla f \cdot \mathbf{j}\right)$$
$$= \frac{1}{\sqrt{2}}\left(D_{\mathbf{i}} f(\mathbf{a}) + D_{\mathbf{j}} f(\mathbf{a})\right) = \frac{1}{\sqrt{2}}\left(\frac{\partial f}{\partial x}(\mathbf{a}) + \frac{\partial f}{\partial y}(\mathbf{a})\right) = 2;$$

i.e., $\partial f/\partial x(\mathbf{a}) + \partial f/\partial y(\mathbf{a}) = 2\sqrt{2}$. Similarly,

$$D_{(\mathbf{i}-\mathbf{j})/\sqrt{2}}f(\mathbf{a}) = \tfrac{1}{\sqrt{2}}\left(\frac{\partial f}{\partial x}(\mathbf{a}) - \frac{\partial f}{\partial y}(\mathbf{a})\right) = 5,$$

and thus $\partial f/\partial x(\mathbf{a}) - \partial f/\partial y(\mathbf{a}) = 5\sqrt{2}$. Adding up the equations, we get $2\partial f/\partial x(\mathbf{a}) = 7\sqrt{2}$, and thus $\partial f/\partial x(\mathbf{a}) = 7\sqrt{2}/2$. From either equation we get $\partial f/\partial y(\mathbf{a}) = -3\sqrt{2}/2$.

3. VECTOR-VALUED FUNCTIONS OF ONE VARIABLE

3.1. World of Curves

1. The velocity of $\mathbf{c}(t)$ is computed to be

$$\mathbf{c}'(t) = \left(\frac{2 - 2t^2}{(1+t^2)^2}, \frac{-4t}{(1+t^2)^2} \right).$$

Thus, the speed is

$$\|\mathbf{c}'(t)\| = \left(\frac{4 - 8t^2 + 4t^4 + 16t^2}{(1+t^2)^4} \right)^{1/2} = \left(\frac{4(1+t^2)^2}{(1+t^2)^4} \right)^{1/2} = \frac{2}{1+t^2}.$$

Since $2/(1+t^2) \leq 2$ for all t, we conclude that the speed is largest (equal to 2) when $t = 0$.

3. We can replace t in parametrization of Exercise 3.1 by t^3, t^5, $\tan t$, etc. (i.e., with any continuous function whose range is $\mathbb{R}$). For instance, the speed of $\mathbf{c}(t) = (2t^3, 1 - t^6)/(1 + t^6)$ is $\|\mathbf{c}'(t)\| = 6t^2/(1 + t^6)$, which is clearly non-constant. Note that the image of all these parametrizations is the circle $x^2 + y^2 = 1$ without the point $(0, -1)$.

The parametrization $\mathbf{c}(t) = (\cos t^3, \sin t^3)$, $t \in [0, \sqrt[3]{2\pi}]$, has non-constant speed, computed to be $\|\mathbf{c}'(t)\| = 3t^2$. Its image is all of the circle $x^2 + y^2 = 1$.

More generally, we can use $\mathbf{c}(t) = (\cos f(t), \sin f(t))$, where the range of $f(t)$ contains $[0, 2\pi]$ (or any interval of length 2π). Since $\mathbf{c}'(t) = (-f'(t) \sin f(t), f'(t) \cos f(t))$, it follows that $\|\mathbf{c}'(t)\| = |f'(t)|$. Thus, to ensure that the speed is not constant, we must also require that $f(t)$ is differentiable and $f'(t) \neq 0$ for some t in its domain.

Note: there are many possible parametrizations.

5. A straightforward calculation shows that

$$x'(t) = -3a_x(1-t)^2 - 6b_x(1-t)t + 3b_x(1-t)^2 - 3c_xt^2 + 6c_x(1-t)t + 3d_xt^2.$$

The expression for $y'(t)$ is obtained by replacing a_x with a_y, b_x with b_y, etc. It follows that $x'(0) = -3a_x + 3b_x$ and $y'(0) = -3a_y + 3b_y$, and thus $\mathbf{c}'(0) = 3(b_x - a_x, b_y - a_y)$. Likewise, $x'(1) = -3c_x + 3d_x$ $y'(1) = -3c_y + 3d_y$ and $\mathbf{c}'(1) = 3(d_x - c_x, d_y - c_y)$.

7. (a) From the fact that $-1 \leq \cos 2t, \sin 3t \leq 1$ for all t, we conclude that the image of $\mathbf{c}(t)$ is contained in the square in the xy-plane defined by $x = -1$, $x = 1$, $y = -1$ and $y = 1$.

Solving $\cos 2t = 1$, $t \in [0, 2\pi]$, we get that $t = 0$, $t = \pi$, or $t = 2\pi$. Thus, the curve touches the vertical line $x = 1$ at the point $\mathbf{c}(0) = \mathbf{c}(\pi) = \mathbf{c}(2\pi) = (1, 0)$. Likewise, $\cos 2t = -1$ implies that $t = \pi/2$ or $t = 3\pi/2$. The curve $\mathbf{c}$ touches the line $x = -1$ at $\mathbf{c}(\pi/2) = (-1, -1)$ and $\mathbf{c}(3\pi/2) = (-1, 1)$.

Similarly, from $\sin 3t = 1$, $t \in [0, 2\pi]$, we get $t = \pi/6$ and $t = 9\pi/6 = 3\pi/2$. Thus, $\mathbf{c}$ touches $y = 1$ at $\mathbf{c}(\pi/6) = (1/2, 1)$ and $\mathbf{c}(3\pi/2) = (-1, 1)$. Finally, $\sin 3t = -1$ implies $t = \pi/2$ and $t = 11\pi/6$, so the given curve touches $y = -1$ at the points $\mathbf{c}(\pi/2) = (-1, -1)$ and $\mathbf{c}(11\pi/6) = (1/2, -1)$.

(b) See Figure 7. Initial point is $\mathbf{c}(0) = (1,0)$, and the terminal point is $\mathbf{c}(2\pi) = (1,0)$. The object traveling along $\mathbf{c}$ is initially at $(1,0)$, then moves in the first quadrant towards the point $\mathbf{c}(\pi/6) = (1/2,1)$. Then it reaches $(-1,1)$ when $t = \pi/2$, and returns to $(1,0)$ when $t = \pi$. From there, the object moves through first and second quadrants to $\mathbf{c}(3\pi/2) = (-1,1)$, reaches $(1/2,-1)$ when $t = 11\pi/6$ and returns to $(1,0)$ when $t = 2\pi$.

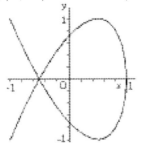

Figure 7

9. (a) Since $\sin 4t = 2\sin 2t \cos 2t$, we get that $\sin^2 4t = 2(1 - \cos^2 2t)\cos^2 2t$, i.e., $y^2 = 4(1-x^2)x^2$. Let $F(x,y) = y^2 - 4x^2(1-x^2)$, $-1 \leq x, y \leq 1$. Then $F(x,y) = 0$ is an implicit equation for $\mathbf{c}(t)$.

Since replacing x by $-x$ or y by $-y$ does not change $F(x,y)$, we conclude that the image of $F(x,y) = 0$ (i.e., the curve $\mathbf{c}$) is symmetric with respect to both x-axis and y-axis.

(b) From $y^2 - 4x^2(1-x^2) = 0$ we get $y = \pm\sqrt{4x^2(1-x^2)}$. Since $(\sqrt{2}/2, 1)$ is in the first quadrant, both x and y are positive, and thus $y = g(x) = 2x\sqrt{1-x^2}$ is a local solution near $(\sqrt{2}/2, 1)$.

(c) The local solution $g(x)$ from (b) is the part of the curve in Figure 9 in the first quadrant.

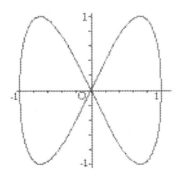

Figure 9

11. We are actually asked to solve the system $x + y - z = 2$ and $2x - 5y + z = 3$ for x, y and z. There are 2 equations and three unknowns, so we are allowed to choose one value — say, $x = t$, $t \in \mathbb{R}$ (in other words, t is a parameter). Adding $y - z = 2 - t$ and $-5y + z = 3 - 2t$ we get $-4y = 5 - 3t$; i.e., $y = -\frac{5}{4} + \frac{3}{4}t$. From $y - z = 2 - t$ it follows that $z = y - 2 + t = -\frac{5}{4} + \frac{3}{4}t - 2 + t = -\frac{13}{4} + \frac{7}{4}t$. So one possible parametrization is $\mathbf{c}(t) = (t, -\frac{5}{4} + \frac{3}{4}t, -\frac{13}{4} + \frac{7}{4}t)$, $t \in \mathbb{R}$.

13. Let $\mathbf{c}(t) = (x(t), y(t), z(t))$ (from now on, drop t from the notation). To satisfy $(x+2)^2 + (z-2)^2 = 4$, we choose $x = 2\cos t - 2$ and $z = 2\sin t + 2$. Since $y = 3$ (that is given), it follows that

$$\mathbf{c}(t) = (2\cos t - 2, 3, 2\sin t + 2), \qquad t \in [0, 2\pi],$$

is one possible parametrization.

Here is an alternative argument: the given cylinder intersects the xz-plane along the curve $(x+2)^2 + (z-2)^2 = 4$, which is the circle of radius 2 centered at $(-2, 0, 2)$. It can be parametrized as $\overline{\mathbf{c}}(t) = (-2, 0, 2) + (2\cos t, 0, 2\sin t) = (-2 + 2\cos t, 0, 2 + 2\sin t)$, $t \in [0, 2\pi]$. Thus, the intersection of the cylinder and the plane $y = 3$ can be described parametrically as $\mathbf{c}(t) = (-2 + 2\cos t, 3, 2 + 2\sin t)$, $t \in [0, 2\pi]$.

15. Choose $x = \sqrt{2}\sin t$, $z = \sqrt{2}\cos t$, so that $x^2 + z^2 = 2$. Then $y = \pm\sqrt{2}\sin t$ (so that $y^2 + z^2 = 2$) and a parametrization is $\mathbf{c}(t) = (\sqrt{2}\sin t, \pm\sqrt{2}\sin t, \sqrt{2}\cos t)$, where $t \in [0, 2\pi]$.

17. (a) Normal vectors to the given planes are $\mathbf{N}_1 = (1, 2, -1)$ and $\mathbf{N}_2 = (2, -1, -1)$. Their cross product is

$$\mathbf{N}_1 \times \mathbf{N}_2 = \begin{vmatrix} \mathbf{i} & \mathbf{j} & \mathbf{k} \\ 1 & 2 & -1 \\ 2 & -1 & -1 \end{vmatrix} = -3\mathbf{i} - \mathbf{j} - 5\mathbf{k}.$$

Since $\mathbf{l}$ belongs to both planes, it must be perpendicular to both $\mathbf{N}_1$ and $\mathbf{N}_2$.

(b) Take, for instance, $z = 0$. Solving $x + 2y = 4$ and $2x - y = 3$ we get $x = 2$ and $y = 1$. Thus, the point $(2, 1, 0)$ belongs to the intersection of the two planes.

(c) A parametric equation of $\mathbf{l}$ is (take $-\mathbf{N}_1 \times \mathbf{N}_2$ as direction vector)

$$\mathbf{l}(t) = (2, 1, 0) + t(3, 1, 5),$$

where $t \in \mathbb{R}$.

Different choice of a point leads to a different parametrization. Substituting, for instance, $x = 0$ into the given equations, we get $2y - z = 4$ and $y + z = -3$. It follows that $y = 1/3$ and $z = -10/3$, and the parametric equation is $\mathbf{l}(t) = (0, 1/3, -10/3) + t(3, 1, 5)$, $t \in \mathbb{R}$.

19. By the Implicit Function Theorem,

$$g'(x) = -\frac{\partial F / \partial x}{\partial F / \partial y} = -\frac{3x^2 - 3y}{3y^2 - 3x} = \frac{y - x^2}{y^2 - x}.$$

Thus, $g'(1) = (1.5066 - 1)/(1.5066^2 - 1) = 0.3989$. Since $g(1) = 1.0566$, we compute the linear approximation of $g(x)$ at $x = 1$ to be

$$L_1(x) = g(1) + g'(1)(x - 1) = 1.5066 + 0.3989(x - 1) = 0.3989x + 1.1077.$$

21. From $\nabla F(2, 4) = (3, -3)$ it follows that $(\partial F/\partial x)(2, 4) = 3$ and $(\partial F/\partial y)(2, 4) = -3$. From $F(2, 4) = 0$ we read $g(2) = 4$. By the Implicit Function Theorem,

$$g'(2) = -\left.\frac{\partial F/\partial x}{\partial F/\partial y}\right|_{(2,4)} = -\frac{3}{-3} = 1.$$

The linear approximation of $g(x)$ at $x = 2$ is $L_2(x) = g(2) + g'(2)(x - 2) = 4 + 1(x - 2)$, i.e., $L_2(x) = x + 2$.

23. (a) Let $F(x, y) = (x - 1)^2 + (y + 4)^2$. Then $F(x, y) = 0$ implies that $x = 1$ and $y = -4$.

We could replace the squares with any positive, even exponents (not necessarily equal). There are other examples, for instance $F(x, y) = (x - 1)^2 + e^{(y+4)^2} - 1 = x^2 - 2x + e^{(y+4)^2}$.

(b) Let $F(x, y) = (x + y)^2 - 1$. Then $F(x, y) = 0$ implies that $(x + y)^2 = 1$, i.e.. $x + y = \pm 1$. So, the set $F(x, y) = 0$ in this case consists of parallel lines and $y = -x + 1$ and and $y = -x - 1$.

More generally, we can use $F(x, y) = (mx + ny)^2 - p = 0$, where $p > 0$ and at least one of m or n is not zero.

(c) Let $F(x, y) = (y - x)y$. Then $F(x, y) = 0$ implies that $y - x = 0$ and $y = 0$. The lines $y = x$ and $y = 0$ intesect at $\pi/4$ radians.

(d) Take, for instance, $F(x, y) = e^{x+y}$. The equation $F(x, y) = e^{x+y} = 0$ has no solutions for x and y.

25. A straightforward calculation shows that

$$x^3 + y^3 - 3xy = \frac{27t^3}{(1+t^3)^3} + \frac{27t^6}{(1+t^3)^3} - 3\frac{9t^3}{(1+t^3)^2} = \frac{27t^3(1+t^3)}{(1+t^3)^3} - \frac{27t^3}{(1+t^3)^2} = 0.$$

27. The angle θ between the line $\mathbf{l}(t)$ and the spiral $\mathbf{c}(t)$ is given by

$$\cos\theta = \frac{\mathbf{c}(t_0) \cdot \mathbf{c}'(t_0)}{\|\mathbf{c}(t_0)\|\,\|\mathbf{c}'(t_0)\|}.$$

From $\mathbf{c}(t_0) = (e^{at_0}\cos t_0, e^{at_0}\sin t_0)$ and

$$\mathbf{c}'(t_0) = (ae^{at_0}\cos t_0 - e^{at_0}\sin t_0, ae^{at_0}\sin t_0 + e^{at_0}\cos t_0)$$

we compute the lengths $\|\mathbf{c}(t_0)\| = e^{at_0}$ and $\|\mathbf{c}'(t_0)\| = \sqrt{a^2e^{2at_0} + e^{2at_0}} = e^{at_0}\sqrt{a^2 + 1}$. Moreover, $\mathbf{c}(t_0) \cdot \mathbf{c}'(t_0) = ae^{2at_0}$, and thus

$$\cos\theta = \frac{ae^{2at_0}}{e^{at_0}e^{at_0}\sqrt{a^2 + 1}} = \frac{a}{\sqrt{a^2 + 1}}.$$

Consequently, the angles that the spiral makes with rays emanating from the origin are all equal.

29. The length $\|\mathbf{c}(t)\|$ represents the distance from the origin to the point $\mathbf{c}(t)$ on the spiral. If $\mathbf{c}(t) = (f(t)\cos t, f(t)\sin t)$, then $\|\mathbf{c}(t)\| = |f(t)|$.

Since $f(t) = t$, $(t \geq 0)$, $f(t) = t^2$ $(t \geq 0)$, $f(t) = \ln t$ $(t > 1)$ all increase, the corresponding spirals will spiral outward. The spiral defined by $f(t) = 1/t$ will spiral inward, since $1/t$ $(t > 0)$ is a decreasing function.

Pick a ray emanating from the origin. The spiral defined by $f(t) = t$ will cross the ray at equidistant points (i.e., that are spaced at equal distance from each other). The windings of the spiral defined by t^2 will move further away from each other as t increases (so, the spiral opens faster and faster), whereas the windings of the spiral defined by $\ln t$ will get closer and closer to each other as t increases. The windings of the spiral defined by $1/t$ will get closer and closer as t increases. The spiral curls around the origin.

Note that the $f(t)$ part of $\mathbf{c}(t) = (f(t)\cos t, f(t)\sin t)$ contributes to the size, but not to the orientation of the spiral. All spirals are oriented counterclockwise.

31. The object moves along the helix on a cylinder of radius 1. The requirement that the trajectory makes the angle of $\pi/4$ radians with respect to horizontal planes implies that the pitch is equal to 2π (since it must be equal to the circumference of the circle of radius equal to the radius of the cylinder). Thus, $\mathbf{c}_1(t) = (\cos t, \sin t, t)$, $t \geq 0$, is a parametrization of the curve along which the object moves. Since $\|\mathbf{c}_1'(t)\| = \sqrt{2}$, $\mathbf{c}_1(t)$ does not have the required speed. To get a path with the speed of 3 units/second, we consider $\mathbf{c}(t) = (\cos \alpha t, \sin \alpha t, t)$, and compute α from the requirement that $\|\mathbf{c}'(t)\| = 3$:

$$\|\mathbf{c}'(t)\| = \|(-\alpha \sin \alpha t, \alpha \cos \alpha t, 1)\| = \sqrt{\alpha^2 + 1} = 3.$$

Thus, $\alpha = \pm\sqrt{8}$, and we pick $\alpha = +\sqrt{8}$ to satisfy the orientation requirement. It follows that the trajectory of the object is

$$\mathbf{c}(t) = (\cos \sqrt{8}t, \sin \sqrt{8}t, t), \quad t \geq 0;$$

the position of the object after 12 seconds is $\mathbf{c}(12) = (-0.8160, 0.5781, 12)$.

33. We need to find a and b in $\mathbf{c}(t) = (t, a\cosh bt)$. Assuming that the lowest point is obtained for $t = 0$, we get $\mathbf{c}(0) = (0, a) = (0, 10)$, so, $a = 10$. Let t be the horizontal distance. Then, from $\mathbf{c}(50) = (50, 10\cosh 50b) = (50, 25)$ we get that $\cosh 50b = 2.5$. Thus, (using formula $\cosh^{-1} x = \ln\left(x + \sqrt{x^2 - 1}\right)$) we get that $b = \frac{1}{50}\ln\left(2.5 + \sqrt{2.5^2 - 1}\right) = 0.031336$. The required parametrization is $\mathbf{c}(t) = (t, 10\cosh 0.031336t)$, where $t \in [-50, 50]$.

3.2. Tangents, Velocity and Acceleration

1. Recall that the cycloid can be parametrized as $\mathbf{c}(\theta) = (\theta - \sin\theta, 1 - \cos\theta)$, where $\theta \in \mathbb{R}$. The velocity is $\mathbf{c}'(\theta) = (1 - \cos\theta, \sin\theta)$, and the speed is $\|\mathbf{c}'(\theta)\| = \sqrt{2 - 2\cos\theta} = \sqrt{2}\sqrt{1 - \cos\theta}$. It follows that the speed is largest (and equal to 2) when $\cos\theta = -1$, i.e., when $\theta = \pi + 2\pi k$ (k is an integer). In words, the speed is largest at the highest points on the cycloid (y-coordinate of $\mathbf{c}(\theta)$ reaches its maximum of 2 when $\theta = \pi + 2\pi k$.)

3. From $\mathbf{c}(t) = (a\cos t, a\sin t, bt)$ we get $\mathbf{c}'(t) = (-a\sin t, a\cos t, b)$, and thus $\|\mathbf{c}'(t)\| = \sqrt{a^2 + b^2}$. Thus, the speed is constant. Since $\mathbf{c}''(t) = (-a\cos t, -a\sin t, 0)$, it follows that $\mathbf{c}'(t) \cdot \mathbf{c}''(t) = 0$. In words, at every point on the helix the velocity vector is perpendicular to the acceleration.

5. The velocity is

$$\mathbf{v}(t) = \int \mathbf{a}(t)\, dt = \int (-1, 1, 0)\, dt = (-t + C_1, t + C_2, C_3).$$

Since $\mathbf{v}(0) = (1, 2, 0)$ it follows that $(C_1, C_2, C_3) = (1, 2, 0)$; i.e., $C_1 = 1$, $C_2 = 2$ and $C_3 = 0$, and hence $\mathbf{v}(t) = (-t + 1, t + 2, 0)$. Now

$$\mathbf{c}(t) = \int \mathbf{v}(t)\, dt = \int (-t + 1, t + 2, 0)\, dt = \left(-\tfrac{1}{2}t^2 + t + D_1, \tfrac{1}{2}t^2 + 2t + D_2, D_3\right).$$

Since $\mathbf{c}(0) = (0, 2, 0)$ it follows that $(D_1, D_2, D_3) = (0, 2, 0)$; i.e., $D_1 = 0$, $D_2 = 2$ and $D_3 = 0$. Therefore, $\mathbf{c}(t) = (t - t^2/2, 2t + 2 + t^2/2, 0)$.

7. The velocity is $\mathbf{v}(t) = \int \mathbf{a}(t)\, dt = (t^2/2 + C_1, t + C_2, t + C_3)$. Since $\mathbf{v}(0) = (0, 1, 0)$ it follows that $(C_1, C_2, C_3) = (0, 1, 0)$; i.e., $C_1 = 0$, $C_2 = 1$ and $C_3 = 0$, so $\mathbf{v}(t) = (t^2/2, t + 1, t)$. Integrating again, we get

$$\mathbf{c}(t) = \int \mathbf{v}(t)\, dt = \left(\tfrac{1}{6}t^3 + D_1, \tfrac{1}{2}t^2 + t + D_2, \tfrac{1}{2}t^2 + D_3\right).$$

Since $\mathbf{c}(0) = (-2, 0, 3)$ it follows that $(D_1, D_2, D_3) = (-2, 0, 3)$; i.e., $D_1 = -2$, $D_2 = 0$ and $D_3 = 3$. Therefore, $\mathbf{c}(t) = (t^3/6 - 2, t^2/2 + t, t^2/2 + 3)$.

9. The velocity is

$$\mathbf{v}(t) = \int \mathbf{a}(t)\, dt = \int (t\mathbf{i} + t^2\mathbf{j} + t\mathbf{k})\, dt = (\tfrac{1}{2}t^2 + C_1)\mathbf{i} + (\tfrac{1}{3}t^3 + C_2)\mathbf{j} + (\tfrac{1}{2}t^2 + C_3)\mathbf{k}.$$

Since $\mathbf{v}(0) = 2\mathbf{j} - 3\mathbf{k}$ it follows that $C_1\mathbf{i} + C_2\mathbf{j} + C_3\mathbf{k} = 2\mathbf{j} - 3\mathbf{k}$; i.e., $C_1 = 0$, $C_2 = 2$ and $C_3 = -3$, and hence $\mathbf{v}(t) = (t^2/2)\mathbf{i} + (2 + t^3/3)\mathbf{j} + (-3 + t^2/2)\mathbf{k}$. Now

$$\mathbf{c}(t) = \int \mathbf{v}(t)\, dt = \left(\tfrac{1}{6}t^3 + D_1\right)\mathbf{i} + \left(2t + \tfrac{1}{12}t^4 + D_2\right)\mathbf{j} + \left(-3t + \tfrac{1}{6}t^3 + D_3\right)\mathbf{k}.$$

Since $\mathbf{c}(0) = 4\mathbf{i} + 2\mathbf{j} - 6\mathbf{k}$, it follows that $D_1\mathbf{i} + D_2\mathbf{j} + D_3\mathbf{k} = 4\mathbf{i} + 2\mathbf{j} - 6\mathbf{k}$; i.e., $D_1 = 4$, $D_2 = 2$ and $D_3 = -6$. So, $\mathbf{c}(t) = (4 + t^3/6)\mathbf{i} + (2 + 2t + t^4/12)\mathbf{j} + (-6 - 3t + t^3/6)\mathbf{k}$.

11. The velocity of the particle is $\mathbf{c}'(t) = (-t^{-2}, 0, 2t)$. We have to find the maximum value of its speed $s(t) = \sqrt{t^{-4} + 4t^2}$ on $[1, 4]$. Recall that the maximum of $s(t)$ on $[1, 4]$ will occur either at the endpoints $t = 1$ and $t = 4$ or at the critical points inside $[1, 4]$. Now $s'(t) = 0$ implies that $(t^{-4} + 4t^2)^{-1/2}(-4t^{-5} + 8t) = 0$ and hence $-4t^{-5} + 8t = 0$ and $t = (0.5)^{1/6}$ or $t = 0$. It follows that there are no critical points inside $[1, 4]$. From $s(1) = \sqrt{5}$ and $s(4) = \sqrt{4^{-4} + 64} \approx 8.0002$ it follows that, on the interval $[1, 4]$, the particle reaches its maximum speed of approximately 8.0002 when $t = 4$.

13. To find the highest position of the particle, we have to maximize the z-coordinate of the position vector. From $\mathbf{v}(t) = \int \mathbf{a}(t)\,dt = \int (3,0,1)\,dt = (3t+C_1, C_2, t+C_3)$ and $\mathbf{v}(0) = (1,3,2)$ it follows that $(C_1, C_2, C_3) = (1,3,2)$, i.e., $\mathbf{v}(t) = (3t+1, 3, t+2)$. Now

$$\mathbf{c}(t) = \int \mathbf{v}(t)\,dt = \int (3t+1, 3, t+2)\,dt = \left(\tfrac{3}{2}t^2 + t + D_1,\ 3t + D_2,\ \tfrac{1}{2}t^2 + 2t + D_3\right).$$

Since $\mathbf{c}(0) = (0,0,0)$, it follows that $D_1 = D_2 = D_3 = 0$ and so $\mathbf{c}(t) = (3t^2/2+t, 3t, t^2/2+2t)$. The height $h(t)$ of the particle is given by $h(t) = t^2/2+2t$. Since $h'(t) = t+2 > 0$ when $t \geq 0$, it follows that $h(t)$ is increasing for $t \geq 0$. Hence its maximum on $[0, 12]$ occurs when $t = 12$.

15. Choose a coordinate system so that the y-axis represents the height and the x-axis the horizontal distance. Since the acceleration is $\mathbf{a}(t) = -g\mathbf{j}$, it follows that $\mathbf{v}(t) = \int \mathbf{a}(t)\,dt = C_1\mathbf{i} + (-gt + C_2)\mathbf{j}$. The initial condition on the velocity is $\|\mathbf{v}(0)\| = 700$ and the angle of $\mathbf{v}(0)$ with respect to the x-axis is $\theta = \pi/3$. Hence $\mathbf{v}(0) = \|\mathbf{v}(0)\|(\cos\theta\mathbf{i} + \sin\theta\mathbf{j}) = 350\mathbf{i} + 350\sqrt{3}\mathbf{j}$. From the formula for $\mathbf{v}(t)$, we get $\mathbf{v}(0) = C_1\mathbf{i} + C_2\mathbf{j}$; so $C_1 = 350$ and $C_2 = 350\sqrt{3}$ and $\mathbf{v}(t) = 350\mathbf{i} + (-gt + 350\sqrt{3})\mathbf{j}$.

Now $\mathbf{c}(t) = (350t + D_1)\mathbf{i} + (-gt^2/2 + 350\sqrt{3}\,t + D_2)\mathbf{j}$, and since $\mathbf{c}(0) = \mathbf{0}$, we get $\mathbf{c}(t) = 350t\mathbf{i} + (-gt^2/2 + 350\sqrt{3}t)\mathbf{j}$. The highest point corresponds to the maximum value of the y-coordinate $y(t) = -gt^2/2 + 350\sqrt{3}\,t$. Since $y'(t) = 350\sqrt{3} - gt = 0$, it follows that $t = 350\sqrt{3}/g$; assuming that $g \approx 9.80$ m/s^2, we get $t \approx 61.86$ s (and this gives a maximum, since $y''(t) = -g < 0$). The maximum height reached is $y(61.86) \approx 18,750$ m. The projectile will hit the ground when $y(t) = -gt^2/2 + 350\sqrt{3}t = 0$; i.e., when $t = 700\sqrt{3}/g \approx 123.72$ s (i.e., double the time needed for the projectile to reach its highest position). The range of the projectile is the value of the x-coordinate at the moment of impact, $x(123.72) = (350)(123.72) \approx 43,302$ m. Finally, the speed at time of impact is $\|\mathbf{v}(123.72)\| = \|350\mathbf{i} - 606.24\mathbf{j}\| = 700$ m/s.

17. We could use the implicit derivative to compute the slope of the tangent. Alternatively, rewrite the equation of the ellipse as $x^2/3 + y^2/(3/4) = 1$, and parametrize it using $\mathbf{c}(t) = (\sqrt{3}\cos t, \sqrt{3/4}\sin t) = (\sqrt{3}\cos t, (\sqrt{3}/2)\sin t)$, $t \in [0, 2\pi]$ (in this case, $x = \sqrt{3}$ is obtained by choosing $t = 0$). It follows that $\mathbf{c}'(0) = (-\sqrt{3}\sin t, (\sqrt{3}/2)\cos t)\big|_{t=0} = (0, \sqrt{3}/2)$; consequently, the tangent at $\mathbf{c}(0) = (\sqrt{3}, 0)$ is vertical. Its parametric equation is (we can take any vector parallel to $\mathbf{c}'(0)$, so take $(0,1)$) $\ell(t) = (\sqrt{3}, 0) + t(0, 1) = (\sqrt{3}, t)$, $t \in \mathbb{R}$.

19. The point $(0, 3, 2\pi)$ is obtained for $t = \pi/2$; so, to get the tangent, we need $\mathbf{c}'(\pi/2)$. From $\mathbf{c}'(t) = (-3\sin t, 3\cos t, 4)$ it follows that $\mathbf{c}'(\pi/2) = (-3, 0, 4)$ and hence

$$\ell(t) = (0, 3, 2\pi) + t(-3, 0, 4) = (-3t, 3, 2\pi + 4t),$$

where $t \in \mathbb{R}$, is a parametrization of the desired tangent line.

21. Since $\cosh 0 = 1$ and $\sinh 0 = 0$, it follows that the point of tangency is $\mathbf{c}(0)$. From $\mathbf{c}'(0) = (-\sinh t, \cosh t)\big|_{t=0} = (0, 1)$, it follows that the tangent is $\ell(t) = (-1, 1) + t(0, 1) = (-1, 1+t)$, $t \in \mathbb{R}$.

23. Using the product rule and the chain rule we get

$$\mathbf{c}' = (r'\cos\theta + r\theta'(-\sin\theta), r'\sin\theta + r\theta'\cos\theta) = r'(\cos\theta, \sin\theta) + r\theta'(-\sin\theta, \cos\theta).$$

Similarly,

$$\begin{aligned}
\mathbf{c}'' &= r''(\cos\theta, \sin\theta) + r'(-\theta'\sin\theta, \theta'\cos\theta) + r'\theta'(-\sin\theta, \cos\theta) \\
&\quad + r\theta''(-\sin\theta, \cos\theta) + r\theta'(-\theta'\cos\theta, -\theta'\sin\theta) \\
&= r''(\cos\theta, \sin\theta) + 2r'\theta'(-\sin\theta, \cos\theta) + r\theta''(-\sin\theta, \cos\theta) - r(\theta')^2(\cos\theta, \sin\theta) \\
&= (r'' - r(\theta')^2)(\cos\theta, \sin\theta) + (2r'\theta' + r\theta'')(-\sin\theta, \cos\theta).
\end{aligned}$$

25. The velocity is

$$\mathbf{c}'(t) = (-e^{-t}\cos t - e^{-t}\sin t, -e^{-t}\sin t + e^{-t}\cos t) = -e^{-t}(\cos t + \sin t, \sin t - \cos t).$$

The velocity is horizontal when its y component $\sin t - \cos t$ is zero. Hence $\sin t = \cos t$; i.e., $\tan t = 1$ and $t = \pi/4$, $t = 5\pi/4$ and $t = 9\pi/4$. The velocity is vertical when the x component of $\mathbf{c}'(t) = 0$; i.e., when $\cos t + \sin t = 0$. In that case, $\tan t = -1$ and $t = 3\pi/4$, $t = 7\pi/4$ and $t = 11\pi/4$. It follows that the velocity is horizontal at $\mathbf{c}(\pi/4) = e^{-\pi/4}(\sqrt{2}/2, \sqrt{2}/2)$, $\mathbf{c}(5\pi/4) = -e^{-5\pi/4}(\sqrt{2}/2, \sqrt{2}/2)$ and $\mathbf{c}(9\pi/4) = e^{-9\pi/4}(\sqrt{2}/2, \sqrt{2}/2)$. It is vertical at the points $\mathbf{c}(3\pi/4) = e^{-3\pi/4}(-\sqrt{2}/2, \sqrt{2}/2)$, $\mathbf{c}(7\pi/4) = e^{-7\pi/4}(\sqrt{2}/2, -\sqrt{2}/2)$ and $\mathbf{c}(11\pi/4) = e^{-11\pi/4}(-\sqrt{2}/2, \sqrt{2}/2)$.

27. The composition $\mathbf{d}(t) = \mathbf{F}(\mathbf{c}(t))$ is the image of the curve $\mathbf{c}$ under $\mathbf{F}$; we need its tangent. By the chain rule, $\mathbf{d}'(t) = D\mathbf{F}(\mathbf{c}(t)) \cdot \mathbf{c}'(t)$ and hence

$$\begin{aligned}
\mathbf{d}'(0) &= D\mathbf{F}(\mathbf{c}(0)) \cdot \mathbf{c}'(0) = D\mathbf{F}(0,0) \cdot \mathbf{c}'(0) \\
&= \begin{bmatrix} 2xy - 3x^2 & x^2 \\ ye^x & e^x \end{bmatrix}_{at\ (0,0)} \cdot \begin{bmatrix} \cos t \\ 2t - 1 \end{bmatrix}_{at\ t=0} = \begin{bmatrix} 0 & 0 \\ 0 & 1 \end{bmatrix} \cdot \begin{bmatrix} 1 \\ -1 \end{bmatrix} = \begin{bmatrix} 0 \\ -1 \end{bmatrix}.
\end{aligned}$$

In words, $\mathbf{d}'(0) = -\mathbf{j}$ is the tangent to the image of $\mathbf{c}$ under $\mathbf{F}$ at $t = 0$.

29. The image $\mathbf{d}(t)$ of $\mathbf{c}(t)$ under $\mathbf{F}$ is given by the composition $\mathbf{d}(t) = \mathbf{F}(\mathbf{c}(t))$, and the corresponding tangents are related by $\mathbf{d}'(t) = D\mathbf{F}(\mathbf{c}(t)) \cdot \mathbf{c}'(t)$.

We have to compute $D\mathbf{F}$. Let $A = \begin{bmatrix} a_{11} & a_{12} \\ a_{21} & a_{22} \end{bmatrix}$ and $\mathbf{x} = \begin{bmatrix} x \\ y \end{bmatrix}$. Then

$$\mathbf{F}(\mathbf{x}) = \begin{bmatrix} a_{11} & a_{12} \\ a_{21} & a_{22} \end{bmatrix} \cdot \begin{bmatrix} x \\ y \end{bmatrix} = \begin{bmatrix} a_{11}x + a_{12}y \\ a_{21}x + a_{22}y \end{bmatrix}$$

and it follows that $D\mathbf{F} = A$. Hence

$$\mathbf{d}'(0) = D\mathbf{F}(\mathbf{c}(0)) \cdot \mathbf{c}'(0) = \begin{bmatrix} a_{11} & a_{12} \\ a_{21} & a_{22} \end{bmatrix}_{at\ (0,0)} \cdot \begin{bmatrix} c_1 \\ c_2 \end{bmatrix} = \begin{bmatrix} a_{11}c_1 + a_{12}c_2 \\ a_{21}c_1 + a_{22}c_2 \end{bmatrix},$$

i.e., $\mathbf{d}'(0) = (a_{11}c_1 + a_{12}c_2)\mathbf{i} + (a_{21}c_1 + a_{22}c_2)\mathbf{j}$.

3.3. Length of a Curve

1. The polygonal path p_5 is obtained by joining $\mathbf{c}(1)$, $\mathbf{c}(1.2)$, $\mathbf{c}(1.4)$, $\mathbf{c}(1.6)$, $\mathbf{c}(1.8)$ and $\mathbf{c}(2)$ with straight line segments; see Figure 1. From $\mathbf{c}'(t) = (1, -t^{-2})$ it follows that $\|\mathbf{c}'(t)\| = \sqrt{1 + t^{-4}}$ and hence, using (3.13),

$$\ell(p_5) \approx \|\mathbf{c}'(1)\|\Delta t_1 + \|\mathbf{c}'(1.2)\|\Delta t_2 + \|\mathbf{c}'(1.4)\|\Delta t_3 + \|\mathbf{c}'(1.6)\|\Delta t_4 + \|\mathbf{c}'(1.8)\|\Delta t_5.$$

Since $\Delta t_1 = \ldots = \Delta t_5 = 0.2$, we get

$$\ell(p_5) \approx (1.4142)(0.2) + (1.2175)(0.2) + (1.1226)(0.2)$$
$$+ (1.0736)(0.2) + (1.0465)(0.2) = 1.17488.$$

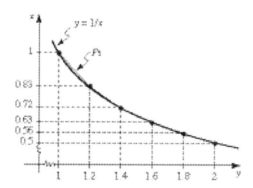

Figure 1

3. (a) Since $\mathbf{c}(t_{i+1}) = (x(t_{i+1}), y(t_{i+1}))$ and $\mathbf{c}(t_i) = (x(t_i), y(t_i))$, we get (using the formula for the distance between two points)

$$\ell(c_i) = \|\mathbf{c}(t_{i+1}) - \mathbf{c}(t_i)\| = \sqrt{\left(x(t_{i+1}) - x(t_i)\right)^2 + \left(y(t_{i+1}) - y(t_i)\right)^2}.$$

(b) Recall the Mean Value Theorem: if $f(t)$ is differentiable on (a,b) and continuous on $[a,b]$, then there exists a number $t_0 \in (a,b)$ such that $(f(b) - f(a))/(b - a) = f'(t_0)$, i.e., $f(b) - f(a) = f'(t_0)(b - a)$.

Applying the Mean Value Theorem to the functions $x(t)$ and $y(t)$ and each of the subintervals $[t_i, t_{i+1}]$, we conclude that there exists t_i^* in $[t_i, t_{i+1}]$ such that $x(t_{i+1}) - x(t_i) = x'(t_i^*)\Delta t$ and t_i^{**} in $[t_i, t_{i+1}]$ such that $y(t_{i+1}) - y(t_i) = y'(t_i^{**})\Delta t$, where $\Delta t = t_{i+1} - t_i$. Combining with (a), we get

$$\ell(c_i) = \sqrt{\left(x'(t_i^*)\right)^2 + \left(y'(t_i^{**})\right)^2}\,\Delta t.$$

(c) Adding up the lengths from (b), we get

$$\ell(p_n) = \sum_{i=1}^{n} \sqrt{\left(x'(t_i^*)\right)^2 + \left(y'(t_i^{**})\right)^2}\,\Delta t.$$

In the limit, we obtain

$$\lim_{n \to \infty} \ell(p_n) = \lim_{n \to \infty} \sum_{i=1}^{n} \sqrt{\left(x'(t_i^*)\right)^2 + \left(y'(t_i^{**})\right)^2} \, \Delta t$$

$$= \int_a^b \sqrt{(x'(t))^2 + (y'(t))^2} \, dt = \int_a^b \|\mathbf{c}'(t)\| \, dt.$$

5. (a) By the definition,

$$\lim_{h \to 0} \frac{y(0+h) - y(0)}{h} = \lim_{h \to 0} \frac{h^{2/3}}{h} = \lim_{h \to 0} \frac{1}{h^{1/3}} = \pm\infty.$$

Thus, $y = x^{2/3}$ is not differentiable at 0, and, consequently, the parametrization $\mathbf{c}(t) = (t, t^{2/3})$, $t \in [-1, 1]$, is not differentiable.

(b) Since $x = \cos^3 t$ and $y = \cos^2 t$, it follows that $x^{2/3} = (\cos^3 t)^{2/3} = \cos^2 t = y$ (so $\mathbf{c}(t) = (\cos^3 t, \cos^2 t)$ does parametrize the given curve). Since both $\cos^3 t$ and $\cos^2 t$ are differentiable on $[-\pi, \pi]$, we conclude that $\mathbf{c}(t) = (\cos^3 t, \cos^2 t)$, $t \in [-\pi, \pi]$, is a differentiable parametrization of the given curve.

(c) The parametrization given in (b) is not smooth since

$$\mathbf{c}'(t) = (-3\cos^2 t \sin t, -2\cos t \sin t) = -\cos t \sin t (3\cos t, 2)$$

is equal to zero (vector) when $t = 0$ (also when $t = -\pi$, $-\pi/2$, $\pi/2$ and π).

7. No integration is needed: $\mathbf{c}(t)$ is the semi-circle in the first and the fourth quadrants of radius 1 (it goes from $\mathbf{c}(0) = (0, 1)$ to $\mathbf{c}(\pi/2) = (0, -1)$). So its length is π.

9. From

$$\mathbf{c}'(t) = (e^t \cos t - e^t \sin t)\mathbf{i} + (e^t \sin t + e^t \cos t)\mathbf{j} = e^t [(\cos t - \sin t)\mathbf{i} + (\sin t + \cos t)\mathbf{j}]$$

it follows that

$$\|\mathbf{c}'(t)\| = e^t \|(\cos t - \sin t)\mathbf{i} + (\sin t + \cos t)\mathbf{j}\| = e^t \sqrt{2}$$

and therefore the length of $\mathbf{c}$ is

$$\ell(\mathbf{c}) = \int_0^\pi e^t \sqrt{2} \, dt = \sqrt{2} \, e^t \Big|_0^\pi = \sqrt{2}(e^\pi - 1) \approx 31.31167.$$

11. From $\mathbf{c}'(t) = (1, 3(1+t)^{1/2}/2)$ it follows that

$$\|\mathbf{c}'(t)\| = \sqrt{1 + \tfrac{9}{4}(1+t)} = \tfrac{1}{2}\sqrt{9t + 13},$$

and therefore the length of $\mathbf{c}$ is

$$\ell(\mathbf{c}) = \int_0^1 \tfrac{1}{2}\sqrt{9t + 13} \, dt = \tfrac{1}{27}(9t + 13)^{3/2}\Big|_0^1 = \tfrac{1}{27}(22^{3/2} - 13^{3/2}) \approx 2.08581.$$

13. From $\mathbf{c}'(t) = (2 - 2t)\mathbf{i} + 4t^{1/2}\mathbf{j}$ it follows that

$$\|\mathbf{c}'(t)\| = \sqrt{4 + 4t^2 - 8t + 16t} = \sqrt{4t^2 + 8t + 4} = \sqrt{(2t + 2)^2} = 2t + 2,$$

since $2t + 2 > 0$ when $1 \leq t \leq 3$. The length of $\mathbf{c}$ is

$$\ell(\mathbf{c}) = \int_1^3 (2t + 2)\, dt = (t^2 + 2t)\big|_1^3 = 12.$$

15. From

$$\mathbf{c}'(t) = (ae^{at}\cos t - e^{at}\sin t, ae^{at}\sin t + e^{at}\cos t)$$

we get $\|\mathbf{c}'(t)\| = \sqrt{a^2 e^{2at} + e^{2at}} = e^{at}\sqrt{a^2 + 1}$. Thus, the length of the logarithmic spiral for $t \geq 0$ is

$$\ell(\mathbf{c}) = \int_0^\infty e^{at}\sqrt{a^2 + 1}\, dt = \frac{\sqrt{a^2 + 1}}{a}e^{at}\Big|_0^\infty = -\frac{\sqrt{a^2 + 1}}{a}.$$

(since $\lim_{t \to \infty} e^{at} = 0$ if $a < 0$). Note that, because $a < 0$, $\ell(\mathbf{c})$ is positive.

17. Using the formula of Example 3.28 we get that the length of the curve $y = \sin x$, $x \in [0, 2\pi]$ (call it $\mathbf{c}_1$) is

$$\ell(\mathbf{c}_1) = \int_0^{2\pi} \sqrt{1 + \cos^2 x}\, dx.$$

Similarly, the length of the curve $y = 2\sin x$, $x \in [0, 2\pi]$ (call it $\mathbf{c}_2$) is

$$\ell(\mathbf{c}_2) = \int_0^{2\pi} \sqrt{1 + 4\cos^2 x}\, dx.$$

We will show that

$$\sqrt{1 + 4\cos^2 x} \leq \sqrt{2.5}\sqrt{1 + \cos^2 x}$$

and therefore (by integrating both sides from 0 to 2π), $\ell(\mathbf{c}_2) \leq \sqrt{2.5}\,\ell(\mathbf{c}_1)$. Since $\sqrt{2.5} < 2$, it follows that $\mathbf{c}_2$ is not twice as long as $\mathbf{c}_1$.

Consider the function $f(t) = (1 + 4t^2)/(1 + t^2)$. From $f'(t) = 6t/(1 + t^2)^2$, it follows that $f(t)$ is increasing for $t \geq 0$. Now let $t = \cos x$. Since $f(t)$ is increasing, its largest value will be obtained for the largest possible value of t, which is 1 (since $\cos x \leq 1$). So

$$f(\cos x) = \frac{1 + 4\cos^2 x}{1 + \cos^2 x} \leq \frac{5}{2},$$

and (by computing the square root) $\sqrt{1 + 4\cos^2 x} \leq \sqrt{5/2}\sqrt{1 + \cos^2 x}$.

19. From $dr/d\theta = -a\sin\theta$, it follows that the length of the curve is (since $a > 0$)

$$\ell = \int_0^{\pi/4} \sqrt{a^2\cos^2\theta + a^2\sin^2\theta}\, d\theta = \int_0^{\pi/4} a\, d\theta = \frac{a\pi}{4}.$$

21. From $dr/d\theta = 6\theta$, it follows that

$$\ell = \int_1^2 \sqrt{9\theta^4 + 36\theta^2}\, d\theta = \int_1^2 \sqrt{9\theta^2(\theta^2 + 4)}\, d\theta = \int_1^2 3\theta\sqrt{\theta^2 + 4}\, d\theta,$$

since $\theta > 0$ (so that $\sqrt{\theta^2} = |\theta| = \theta$). We proceed by substitution $u = \theta^2 + 4$, thus getting

$$\ell = \frac{3}{2}\int_5^8 \sqrt{u}\, du = u^{3/2}\Big|_5^8 = 8^{3/2} - 5^{3/2} \approx 11.447.$$

23. Since $r^2 + (dr/d\theta)^2 = (1 + \sin\theta)^2 + \cos^2\theta = 2 + 2\sin\theta$, it follows that

$$\ell = \int_{-\pi/2}^{\pi/2} \sqrt{2}\sqrt{1 + \sin\theta}\, d\theta = \sqrt{2}\int_{-\pi/2}^{\pi/2} \sqrt{1 + \sin\theta}\,\frac{\sqrt{1 - \sin\theta}}{\sqrt{1 - \sin\theta}}\, d\theta = \sqrt{2}\int_{-\pi/2}^{\pi/2} \frac{\cos\theta}{\sqrt{1 - \sin\theta}}\, d\theta$$

(we used the fact that $\sqrt{1 - \sin^2\theta} = \sqrt{\cos^2\theta} = |\cos\theta| = \cos\theta$, since $-\pi/2 \le \theta \le \pi/2$). Proceed by the substitution $u = 1 - \sin\theta$ to get

$$\ell = \sqrt{2}(-2)(1 - \sin\theta)^{1/2}\Big|_{-\pi/2}^{\pi/2} = -2\sqrt{2}(-\sqrt{2}) = 4.$$

25. From $\mathbf{c}'(t) = (\sin 2t + 2t\cos 2t, \cos 2t - 2t\sin 2t, 2t^{1/2})$ we get

$$\|\mathbf{c}'(t)\| = \sqrt{\sin^2 2t + 4t^2\cos^2 2t + \cos^2 2t + 4t^2\sin^2 2t + 4t}$$
$$= \sqrt{1 + 4t + 4t^2} = \sqrt{(2t + 1)^2} = 2t + 1,$$

(since $t \ge 0$ implies $2t + 1 \ge 0$ and so $\sqrt{(2t + 1)^2} = |2t + 1| = 2t + 1$). Consequently, the arc-length function $s(t)$ is given by

$$s(t) = \int_0^t \|\mathbf{c}'(\tau)\|\, d\tau = \int_0^t (2\tau + 1)\, d\tau = t^2 + t,$$

where $0 \le t \le 2\pi$.

27. From $\mathbf{c}'(t) = -5\sin t\,\mathbf{i} + 5\cos t\,\mathbf{j} + 12\mathbf{k}$ we get $\|\mathbf{c}'(t)\| = 13$ and the arc-length function of $\mathbf{c}$ is

$$s(t) = \int_0^t \|\mathbf{c}'(\tau)\|\, d\tau = \int_0^t 13\, d\tau = 13t,$$

where $t \in [0, \pi/4]$. The path $\mathbf{c}$, parametrized by the arc-length function s is (from $s = 13t$ it follows that $t = s/13$)

$$\mathbf{c}(s) = 5\cos(s/13)\,\mathbf{i} + 5\sin(s/13)\,\mathbf{j} + \tfrac{12}{13}s\mathbf{k}, \qquad \text{where} \qquad s \in [0, 13\pi/4].$$

29. From $\mathbf{c}(t) = e^t\cos t\,\mathbf{i} + e^t\sin t\,\mathbf{j}$ we get

$$\mathbf{c}'(t) = (e^t\cos t - e^t\sin t)\mathbf{i} + (e^t\sin t + e^t\cos t)\mathbf{j} = e^t[(\cos t - \sin t)\mathbf{i} + (\sin t + \cos t)\mathbf{j}];$$

it follows that

$$\|\mathbf{c}'(t)\| = e^t\|(\cos t - \sin t)\mathbf{i} + (\sin t + \cos t)\mathbf{j}\| = e^t\sqrt{2},$$

and therefore

$$s(t) = \int_0^t \|\mathbf{c}'(\tau)\|\, d\tau = \int_0^t \sqrt{2}\,e^\tau\, d\tau = \sqrt{2}(e^t - 1), \qquad \text{where} \qquad 0 \le t \le 1.$$

Solving $s = \sqrt{2}(e^t - 1)$ for t, we get $t = \ln(1 + s/\sqrt{2})$ and the parametrization by the arc-length function is (since $e^t = s/\sqrt{2} + 1$)

$$\mathbf{c}(s) = \left(\frac{s}{\sqrt{2}} + 1\right)\cos\left(\ln(1 + s/\sqrt{2})\right)\mathbf{i} + \left(\frac{s}{\sqrt{2}} + 1\right)\sin\left(\ln(1 + s/\sqrt{2})\right)\mathbf{j},$$

where $s \in [0, \sqrt{2}(e - 1)]$.

31. Since $\mathbf{c}_2(t) = \mathbf{c}_1(t/2)$ and $\mathbf{c}_3(t) = \mathbf{c}_1(-t)$, it follows that the paths $\mathbf{c}_1$, $\mathbf{c}_2$ and $\mathbf{c}_3$ are reparametrizations of each other, so they have the same image (i.e., represent the same curve).

Computing $\mathbf{c}_1'(t) = (-2\sin 2t, 2\cos 2t, 1)$, we conclude that $\mathbf{c}_1'(t) \neq \mathbf{0}$. Similarly, $\mathbf{c}_2'(t) = (-\sin t, \cos t, 1/2)$ and $\mathbf{c}_3'(t) = (-2\sin 2t, -2\cos 2t, -1)$ are non-zero for all t. Moreover, $\mathbf{c}_1$ maps distinct points to distinct points (for two different values of t the z-coordinates are different), so $\mathbf{c}_1$ is a smooth parametrization. For the same reasons $\mathbf{c}_2$ and $\mathbf{c}_3$ are smooth as well. Consequently, it does not matter what parametrization is used in the computation of the length. Using $\mathbf{c}_1$, we get

$$\ell = \int_0^\pi \|\mathbf{c}_1'(t)\| \, dt = \int_0^\pi \sqrt{5} \, dt = \pi\sqrt{5}.$$

33. From $\mathbf{c}'(t) = (-t^{-2} - 1, -t^{-2} + 1, 0)$, we get

$$\mathbf{T}(1) = \mathbf{c}'(1)/\|\mathbf{c}'(1)\| = (-2, 0, 0)/2 = (-1, 0, 0) = -\mathbf{i}.$$

35. The curve $\mathbf{c}(t)$ is not smooth, since $\mathbf{c}(\pi) = (0, 0) = \mathbf{c}(0)$; i.e., $\mathbf{c}$ intersects itself.

3.4. Acceleration and Curvature

1. From $\mathbf{c}(t) = (t^2, t, t^2)$ we get $\mathbf{c}'(t) = (2t, 1, 2t)$, $\|\mathbf{c}'(t)\| = \sqrt{8t^2 + 1}$ and the acceleration $\mathbf{a}(t) = (2, 0, 2)$. From $ds/dt = \|\mathbf{c}'(t)\| = \sqrt{8t^2 + 1}$ it follows that

$$\frac{d^2 s}{dt^2} = \frac{1}{2\sqrt{8t^2 + 1}} 16t = \frac{8t}{\sqrt{8t^2 + 1}};$$

so the tangential component is (recall that $\mathbf{T}(t) = \mathbf{c}'(t)/\|\mathbf{c}'(t)\|$)

$$\mathbf{a}_T(t) = \frac{d^2 s}{dt^2} \mathbf{T}(t) = \frac{8t}{\sqrt{8t^2 + 1}} \frac{(2t, 1, 2t)}{\sqrt{8t^2 + 1}} = \frac{(16t^2, 8t, 16t^2)}{8t^2 + 1}.$$

The normal component is

$$\mathbf{a}_N(t) = \mathbf{a} - \mathbf{a}_T = (2, 0, 2) - \frac{(16t^2, 8t, 16t^2)}{8t^2 + 1}$$
$$= \frac{(16t^2 + 2, 0, 16t^2 + 2)}{8t^2 + 1} - \frac{(16t^2, 8t, 16t^2)}{8t^2 + 1} = \frac{(2, -8t, 2)}{8t^2 + 1}.$$

3. From $\mathbf{c}(t) = 5t\,\mathbf{i} + 12\sin t\,\mathbf{j} + 12\cos t\,\mathbf{k}$ we get $\mathbf{c}'(t) = 5\mathbf{i} + 12\cos t\,\mathbf{j} - 12\sin t\,\mathbf{k}$, $\|\mathbf{c}'(t)\| = \sqrt{5^2 + 12^2} = 13$ and the acceleration $\mathbf{a}(t) = -12\sin t\,\mathbf{j} - 12\cos t\,\mathbf{k}$. Since $ds/dt = \|\mathbf{c}'(t)\| = 13$ it follows that $d^2 s/dt^2 = 0$ and the tangential component $\mathbf{a}_T = (d^2 s/dt^2)\mathbf{T}(t)$ of $\mathbf{a}$ is zero. Therefore $\mathbf{a}_N = \mathbf{a} = -12\sin t\,\mathbf{j} - 12\cos t\,\mathbf{k}$.

5. From $\mathbf{c}'(t) = (1 - \cos t)\mathbf{i} + \sin t\,\mathbf{j}$ it follows that $\|\mathbf{c}'(t)\| = \sqrt{2 - 2\cos t}$ and $\mathbf{a}(t) = \sin t\,\mathbf{i} + \cos t\,\mathbf{j}$. Next, $ds/dt = \|\mathbf{c}'(t)\| = \sqrt{2 - 2\cos t}$ and

$$\frac{d^2 s}{dt^2} = \frac{2\sin t}{2\sqrt{2 - 2\cos t}} = \frac{\sin t}{\sqrt{2 - 2\cos t}}.$$

The unit tangent is

$$\mathbf{T}(t) = \frac{\mathbf{c}'(t)}{\|\mathbf{c}'(t)\|} = \frac{(1 - \cos t)\mathbf{i} + \sin t\,\mathbf{j}}{\sqrt{2 - 2\cos t}}$$

and so

$$\mathbf{a}_T = \frac{d^2 s}{dt^2}\mathbf{T}(t) = \frac{\sin t}{\sqrt{2 - 2\cos t}}\frac{(1 - \cos t)\mathbf{i} + \sin t\mathbf{j}}{\sqrt{2 - 2\cos t}} = \frac{\sin t(1 - \cos t)\mathbf{i} + \sin^2 t\mathbf{j}}{2 - 2\cos t}.$$

The normal component is

$$\mathbf{a}_N = \mathbf{a} - \mathbf{a}_T = \sin t\mathbf{i} + \cos t\mathbf{j} - \frac{\sin t(1 - \cos t)\mathbf{i} + \sin^2 t\mathbf{j}}{2 - 2\cos t}$$

$$= \frac{\sin t(1 - \cos t)\mathbf{i} + (2\cos t - 2\cos^2 t - \sin^2 t)\mathbf{j}}{2 - 2\cos t}.$$

7. Drop t from the notation. From $\mathbf{c}' = (x', y')$ we get $\|\mathbf{c}'\| = \sqrt{(x')^2 + (y')^2}$ and $\mathbf{c}'' = (x'', y'')$. The unit tangent vector is

$$\mathbf{T} = \mathbf{c}'/\|\mathbf{c}'\| = (x'/\sqrt{(x')^2 + (y')^2}, y'/\sqrt{(x')^2 + (y')^2}).$$

Now

$$\frac{d}{dt}\left(\frac{x'}{\sqrt{(x')^2 + (y')^2}}\right) = \frac{x''\sqrt{(x')^2 + (y')^2} - x'\frac{1}{2}((x')^2 + (y')^2)^{-1/2}(2x'x'' + 2y'y'')}{(x')^2 + (y')^2}$$

$$= \frac{x''((x')^2 + (y')^2) - (x')^2 x'' - x'y'y''}{((x')^2 + (y')^2)^{3/2}} = \frac{x''(y')^2 - x'y'y''}{((x')^2 + (y')^2)^{3/2}}.$$

By symmetry,

$$\frac{d}{dt}\left(\frac{y'}{\sqrt{(x')^2 + (y')^2}}\right) = \frac{y''(x')^2 - x'y'x''}{((x')^2 + (y')^2)^{3/2}},$$

and therefore

$$\mathbf{T}' = \left(\frac{x''(y')^2 - x'y'y''}{((x')^2 + (y')^2)^{3/2}}, \frac{y''(x')^2 - x'y'x''}{((x')^2 + (y')^2)^{3/2}}\right)$$

$$= ((x')^2 + (y')^2)^{-3/2}\left(x''(y')^2 - x'y'y'', y''(x')^2 - x'y'x''\right).$$

The norm of $\mathbf{T}'$ is

$$\|\mathbf{T}'\| = ((x')^2 + (y')^2)^{-3/2}\big((x'')^2(y')^4 + (x')^2(y')^2(y'')^2 - 2x'(y')^3 x''y''$$
$$+ (y'')^2(x')^4 + (x')^2(y')^2(x'')^2 - 2(x')^3 y'x''y''\big)^{1/2}$$

$$= ((x')^2 + (y')^2)^{-3/2}\big((x'')^2(y')^2((x')^2 + (y')^2) + (x')^2(y'')^2((x')^2 + (y')^2)$$
$$- 2x'y'x''y''((x')^2 + (y')^2)\big)^{1/2}$$

$$= ((x')^2 + (y')^2)^{-3/2}((x')^2 + (y')^2)^{1/2}\big((x'')^2(y')^2 + (x')^2(y'')^2 - 2x'y'x''y''\big)^{1/2}$$

$$= ((x')^2 + (y')^2)^{-1}\big((x''y' - x'y'')^2\big)^{1/2} = ((x')^2 + (y')^2)^{-1}|x''y' - x'y''|.$$

Finally,

$$\kappa = \frac{\|\mathbf{T}'\|}{\|\mathbf{c}'\|} = \frac{((x')^2 + (y')^2)^{-1}|x''y' - x'y''|}{\sqrt{(x')^2 + (y')^2}} = \frac{|x''y' - x'y''|}{((x')^2 + (y')^2)^{3/2}}.$$

9. We use the formula of Exercise 7. From $x(t) = 2 - 2t^3$ we get $x'(t) = -6t^2$ and $x''(t) = -12t$; similarly, $y(t) = t^3 + 1$ implies $y'(t) = 3t^2$ and $y''(t) = 6t$. Hence

$$\kappa(t) = \frac{|x'(t)y''(t) - x''(t)y'(t)|}{((x'(t))^2 + (y'(t))^2)^{3/2}} = \frac{|(-6t^2)6t - (-12t)3t^2|}{(36t^4 + 9t^4)^{3/2}} = 0.$$

We can write $\mathbf{c}(t)$ as $\mathbf{c}(t) = (2, 1) + t^3(-2, 1)$, $t \in \mathbb{R}$. In words, $\mathbf{c}(t)$ is the line through $(2, 1)$ in the direction of the vector $(-2, 1)$. The curvature of a line is zero, and that confirms our calculation.

11. We will use the formula of Exercise 7. From $x(t) = t^2$ we get $x'(t) = 2t$ and $x''(t) = 2$; similarly, $y(t) = 3 - t$ implies $y'(t) = -1$ and $y''(t) = 0$. Hence

$$\kappa(t) = \frac{|x'(t)y''(t) - x''(t)y'(t)|}{((x'(t))^2 + (y'(t))^2)^{3/2}} = \frac{|2t(0) - 2(-1)|}{(4t^2 + 1)^{3/2}} = \frac{2}{(4t^2 + 1)^{3/2}}.$$

The curvature is largest when the denominator is smallest; i.e., when $t = 0$. It follows that the maximum curvature $\kappa(0) = 2$ occurs at $\mathbf{c}(0) = (0, 3)$. As $t \to \infty$, $\kappa(t) = 2/(4t^2 + 1)^{3/2} \to 0$.

13. From $\mathbf{c}(t) = (e^t \sin t, 0, e^t \cos t)$ we get

$$\mathbf{c}'(t) = (e^t \sin t + e^t \cos t, 0, e^t \cos t - e^t \sin t) = e^t(\sin t + \cos t, 0, \cos t - \sin t)$$

and hence

$$\|\mathbf{c}'(t)\| = e^t\|(\sin t + \cos t, 0, \cos t - \sin t)\| = e^t\sqrt{2}.$$

The unit tangent vector is

$$\mathbf{T}(t) = \frac{\mathbf{c}'(t)}{\|\mathbf{c}'(t)\|} = \frac{1}{\sqrt{2}}(\sin t + \cos t, 0, \cos t - \sin t).$$

Now

$$\mathbf{T}'(t) = \frac{1}{\sqrt{2}}(\cos t - \sin t, 0, -\sin t - \cos t),$$

and since

$$\|\mathbf{T}'(t)\| = \frac{1}{\sqrt{2}}\|(\cos t - \sin t, 0, -\sin t - \cos t)\| = \frac{1}{\sqrt{2}}\sqrt{2} = 1,$$

it follows that the principal unit normal is

$$\mathbf{N}(t) = \mathbf{T}'(t) = \frac{1}{\sqrt{2}}(\cos t - \sin t, 0, -\sin t - \cos t).$$

The curvature of $\mathbf{c}$ is

$$\kappa(t) = \frac{\|\mathbf{T}'(t)\|}{\|\mathbf{c}'(t)\|} = \frac{1}{\sqrt{2}e^t},$$

and

$$\mathbf{a}_N = \|\mathbf{c}'(t)\|\mathbf{T}'(t) = \sqrt{2}e^t \frac{1}{\sqrt{2}}(\cos t - \sin t, 0, -\sin t - \cos t)$$
$$= e^t(\cos t - \sin t, 0, -\sin t - \cos t)$$

is the normal component of the acceleration.

15. From $\mathbf{c}(t) = (t, \cos t, 1 - \sin t)$ we get $\mathbf{c}'(t) = (1, -\sin t, -\cos t)$ and hence $\|\mathbf{c}'(t)\| = \sqrt{2}$. The unit tangent vector is

$$\mathbf{T}(t) = \frac{\mathbf{c}'(t)}{\|\mathbf{c}'(t)\|} = \frac{1}{\sqrt{2}}(1, -\sin t, -\cos t).$$

Now

$$\mathbf{T}'(t) = \frac{1}{\sqrt{2}}(0, -\cos t, \sin t),$$

and since $\|\mathbf{T}'(t)\| = 1/\sqrt{2}$, it follows that the principal unit normal is

$$\mathbf{N}(t) = \frac{\mathbf{T}'(t)}{\|\mathbf{T}'(t)\|} = (0, -\cos t, \sin t).$$

The curvature of $\mathbf{c}$ is

$$\kappa(t) = \frac{\|\mathbf{T}'(t)\|}{\|\mathbf{c}'(t)\|} = \frac{1/\sqrt{2}}{\sqrt{2}} = \frac{1}{2},$$

and

$$\mathbf{a}_N = \|\mathbf{c}'(t)\|\mathbf{T}'(t) = (0, -\cos t, \sin t)$$

is the normal component of the acceleration.

17. From $\mathbf{c}(t) = (e^{-t}\cos t, e^{-t}\sin t, e^{-t})$ we get $\mathbf{c}'(t) = e^{-t}(-\cos t - \sin t, -\sin t + \cos t, -1)$ and hence $\|\mathbf{c}'(t)\| = e^{-t}\sqrt{3}$. The unit tangent vector is

$$\mathbf{T}(t) = \frac{\mathbf{c}'(t)}{\|\mathbf{c}'(t)\|} = \frac{1}{\sqrt{3}}(-\cos t - \sin t, -\sin t + \cos t, -1).$$

Now

$$\mathbf{T}'(t) = \frac{1}{\sqrt{3}}(\sin t - \cos t, -\sin t - \cos t, 0),$$

and since $\|\mathbf{T}'(t)\| = (1/\sqrt{3})\sqrt{2} = \sqrt{2/3}$, it follows that the principal unit normal is

$$\mathbf{N}(t) = \frac{\mathbf{T}'(t)}{\|\mathbf{T}'(t)\|} = \frac{1}{\sqrt{2}}(\sin t - \cos t, -\sin t - \cos t, 0).$$

The curvature of $\mathbf{c}$ is

$$\kappa(t) = \frac{\|\mathbf{T}'(t)\|}{\|\mathbf{c}'(t)\|} = \frac{\sqrt{2}}{3}e^t,$$

and

$$\mathbf{a}_N = \|\mathbf{c}'(t)\|\mathbf{T}'(t) = e^{-t}(\sin t - \cos t, -\sin t - \cos t, 0)$$

is the normal component of the acceleration.

19. We will use the formula in Exercise 7 to compute the curvature. From $x(t) = t^3$ we get $x'(t) = 3t^2$ and $x''(t) = 6t$; similarly, $y(t) = t$ implies $y'(t) = 1$ and $y''(t) = 0$. Hence

$$\kappa(t) = \frac{|x'(t)y''(t) - x''(t)y'(t)|}{((x'(t))^2 + (y'(t))^2)^{3/2}} = \frac{|3t^2(0) - 6t(1)|}{(9t^4 + 1)^{3/2}} = \frac{6|t|}{(9t^4 + 1)^{3/2}}.$$

It follows that the radius of the osculating circle at $(8,2)$ is $1/\kappa(2) = (145)^{3/2}/12 \approx 145.5026$. Since $\mathbf{c}(t)$ lies in the xy-plane, the osculating circle lies in it as well. Its center is on the line normal to $\mathbf{c}$ at $\mathbf{c}(2) = (8,2)$. To get a direction normal to $\mathbf{c}$, take any vector perpendicular to $\mathbf{c}' = (3t^2, 1)$; for example, $(1, -3t^2)$. Hence a normal direction at $t = 2$ is

$$\mathbf{N} = (1, -3t^2)\Big|_{t=2} = (1, -12).$$

The equation of the normal line at $(8,2)$ is

$$\ell(u) = (8,2) + u(1, -12) = (8 + u, 2 - 12u),$$

where $u \in \mathbb{R}$. To locate the center of the osculating circle, we have to find the point on that line that is $1/\kappa(2) = (145)^{3/2}/12$ units away from $(8,2)$ in the direction towards which $\overline{\mathbf{N}}$ points (that is, with $u > 0$). The distance between $(8,2)$ and $(8 + u, 2 - 12u)$ is

$$\sqrt{u^2 + 144u^2} = \sqrt{145u^2},$$

so $\sqrt{145u^2} = (145)^{3/2}/12$ implies that $145u^2 = 145^3/144$ or $u = 145/12$ ($u = -145/12$ gives a point on the wrong side). Hence

$$\ell(145/12) = (8 + (145/12), 2 - 12(145/12)) = (241/12, -143)$$

is the center of the osculating circle. Its equation is $(x - 241/12)^2 + (y + 143)^2 = 145^3/144$.

21. The osculating plane goes through $\mathbf{c}(\pi/2) = (0, 2, \pi/2)$ and is spanned by the vectors $\mathbf{T}(\pi/2)$ and $\mathbf{N}(\pi/2)$. From $\mathbf{c}'(t) = (-2\sin t, 2\cos t, 1)$ it follows that $\|\mathbf{c}'(t)\| = \sqrt{5}$ and

$$\mathbf{T}(t) = \frac{\mathbf{c}'(t)}{\|\mathbf{c}'(t)\|} = \left(-\frac{2}{\sqrt{5}}\sin t, \frac{2}{\sqrt{5}}\cos t, \frac{1}{\sqrt{5}}\right).$$

The derivative of $\mathbf{T}$ is

$$\mathbf{T}'(t) = \left(-\frac{2}{\sqrt{5}}\cos t, -\frac{2}{\sqrt{5}}\sin t, 0\right),$$

its norm is $\|\mathbf{T}'(t)\| = 2/\sqrt{5}$ and so the principal unit normal is $\mathbf{N}(t) = \mathbf{T}'(t)/\|\mathbf{T}'(t)\| = (-\cos t, -\sin t, 0)$. To get a normal to the plane, we compute the cross product of the tangent vector $\mathbf{T}(t)$ and the vector $\mathbf{N}(t)$ at $t = \pi/2$:

$$\mathbf{T}(\pi/2) \times \mathbf{N}(\pi/2) = \begin{vmatrix} \mathbf{i} & \mathbf{j} & \mathbf{k} \\ -2/\sqrt{5} & 0 & 1/\sqrt{5} \\ 0 & -1 & 0 \end{vmatrix} = \frac{1}{\sqrt{5}}\mathbf{i} + \frac{2}{\sqrt{5}}\mathbf{k}.$$

The equation of the osculating plane is

$$\frac{1}{\sqrt{5}}(x - 0) + 0(y - 2) + \frac{2}{\sqrt{5}}(z - \frac{\pi}{2}) = 0,$$

i.e., $x + 2z - \pi = 0$.

23. Using the formula of Exercise 22 we get ($y = x^2$, so $y' = 2x$ and $y'' = 2$)

$$\kappa(x_0) = \frac{|2|}{(1 + 4x^2)^{3/2}}\bigg|_{x=x_0} = \frac{2}{(1 + 4x_0^2)^{3/2}}.$$

25. Using the formula of Exercise 22 we get ($y = \ln x$, so $y' = 1/x$ and $y'' = -1/x^2$)

$$\kappa(x) = \frac{|-1/x^2|}{(1 + (1/x)^2)^{3/2}} = \frac{1/x^2}{(x^2 + 1)^{3/2}/x^3} = \frac{x}{(x^2 + 1)^{3/2}}.$$

From

$$\kappa'(x) = \frac{(x^2 + 1)^{3/2} - x\frac{3}{2}(x^2 + 1)^{1/2}2x}{(x^2 + 1)^3} = \frac{1 - 2x^2}{(x^2 + 1)^{5/2}} = 0$$

it follows that $x = 1/\sqrt{2}$ ($x = -1/\sqrt{2}$ makes no sense, since $\ln x$ is not defined for negative x). If $0 < x < 1/\sqrt{2}$, then $\kappa'(x) > 0$ and if $x > 1/\sqrt{2}$, then $\kappa'(x) < 0$. Therefore $x = 1/\sqrt{2}$ is the point where maximum curvature occurs. Its value is

$$\kappa(1/\sqrt{2}) = \frac{1/\sqrt{2}}{(3/2)^{3/2}} = \frac{2}{3\sqrt{3}} \approx 0.3849.$$

27. The osculating plane goes through $\mathbf{c}(1) = (1, 0, 2)$ and is spanned by the vectors $\mathbf{T}(1)$ and $\mathbf{N}(1)$. From $\mathbf{c}'(t) = (1, -2t, 4t)$ it follows that $\|\mathbf{c}'(t)\| = \sqrt{1 + 20t^2}$ and

$$\mathbf{T}(t) = \frac{\mathbf{c}'(t)}{\|\mathbf{c}'(t)\|} = \frac{(1, -2t, 4t)}{\sqrt{1 + 20t^2}}.$$

The derivative of $\mathbf{T}$ is

$$\mathbf{T}'(t) = \frac{(0, -2, 4)(1 + 20t^2)^{1/2} - (1, -2t, 4t)\frac{1}{2}(1 + 20t^2)^{-1/2}40t}{1 + 20t^2} = \frac{(-20t, -2, 4)}{(1 + 20t^2)^{3/2}}.$$

We do not really need the unit tangent and the principal unit normal vectors — we only need their directions. So instead of $\mathbf{T}(1)$ take $\mathbf{c}'(1) = (1, -2, 4)$ and instead of $\mathbf{N}(1)$ or $\mathbf{T}'(1)$ take the numerator of $\mathbf{T}'(1)$, which is $(-20, -2, 4)$. From

$$\begin{vmatrix} \mathbf{i} & \mathbf{j} & \mathbf{k} \\ 1 & -2 & 4 \\ -20 & -2 & 4 \end{vmatrix} = -84\mathbf{j} - 42\mathbf{k} = -42(2\mathbf{j} + \mathbf{k})$$

it follows that we can take a normal vector to be $(0, 2, 1)$; the equation of the osculating plane is $0(x - 1) + 2(y - 0) + 1(z - 2) = 0$; i.e., $2y + z - 2 = 0$.

29. Since $y' = \cos x$ and $y'' = -\sin x$, it follows from Exercise 22 that $\kappa(x) = |-\sin x|/(1 + \cos^2 x)^{3/2}$ and so $\kappa(\pi/2) = 1$. It follows that the osculating circle at $(\pi/2, 1)$ has radius $1/\kappa(\pi/2) = 1$. The graph of $y = \sin x$ has a horizontal tangent at $x = \pi/2$, and so its normal is the vertical line $x = \pi/2$. So the osculating circle has a center at $(\pi/2, 0)$ (i.e., along the normal, 1 unit away from $(\pi/2, 1)$) and is of radius 1; its equation is $(x - \pi/2)^2 + y^2 = 1$.

3.5. Introduction to Differential Geometry of Curves

1. By the product rule and (3.19)

$$\frac{d}{ds}(\mathbf{T}(s) \cdot \mathbf{T}(s)) = \frac{d\mathbf{T}(s)}{ds} \cdot \mathbf{T}(s) + \mathbf{T}(s) \cdot \frac{d\mathbf{T}(s)}{ds}$$

$$= 2\mathbf{T}(s) \cdot \frac{d\mathbf{T}(s)}{ds} = 2\mathbf{T}(s) \cdot \kappa(s)\mathbf{N}(s) = 2\kappa(s)\mathbf{T}(s) \cdot \mathbf{N}(s) = 0,$$

since $\mathbf{T}(s)$ and $\mathbf{N}(s)$ are orthogonal. Next,

$$\frac{d\mathbf{c}(t)}{dt} \cdot \mathbf{T}(t) = \mathbf{c}'(t) \cdot \frac{\mathbf{c}'(t)}{\|\mathbf{c}'(t)\|} = \|\mathbf{c}'(t)\|,$$

since $\mathbf{c}'(t) \cdot \mathbf{c}'(t) = \|\mathbf{c}'(t)\|^2$. Using (3.20), we get

$$\frac{d\mathbf{N}(s)}{ds} \cdot \mathbf{B}(s) = (-\kappa(s)\mathbf{T}(s) + \tau(s)\mathbf{B}(s)) \cdot \mathbf{B}(s) = \tau(s),$$

since $\mathbf{T}(s) \cdot \mathbf{B}(s) = 0$ and $\mathbf{B}(s) \cdot \mathbf{B}(s) = \|\mathbf{B}(s)\|^2 = 1$.

3. The dot product of $\mathbf{c}''$ and $\mathbf{B}$ is equal to $\mathbf{c}'' \cdot \mathbf{B} = \mathbf{c}'' \cdot (\mathbf{T} \times \mathbf{N})$; i.e., it is equal to the scalar triple product of $\mathbf{c}''$, $\mathbf{T}$ and $\mathbf{N}$. Since $\mathbf{T} = \mathbf{c}'/\|\mathbf{c}'\|$, it follows that

$$\mathbf{N} = \frac{1}{\|\mathbf{T}'\|}\mathbf{T}' = \frac{1}{\|\mathbf{T}'\|}\frac{\|\mathbf{c}'\|\mathbf{c}'' - \|\mathbf{c}'\|'\mathbf{c}'}{\|\mathbf{c}'\|^2} = \frac{1}{\|\mathbf{c}'\|\,\|\mathbf{T}'\|}\mathbf{c}'' - \frac{\|\mathbf{c}'\|'}{\|\mathbf{c}'\|^2\,\|\mathbf{T}'\|}\mathbf{c}'.$$

In words, $\mathbf{N}$ is a linear combination of $\mathbf{c}'$ and $\mathbf{c}''$. So the three vectors $\mathbf{c}''$, $\mathbf{T}$ and $\mathbf{N}$ lie in the plane spanned by $\mathbf{c}'$ and $\mathbf{c}''$. Consequently, their scalar triple product is zero, and so $\mathbf{c}''$ is perpendicular to $\mathbf{B}$.

5. From $\mathbf{c}(t) = e^t(\cos t\,\mathbf{i} + \sin t\,\mathbf{k})$ we get (using the product rule)

$$\mathbf{c}'(t) = e^t(\cos t\,\mathbf{i} + \sin t\,\mathbf{k}) + e^t(-\sin t\,\mathbf{i} + \cos t\,\mathbf{k}) = e^t((\cos t - \sin t)\mathbf{i} + (\sin t + \cos t)\mathbf{k})$$

and $\|\mathbf{c}'(t)\| = e^t\sqrt{2}$. The unit tangent vector is

$$\mathbf{T}(t) = \frac{(\cos t - \sin t)\mathbf{i} + (\sin t + \cos t)\mathbf{k}}{\sqrt{2}}.$$

From

$$\mathbf{T}'(t) = \frac{(-\sin t - \cos t)\mathbf{i} + (\cos t - \sin t)\mathbf{k}}{\sqrt{2}}$$

and $\|\mathbf{T}'(t)\| = 1$ it follows that the principal unit normal is

$$\mathbf{N}(t) = \mathbf{T}'(t) = \frac{(-\sin t - \cos t)\mathbf{i} + (\cos t - \sin t)\mathbf{k}}{\sqrt{2}}.$$

The binormal vector is

$$\mathbf{B}(t) = \mathbf{T}(t) \times \mathbf{N}(t) = \frac{1}{\sqrt{2}}\frac{1}{\sqrt{2}}\begin{vmatrix} \mathbf{i} & \mathbf{j} & \mathbf{k} \\ \cos t - \sin t & 0 & \sin t + \cos t \\ -\sin t - \cos t & 0 & \cos t - \sin t \end{vmatrix}$$

$$= -\tfrac{1}{2}((\cos t - \sin t)^2 + (\sin t + \cos t)^2)\mathbf{j} = -\mathbf{j}.$$

At $t = \pi/2$, $\mathbf{T}(\pi/2) = (-\mathbf{i} + \mathbf{k})/\sqrt{2}$, $\mathbf{N}(\pi/2) = (-\mathbf{i} - \mathbf{k})/\sqrt{2}$, $\mathbf{B}(\pi/2) = -\mathbf{j}$.

7. From $\mathbf{c}(t) = (\sin t, \sin t, \sqrt{2}\cos t)$ we get $\mathbf{c}'(t) = (\cos t, \cos t, -\sqrt{2}\sin t)$ and $\|\mathbf{c}'(t)\| = \sqrt{2}$. The unit tangent vector is

$$\mathbf{T}(t) = \frac{\mathbf{c}'(t)}{\|\mathbf{c}'(t)\|} = \left(\frac{\cos t}{\sqrt{2}}, \frac{\cos t}{\sqrt{2}}, -\sin t\right).$$

From

$$\mathbf{T}'(t) = \left(\frac{-\sin t}{\sqrt{2}}, \frac{-\sin t}{\sqrt{2}}, -\cos t\right)$$

and $\|\mathbf{T}'(t)\| = 1$ it follows that the principal unit normal is

$$\mathbf{N}(t) = \mathbf{T}'(t) = \left(\frac{-\sin t}{\sqrt{2}}, \frac{-\sin t}{\sqrt{2}}, -\cos t\right).$$

The binormal vector is

$$\mathbf{B}(t) = \mathbf{T}(t) \times \mathbf{N}(t) = \begin{vmatrix} \mathbf{i} & \mathbf{j} & \mathbf{k} \\ \cos t/\sqrt{2} & \cos t/\sqrt{2} & -\sin t \\ -\sin t/\sqrt{2} & -\sin t/\sqrt{2} & -\cos t \end{vmatrix} = \frac{-\mathbf{i} + \mathbf{j}}{\sqrt{2}} = \left(-\frac{1}{\sqrt{2}}, \frac{1}{\sqrt{2}}, 0\right).$$

9. Differentiating $\mathbf{N} \cdot \mathbf{N} = \|\mathbf{N}\|^2 = 1$ we get $(d\mathbf{N}/ds) \cdot \mathbf{N} + \mathbf{N} \cdot (d\mathbf{N}/ds) = 0$, and hence $(d\mathbf{N}/ds) \cdot \mathbf{N} = 0$. From

$$\left(\frac{d\mathbf{N}}{ds} + \kappa\mathbf{T} \right) \cdot \mathbf{N} = \frac{d\mathbf{N}}{ds} \cdot \mathbf{N} + \kappa\mathbf{T} \cdot \mathbf{N} = 0$$

we conclude that $d\mathbf{N}/ds + \kappa\mathbf{T}$ is perpendicular to $\mathbf{N}$.

Differentiating $\mathbf{T} \cdot \mathbf{N} = 0$ we get $(d\mathbf{T}/ds) \cdot \mathbf{N} + \mathbf{T} \cdot (d\mathbf{N}/ds) = 0$, and therefore

$$\left(\frac{d\mathbf{N}}{ds} + \kappa\mathbf{T} \right) \cdot \mathbf{T} = \frac{d\mathbf{N}}{ds} \cdot \mathbf{T} + \kappa\|\mathbf{T}\|^2 = -\frac{d\mathbf{T}}{ds} \cdot \mathbf{N} + \kappa,$$

since $\|\mathbf{T}\| = 1$. By the chain rule,

$$\frac{d\mathbf{T}}{ds} = \frac{d\mathbf{T}}{dt} \frac{dt}{ds} = \frac{\mathbf{T}'(t)}{\|\mathbf{c}'(t)\|},$$

and

$$\frac{d\mathbf{T}}{ds} \cdot \mathbf{N} = \frac{\mathbf{T}'(t)}{\|\mathbf{c}'(t)\|} \cdot \frac{\mathbf{T}'(t)}{\|\mathbf{T}'(t)\|} = \frac{\|\mathbf{T}'(t)\|}{\|\mathbf{c}'(t)\|} = \kappa.$$

It follows that $(d\mathbf{N}/ds + \kappa\mathbf{T}) \cdot \mathbf{T} = 0$. Since $d\mathbf{N}/ds + \kappa\mathbf{T}$ is perpendicular to both $\mathbf{N}$ and $\mathbf{T}$, it must be parallel to $\mathbf{B}$ (i.e., it is a multiple of $\mathbf{B}$). So we can define τ to be that multiple; that is, τ is the real number such that $d\mathbf{N}/ds + \kappa\mathbf{T} = \tau\mathbf{B}$.

Comment: in the above computation, we used the fact that

$$\mathbf{N}(s) = \frac{\mathbf{T}'(s)}{\|\mathbf{T}'(s)\|} = \frac{\mathbf{T}'(t)/\|\mathbf{c}'(t)\|}{\|\mathbf{T}'(t)\|/\|\mathbf{c}'(t)\|} = \frac{\mathbf{T}'(t)}{\|\mathbf{T}'(t)\|} = \mathbf{N}(t).$$

11. From $\mathbf{T}(t) = \mathbf{c}'(t)/\|\mathbf{c}'(t)\|$ we get $\mathbf{c}'(t) = \|\mathbf{c}'(t)\| \, \mathbf{T}(t) = (ds/dt)\mathbf{T}(t)$; differentiate $\mathbf{c}'(t) = (ds/dt)\mathbf{T}(t)$ with respect to t:

$$\mathbf{c}''(t) = \frac{d^2s}{dt^2}\mathbf{T} + \frac{ds}{dt}\frac{d\mathbf{T}}{dt} = \frac{d^2s}{dt^2}\mathbf{T} + \frac{ds}{dt}\frac{d\mathbf{T}}{ds}\frac{ds}{dt} = \frac{d^2s}{dt^2}\mathbf{T} + \left(\frac{ds}{dt} \right)^2 \frac{d\mathbf{T}}{ds}.$$

(We used the chain rule in computing $d\mathbf{T}/dt$.) By (3.19), $d\mathbf{T}(s)/ds = \kappa(s)\mathbf{N}(s)$. Since $\kappa(s) = \kappa(t)$ (see the text immediately following the definition of the curvature in Section 3.4) and $\mathbf{N}(s) = \mathbf{N}(t)$ (see the comment at the end of Exercise 9) it follows that

$$\frac{d\mathbf{T}(s)}{ds} = \kappa(s)\mathbf{N}(s) = \kappa(t)\mathbf{N}(t)$$

and therefore

$$\mathbf{c}''(t) = \frac{d^2s}{dt^2}\mathbf{T}(t) + \left(\frac{ds}{dt} \right)^2 \kappa(t)\mathbf{N}(t).$$

13. From Exercise 11 we get

$$\mathbf{c}'(t) \times \mathbf{c}''(t) = \frac{ds}{dt}\mathbf{T}(t) \times \left(\frac{d^2s}{dt^2}\mathbf{T}(t) + \kappa(t) \left(\frac{ds}{dt} \right)^2 \mathbf{N}(t) \right).$$

From the distributivity of the cross product and the fact that $\mathbf{T}(t) \times \mathbf{T}(t) = \mathbf{0}$, it follows that

$$\mathbf{c}'(t) \times \mathbf{c}''(t) = \frac{ds}{dt}\kappa(t) \left(\frac{ds}{dt} \right)^2 \mathbf{T}(t) \times \mathbf{N}(t) = \kappa(t) \left(\frac{ds}{dt} \right)^3 \mathbf{B}(t),$$

by the definition of $\mathbf{B}(t)$.

15. From $\mathbf{c}(t) = a\cos t\mathbf{i} + a\sin t\mathbf{j} + bt\mathbf{k}$ we get $\mathbf{c}'(t) = -a\sin t\mathbf{i} + a\cos t\mathbf{j} + b\mathbf{k}$, $\mathbf{c}''(t) = -a\cos t\mathbf{i} - a\sin t\mathbf{j}$ and $\mathbf{c}'''(t) = a\sin t\mathbf{i} - a\cos t\mathbf{j}$ and hence

$$\mathbf{c}'(t) \times \mathbf{c}''(t) = \begin{vmatrix} \mathbf{i} & \mathbf{j} & \mathbf{k} \\ -a\sin t & a\cos t & b \\ -a\cos t & -a\sin t & 0 \end{vmatrix} = ab\sin t\mathbf{i} - ab\cos t\mathbf{j} + a^2\mathbf{k}.$$

So

$$\|\mathbf{c}'(t) \times \mathbf{c}''(t)\| = \sqrt{a^2b^2\sin^2 t + a^2b^2\cos^2 t + a^4} = \sqrt{a^2(a^2 + b^2)} = a\sqrt{a^2 + b^2},$$

since $a > 0$. Finally,

$$(\mathbf{c}'(t) \times \mathbf{c}''(t)) \cdot \mathbf{c}'''(t) = (ab\sin t\mathbf{i} - ab\cos t\mathbf{j} + a^2\mathbf{k}) \cdot (a\sin t\mathbf{i} - a\cos t\mathbf{j}) = a^2 b$$

and

$$\tau(t) = \frac{(\mathbf{c}'(t) \times \mathbf{c}''(t)) \cdot \mathbf{c}'''(t)}{\|\mathbf{c}'(t) \times \mathbf{c}''(t)\|^2} = \frac{a^2 b}{a^2(a^2 + b^2)} = \frac{b}{(a^2 + b^2)}.$$

17. The curve $\mathbf{c}(t)$ is a plane curve, so its torsion is zero. We use the formula of Exercise 28 in Section 3.4 (we could also use the formula of Exercise 7 in the same section). From $\mathbf{c}(t) = at^2\mathbf{i} + 2at\mathbf{j}$ we get $\mathbf{c}'(t) = 2at\mathbf{i} + 2a\mathbf{j}$ and $\mathbf{c}''(t) = 2a\mathbf{i}$ and hence

$$\mathbf{c}'(t) \times \mathbf{c}''(t) = \begin{vmatrix} \mathbf{i} & \mathbf{j} & \mathbf{k} \\ 2at & 2a & 0 \\ 2a & 0 & 0 \end{vmatrix} = -4a^2\mathbf{k}.$$

Therefore

$$\kappa(t) = \frac{\|\mathbf{c}'(t) \times \mathbf{c}''(t)\|}{\|\mathbf{c}'(t)\|^3} = \frac{4a^2}{(4a^2t^2 + 4a^2)^{3/2}} = \frac{1}{2a(t^2 + 1)^{3/2}}.$$

The curvature is largest when the denominator is smallest; i.e., when $t = 0$. So $\mathbf{c}(t)$ has maximum curvature at $\mathbf{c}(0) = (0, 0)$.

19. From $\mathbf{c}(t) = (t + \cos t, t - \sin t, t)$ we get $\mathbf{c}'(t) = (1 - \sin t, 1 - \cos t, 1)$, $\mathbf{c}''(t) = (-\cos t, \sin t, 0)$ and $\mathbf{c}'''(t) = (\sin t, \cos t, 0)$. So

$$\mathbf{c}'(t) \times \mathbf{c}''(t) = \begin{vmatrix} \mathbf{i} & \mathbf{j} & \mathbf{k} \\ 1 - \sin t & 1 - \cos t & 1 \\ -\cos t & \sin t & 0 \end{vmatrix} = -\sin t\mathbf{i} - \cos t\mathbf{j} + (\sin t + \cos t - 1)\mathbf{k}.$$

Since

$$\|\mathbf{c}'(t) \times \mathbf{c}''(t)\| = \sqrt{\sin^2 t + \cos^2 t + \sin^2 t + \cos^2 t + 1 + 2\sin t\cos t - 2\sin t - 2\cos t}$$
$$= \sqrt{3 + 2(\sin t\cos t - \sin t - \cos t)}$$

and

$$\|\mathbf{c}'(t)\| = \sqrt{1 + \sin^2 t - 2\sin t + 1 + \cos^2 t - 2\cos t + 1}$$
$$= \sqrt{4 - 2\sin t - 2\cos t},$$

it follows from the formula of Exercise 28 in Section 3.4 that

$$\kappa(t) = \frac{\|\mathbf{c}'(t) \times \mathbf{c}''(t)\|}{\|\mathbf{c}'(t)\|^3} = \frac{(3 + 2\sin t \cos t - 2\sin t - 2\cos t)^{1/2}}{(4 - 2\sin t - 2\cos t)^{3/2}}.$$

From

$$(\mathbf{c}'(t) \times \mathbf{c}''(t)) \cdot \mathbf{c}'''(t) = (-\sin t\,\mathbf{i} - \cos t\,\mathbf{j} + (\sin t + \cos t - 1)\mathbf{k}) \cdot (\sin t\,\mathbf{i} + \cos t\,\mathbf{j}) = -1,$$

we get (see Exercise 14 in this section)

$$\tau(t) = \frac{(\mathbf{c}'(t) \times \mathbf{c}''(t)) \cdot \mathbf{c}'''(t)}{\|\mathbf{c}'(t) \times \mathbf{c}''(t)\|^2} = \frac{-1}{3 + 2\sin t \cos t - 2\sin t - 2\cos t}.$$

Chapter Review

True/false Quiz

1. False. By the Implicit Function Theorem, $y' = -(\partial F/\partial x)/(\partial F/\partial y)$.

3. True. See Example 3.38.

5. True. From $\mathbf{c} \cdot \mathbf{c} = $ constant we get $\mathbf{c} \cdot \mathbf{c}' + \mathbf{c}' \cdot \mathbf{c} = 0$, i.e., $\mathbf{c} \cdot \mathbf{c}' = 0$. So, $\mathbf{c}$ and $\mathbf{c}'$ are perpendicular to each other.

7. False. $\mathbf{c}(t)$ wraps around the circle of radius 1 four times. Thus, its length is 8π. Alternatively, $\int_{-\pi}^{\pi} \|\mathbf{c}'(t)\|\, dt = \int_{-\pi}^{\pi} 4\, dt = 8\pi$.

Review Exercises and Problems

1. Initial point is $(0, 2)$, and the terminal point is $(3, 2)$. From $t = 0$ to $t = 1$, the x coordinate changes linearly from 0 to 1 while the y coordinate remains fixed at 2. Thus, the object moves at constant speed (equal to 1), and when $t = 1$, it is located at $\mathbf{c}(1) = (1, 2)$.

Between $t = 1$ and $t = 2$, the x coordinate increases from 1 to 2 and the y coordinate decreases from 2 to 1. Thus, From $t = 1$ to $t = 2$ the object travels at constant speed (equal to $\sqrt{2}$), and when $t = 2$, it is located at $\mathbf{c}(2) = (2, 1)$.

From $t = 2$ to $t = 3$: constant speed (equal to 1), $\mathbf{c}(3) = (3, 1)$. From $t = 3$ to $t = 4$: object rests at $(3, 1)$. From $t = 4$ to $t = 5$: constant speed (equal to 1), object moves vertically to $\mathbf{c}(5) = (3, 2)$.

3. Substituting $x = 3-t$, $y = 2+2t$ and $z = 4$ into the equation of the cone $x^2 + y^2 = z^2$, we get $(3-t)^2 + (2+2t)^2 = 16$, i.e., $5t^2 + 2t - 3 = 0$. Solving this equation, we get $t = -1$ and $t = 3/5$. So, $\mathbf{l}(t)$ intersects the cone at two points, $\mathbf{l}(-1) = (4, 0, 4)$ and $\mathbf{l}(3/5) = (12/5, 16/5, 4)$.

5. Since $\mathbf{T}$, $\mathbf{N}$ and $\mathbf{B}$ are mutually orthogonal vectors, we can write $\mathbf{W} = a\mathbf{T} + b\mathbf{N} + c\mathbf{B}$, where a, b and c are scalars. It follows that

$$\mathbf{W} \times \mathbf{T} = a\mathbf{T} \times \mathbf{T} + b\mathbf{N} \times \mathbf{T} + c\mathbf{B} \times \mathbf{T} = -b\mathbf{B} + c\mathbf{N},$$

since $\mathbf{N} \times \mathbf{T} = -\mathbf{T} \times \mathbf{N} = -\mathbf{B}$ and $\mathbf{B} \times \mathbf{T} = \mathbf{N}$ (the TNB frame is a right coordinate system, so $\mathbf{T} \times \mathbf{N} = \mathbf{B}$, $\mathbf{N} \times \mathbf{B} = \mathbf{T}$ and $\mathbf{B} \times \mathbf{T} = \mathbf{N}$). Substituting $\mathbf{W} \times \mathbf{T} = -b\mathbf{B} + c\mathbf{N}$ and $d\mathbf{T}/ds = \kappa\mathbf{N}$ into $d\mathbf{T}/ds = \mathbf{W} \times \mathbf{T}$, we get $-b\mathbf{B} + c\mathbf{N} = \kappa\mathbf{N}$ and so $b = 0$ and $c = \kappa$. From

$$\mathbf{W} \times \mathbf{N} = (a\mathbf{T} + b\mathbf{N} + c\mathbf{B}) \times \mathbf{N} = a\mathbf{B} - c\mathbf{T}$$

and $d\mathbf{N}/ds = -\kappa\mathbf{T} + \tau\mathbf{B}$ it follows that $-\kappa\mathbf{T} + \tau\mathbf{B} = a\mathbf{B} - c\mathbf{T}$ and so $a = \tau$ and $c = \kappa$. Therefore, $\mathbf{W} = \tau\mathbf{T} + \kappa\mathbf{B}$. We found $\mathbf{W}$ — but still have to check that it satisfies the third equation. From the Serret-Frenet fromula (3.21) we get

$$\mathbf{W} \times \mathbf{B} = (\tau\mathbf{T} + \kappa\mathbf{B}) \times \mathbf{B} = \tau\mathbf{T} \times \mathbf{B} = -\tau\mathbf{N} = \frac{d\mathbf{B}(s)}{ds}.$$

7. Recall that $\kappa(s) = \|d\mathbf{T}(s)/ds\|$, where s is the arc-length parameter of $\mathbf{c}$. Consider the tangent vectors $\mathbf{T}(s)$ at $\mathbf{c}(s)$ and $\mathbf{T}(s + \Delta s)$ at $\mathbf{c}(s + \Delta s)$ and let $\theta(s)$ be the angle between $\mathbf{T}(s)$ and the positive x-axis and $\theta(s + \Delta s)$ the angle between $\mathbf{T}(s + \Delta s)$ and the positive x-axis. Now

$$\kappa(s) = \left\| \frac{d\mathbf{T}(s)}{ds} \right\| = \lim_{\Delta s \to 0} \frac{\|\mathbf{T}(s + \Delta s) - \mathbf{T}(s)\|}{|\Delta s|}.$$

Since $\mathbf{T}(s)$ and $\mathbf{T}(s + \Delta s)$ are unit vectors, $\|\mathbf{T}(s + \Delta s) - \mathbf{T}(s)\|$ is the length of the base of an isosceles triangle with sides of length 1 and angle $|\theta(s + \Delta s) - \theta(s)|$ opposite to the base; see Figure 7. Hence $\|\mathbf{T}(s + \Delta s) - \mathbf{T}(s)\| = 2\sin(|\theta(s + \Delta s) - \theta(s)|/2)$ and therefore

$$\left\| \frac{d\mathbf{T}(s)}{ds} \right\| = \lim_{\Delta s \to 0} \frac{\left| 2\sin\left((\theta(s + \Delta s) - \theta(s))/2\right) \right|}{|\Delta s|}.$$

Using the linear approximation for $\sin x$ at zero, we get $\sin x \approx x$ (for x close to 0). Hence $2\sin((\theta(s + \Delta s) - \theta(s))/2) \approx 2(\theta(s + \Delta s) - \theta(s))/2 = \theta(s + \Delta s) - \theta(s)$ and

$$\left\| \frac{d\mathbf{T}(s)}{ds} \right\| = \lim_{\Delta s \to 0} \frac{|\theta(s + \Delta s) - \theta(s)|}{|\Delta s|} = \lim_{\Delta s \to 0} \left| \frac{\theta(s + \Delta s) - \theta(s)}{\Delta s} \right| = \left| \frac{d\theta}{ds} \right|.$$

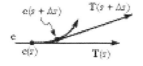 translate $\mathbf{T}(s)$ and $\mathbf{T}(s + \Delta s)$ so that their tails coincide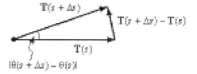

Figure 7

9. Let $\mathbf{c}(t) = (r(t), \theta(t))$ be the trajectory of the particle. We will show that

$$\frac{d}{dt}\left(r^2 \frac{d\theta}{dt} \right) = 2r \frac{dr}{dt} \frac{d\theta}{dt} + r^2 \frac{d^2\theta}{dt^2} = r(2r'\theta' + r\theta'') = 0.$$

From Newton's Second Law $\mathbf{F} = m\mathbf{a}$ and Exercise 23 in Section 3.2 it follows that

$$(f(r)\cos\theta, f(r)\sin\theta) = m\left[(r'' - r(\theta')^2)(\cos\theta, \sin\theta) + (2r'\theta' + r\theta'')(-\sin\theta, \cos\theta) \right]$$

and

$$(f(r)\cos\theta, f(r)\sin\theta) = \big(m(r'' - r(\theta')^2)\cos\theta - (2r'\theta' + r\theta'')\sin\theta,$$
$$m(r'' - r(\theta')^2)\sin\theta + (2r'\theta' + r\theta'')\cos\theta\big).$$

Comparing either the x-coordinates or the y-coordinates we get $2r'\theta' + r\theta'' = 0$, and so $(d/dt)(r^2\theta') = 0$; i.e., $r^2\theta'$ is constant.

11. From $\mathbf{c}(t) = (e^t\cos t, e^t\sin t, e^t)$ we get

$$\mathbf{v}(t) = (e^t\cos t - e^t\sin t, e^t\sin t + e^t\cos t, e^t) = e^t(\cos t - \sin t, \sin t + \cos t, 1);$$

so the speed of the particle, $\|\mathbf{v}(t)\| = e^t\sqrt{3}$, is increasing (and is hence non-constant). The acceleration is

$$\mathbf{a}(t) = \mathbf{v}'(t) = e^t(\cos t - \sin t, \sin t + \cos t, 1) + e^t(-\sin t - \cos t, \cos t - \sin t, 0)$$
$$= e^t(-2\sin t, 2\cos t, 1).$$

The magnitude of $\mathbf{a}(t)$ is non-constant, since $\|\mathbf{a}(t)\| = e^t\sqrt{5}$.

As a matter of fact, there was no need to compute the acceleration. The amount of increase in speed per unit time increases (since the speed $\|\mathbf{v}(t)\|$ depends exponentially on t), and so acceleration cannot be constant.

13. (a) Let us compute $(A(t) - A(t_0))/(t - t_0)$. (For convenience, assume that $t > t_0$.) The quantity $A(t) - A(t_0)$ is equal to the area of the ellipse between $\mathbf{r}(t)$ and $\mathbf{r}(t_0)$; see Figure 3.38 ($\overrightarrow{OP_1}$ could represent $\mathbf{r}(t)$, and $\overrightarrow{OP_2}$ could represent $\mathbf{r}(t_0)$). Assuming that the region between $\mathbf{r}(t)$ and $\mathbf{r}(t_0)$ is a sector of a disk, we get that

$$A(t) - A(t_0) = \tfrac{1}{2}\|\mathbf{r}(t)\|^2(\theta(t) - \theta(t_0)).$$

(Recall that the area of a sector of a disk of radius r and central angle θ (in radians) is $r^2\theta/2$.) In the above, we assumed that the elliptic sector is actually a sector of a disk of radius $\|\mathbf{r}(t)\|$.

Alternatively, we could arrive at the equivalent formula

$$A(t) - A(t_0) = \tfrac{1}{2}\|\mathbf{r}(t)\|\,\|\mathbf{r}(t_0)\|(\theta(t) - \theta(t_0))$$

by computing $1/2$ of the length of the cross product of $\mathbf{r}(t)$ and $\mathbf{r}(t_0)$ and using the fact that $\sin(\theta(t) - \theta(t_0)) \approx \theta(t) - \theta(t_0)$ when t and t_0 are close to each other. Thus,

$$A'(t_0) = \lim_{t \to t_0} \frac{A(t) - A(t_0)}{t - t_0} = \lim_{t \to t_0} \frac{1}{2}\frac{\|\mathbf{r}(t)\|^2(\theta(t) - \theta(t_0))}{t - t_0}$$
$$= \frac{1}{2}\|\mathbf{r}(t_0)\|^2 \lim_{t \to t_0} \frac{\theta(t) - \theta(t_0)}{t - t_0} = \frac{1}{2}\|\mathbf{r}(t_0)\|^2\theta'(t_0).$$

(b) Let $\mathbf{r} = \mathbf{r}(t) = (\|\mathbf{r}(t)\|\cos\theta(t), \|\mathbf{r}(t)\|\sin\theta(t))$. Then (drop t to keep notation simple, but keep in mind that $\mathbf{r}$ and θ are functions of t)

$$\mathbf{v} = \mathbf{r}' = (\|\mathbf{r}\|'\cos\theta - \|\mathbf{r}\|\theta'\sin\theta, \|\mathbf{r}\|'\sin\theta + \|\mathbf{r}\|\theta'\cos\theta),$$

and thus ($\mathbf{k}$ components of $\mathbf{r}$ and $\mathbf{v}$ are zero)

$$\mathbf{d} = \mathbf{r} \times \mathbf{v} = (\|\mathbf{r}\| \, \|\mathbf{r}\|' \cos\theta \sin\theta + \|\mathbf{r}\|^2 \theta' \cos^2\theta - \|\mathbf{r}\| \, \|\mathbf{r}\|' \cos\theta \sin\theta$$
$$+ \|\mathbf{r}\|^2 \theta' \sin^2\theta)\mathbf{k} = \|\mathbf{r}\|^2 \theta'\mathbf{k}.$$

Since $\mathbf{d} \cdot \mathbf{k} = \|\mathbf{d}\| \, \|\mathbf{k}\| \cos(\pi/2) = \|\mathbf{d}\|$, we get that $\|\mathbf{d}\| = \|\mathbf{r}(t)\|^2 \theta'(t)$.

(c) From (a) and (b), we conclude that $A'(t) = \|\mathbf{d}\|/2$. I.e., the time rate at which the area is swept by $\mathbf{r}(t)$ is constant.

15. We are asked to compute a from the equation $T^2 = 4\pi^2 a^3/GM$, where $T = 365.25$ days $= (365.25)(86400)$ seconds, $G = 6.67 \cdot 10^{-11}$ Nm^{2}kg^{-2}, and $M = 2 \cdot 10^{30}$ kg. It follows that

$$a^3 = \frac{GMT^2}{4\pi^2} = 0.3365 \cdot 10^{34}.$$

Thus, $a = 0.1499 \cdot 10^{12}$ metres.

17. Clearly, $\mathbf{c}_1(0) = (0,0)$, $\mathbf{c}_1(1) = (1,1)$; $\mathbf{c}_2(0) = (0,0)$, $\mathbf{c}_2(\pi) = (1,1)$; $\mathbf{c}_3(3\pi/2) = (0,0)$, $\mathbf{c}_3(2\pi) = (1,1)$; and $\mathbf{c}_4(0) = (0,1)$, $\mathbf{c}_4(1) = (1,1)$.

Next, we compute the lengths $\ell(\mathbf{c}_i)$ of $\mathbf{c}_i$ and the corresponding times T_i, $i = 1,2,3,4$.

The curve $\mathbf{c}_1$ is a straight line segment joining $(0,0)$ and $(1,1)$; so $\ell(\mathbf{c}_1) = \sqrt{2} \approx 1.41421$. The time T_1 for $\mathbf{c}_1$ is

$$T_1 = \frac{1}{\sqrt{2g}} \int_0^1 \sqrt{\frac{2}{t}} \, dt = \frac{1}{\sqrt{g}} \int_0^1 t^{-1/2} \, dt = 2\frac{1}{\sqrt{g}}.$$

From $\mathbf{c}_2'(t) = ((1 - \cos t)/\pi, \sin t/2)$ we get (using a computer)

$$\ell(\mathbf{c}_2) = \int_0^1 \left(\frac{(1-\cos t)^2}{\pi^2} + \frac{\sin^2 t}{4} \right)^{1/2} dt \approx 1.53775;$$

the time is

$$T_2 = \frac{1}{\sqrt{2g}} \int_0^1 \left(\frac{(1-\cos t)^2}{\pi^2} + \frac{\sin^2 t}{4} \right)^{1/2} \Big/ \left(\frac{(1-\cos t)}{2} \right)^{1/2} dt \approx 1.84030 \frac{1}{\sqrt{g}}.$$

The curve $\mathbf{c}_3$ is a quarter-circle of radius 1, so $\ell(\mathbf{c}_3) = \pi/2 \approx 1.57080$. The corresponding time T_3 is

$$T_3 = \frac{1}{\sqrt{2g}} \int_0^{\pi/2} \sqrt{\frac{1}{\cos t}} \, dt \approx 1.85407 \frac{1}{\sqrt{g}}.$$

From $\mathbf{c}_4'(t) = (1, (\pi/2)\cos(\pi t/2))$ we get (using a computer)

$$\ell(\mathbf{c}_4) = \int_0^1 \left(1 + \frac{\pi^2}{4}\cos^2(\pi t/2) \right)^{1/2} dt \approx 1.46370$$

and

$$T_4 = \frac{1}{\sqrt{2g}} \int_0^1 \left(\frac{1 + (\pi^2/4)\cos^2(\pi t/2)}{\sin(\pi t/2)} \right)^{1/2} dt \approx 1.88734 \frac{1}{\sqrt{g}}.$$

Contrary to (maybe) our expectations, the line is not the "fastest" curve (although it is the shortest); see Figure 17. The "fastest" curve is $\mathbf{c}_2$.

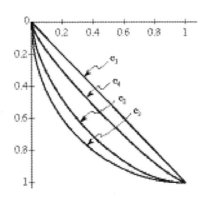

Figure 17

4. SCALAR AND VECTOR FIELDS

4.1. Higher Order Partial Derivatives

1. If we move away from P in the direction of the positive x-axis, we notice that the values of f decrease. Approaching P from the left along the horizontal line, we see that the values of f decrease as well. Thus, $f_x(P) < 0$. Since f does not change in vertical directions, $f_y(P) = 0$.

Recall that $f_{xx} = (f_x)_x$ gives information about the rate of change of changes in f in the direction of the x-axis. Since f decreases at a faster and faster rate, it follows that f_x is a decreasing function, and thus $f_{xx}(P) < 0$. As f does not change in the y-direction (f is a constant function when restricted to vertical lines), $f_{yy}(P) = 0$. Since f_x does not change in the y-direction, we conclude that $f_{xy}(P) = 0$.

3. Since f increases both in x- and y-directions, we conclude that $f_x(P) > 0$ and $f_y(P) > 0$. Both amounts of increase are constant, thus $f_{xx}(P) = 0$ and $f_{yy}(P) = 0$. The amounts of increase in f_x are constant along vertical lines, i.e., $f_{xy}(P) = 0$.

5. Write z as $z = e^{xy} + 2\ln x + 3\ln y$. Then $z_x = ye^{xy} + 2/x$, $z_y = xe^{xy} + 3/y$, $z_{xx} = y^2 e^{xy} - 2/x^2$, $z_{yy} = x^2 e^{xy} - 3/y^2$, and $z_{xy} = z_{yx} = e^{xy}(1 + xy)$.

7. Using the chain rule and the product rule over and over again, we get

$$z_x = \tfrac{5}{2}(x^2 + y^2)^{3/2}\, 2x = 5x(x^2 + y^2)^{3/2}$$

and

$$z_{xx} = 5(x^2 + y^2)^{3/2} + 5x\,\tfrac{3}{2}(x^2 + y^2)^{1/2}\, 2x$$
$$= 5(x^2 + y^2)^{3/2} + 15x^2(x^2 + y^2)^{1/2} = 5(x^2 + y^2)^{1/2}(4x^2 + y^2).$$

By symmetry (interchanging x and y),

$$z_y = 5y(x^2 + y^2)^{3/2} \quad\text{and}\quad z_{yy} = 5(x^2 + y^2)^{1/2}(x^2 + 4y^2).$$

Finally,

$$z_{xy} = z_{yx} = 5x\,\tfrac{3}{2}(x^2 + y^2)^{1/2}\, 2y = 15xy(x^2 + y^2)^{1/2}.$$

9. Using the chain rule and the product rule, we get

$$z_x = 2\sin(x + y)\cos(x + y) = \sin(2x + 2y),$$

using the formula $2\sin\theta\cos\theta = \sin 2\theta$ with $\theta = x + y$. So $z_{xx} = 2\cos(2x + 2y)$, and, by symmetry, $z_{yy} = 2\cos(2x + 2y)$. Finally, $z_{xy} = z_{yx} = 2\cos(2x + 2y)$.

11. Using $'$ to denote the derivatives of f and g, we get $z_x = f'(ax + by)\,a + g'(ax/y)\,(a/y)$ and

$$z_{xx} = f''(ax + by)\,a\,a + g''(ax/y)\frac{a}{y}\frac{a}{y} = a^2\left(f''(ax + by) + \frac{g''(ax/y)}{y^2}\right).$$

Similarly, $z_y = f'(ax+by)\,b + g'(ax/y)\,(-ax/y^2)$ and

$$z_{yy} = f''(ax+by)\,b\,b + g''(ax/y)\frac{-ax}{y^2}\frac{-ax}{y^2} + g'(ax/y)\frac{2ax}{y^3}$$

$$= b^2 f''(ax+by) + \frac{a^2 x^2 g''(ax/y)}{y^4} + \frac{2axg'(ax/y)}{y^3}.$$

Starting with z_x, and computing the derivative with respect to y, we get

$$z_{xy} = f''(ax+by)\,b\,a + g''(ax/y)\frac{-ax}{y^2}\frac{a}{y} + g'(ax/y)\frac{-a}{y^2}$$

$$= abf''(ax+by) - \frac{a^2 x g''(ax/y)}{y^3} - \frac{ag'(ax/y)}{y^2}.$$

Similarly, $z_{yx} = z_{xy}$.

13. The function w is of class C^k, for any k, so we can compute partial derivatives in any order. Since the first term has no z, start with $w_z = 4x^3 y^2 z^3$. Now $w_{zx} = 12x^2 y^2 z^3$, $w_{zxx} = 24xy^2 z^3$ and $w_{xyzx} = w_{zxxy} = 48xyz^3$.

15. A straightforward computation gives $z_x = e^y + ye^x$, $z_{xx} = ye^x$ and $z_{xxx} = ye^x$. Similarly, $z_y = xe^y + e^x$, $z_{yy} = z_{yyy} = xe^y$, and it follows that $z_{xxx} + z_{yyy} = xe^y + ye^x = z$. Next, we compute the right side. From $z_{xy} = z_{yx} = e^y + e^x$ it follows that $z_{xyy} = e^y$, $z_{yxx} = e^x$ and

$$xz_{xyy} + yz_{yxx} = xe^y + ye^x = z;$$

so both sides are equal to z.

17. Let $f(x,y,z)$ be a C^2 function. It has three partial derivatives in which both derivatives are done with respect to the same variable: f_{xx}, f_{yy} and f_{zz}. There are three other distinct partial derivatives: $f_{xy} = f_{yx}$, $f_{xz} = f_{zx}$ and $f_{yz} = f_{zy}$. Therefore, a C^2 function of three variables has six different second-order partial derivatives.

In general, let $f(x_1, \ldots, x_m)$ be a C^2 function. Organize all second-order partial derivatives in a matrix (call it Hf), so that the ij-th entry of Hf is $\partial^2 f/\partial x_i \partial x_j$. The entries on the main diagonal of Hf are $\partial^2 f/\partial x_1^2, \ldots, \partial^2 f/\partial x_m^2$. If $i \neq j$, then $\partial^2 f/\partial x_i \partial x_j = \partial^2 f/\partial x_j \partial x_i$, since f is C^2. Notice that one of these two derivatives is above the main diagonal, and the other (symmetric with respect to the diagonal) is below it. As a matter of fact, every $\partial^2 f/\partial x_i \partial x_j$ above the diagonal has its matching pair $\partial^2 f/\partial x_j \partial x_i$ below the diagonal (and vice versa). Hence all distinct partial derivatives are either on the diagonal or above it (or on the diagonal and below it). There are m entries on the diagonal, and a total of $m^2 - m$ non-diagonal entries. Half of them are above (and half below) the diagonal; hence the total number of distinct second-order partial derivatives is

$$m + \tfrac{1}{2}(m^2 - m) = \tfrac{1}{2}m^2 + \tfrac{1}{2}m = \tfrac{1}{2}m(m+1).$$

19. Let f', f'', g' and g'' denote the derivatives of f and g. Then $u_x = f + xf' + yg'$, where (here and in what follows) f, g and their derivatives are computed at $x+y$). Computing the derivatives of u_x with respect to x and y, we get $u_{xx} = f' + f' + xf'' + yg'' = 2f' + xf'' + yg''$

and $u_{xy} = f' + xf'' + g' + yg''$. Similarly, $u_y = xf' + g + yg'$ and $u_{yy} = xf'' + g' + g' + yg'' = 2g' + xf'' + yg''$. It follows that

$$u_{xx} - 2u_{xy} + u_{yy} = 2f' + xf'' + yg'' - 2(f' + xf'' + g' + yg'') + 2g' + xf'' + yg'' = 0.$$

21. Starting with $u_{xx} = -4k^4 \left(-3\cosh^{-4}A \cdot \sinh^2 A + \cosh^{-2}A\right)$, we get

$$u_{xxx} = -4k^4 \left(12\cosh^{-5}A \cdot \sinh A \cdot k \sinh^2 A\right.$$
$$-3\cosh^{-4}A \cdot 2\sinh A \cdot \cosh A \cdot k - 2\cosh^{-3}A \cdot \sinh A \cdot k\right)$$
$$= -4k^5 \left(12\cosh^{-5}A \cdot \sinh^3 A - 8\cosh^{-3}A \cdot \sinh A\right).$$

Therefore

$$u_t + 6uu_x + u_{xxx} = k^5 \left(16\cosh^{-3}A \cdot \sinh A - 48\cosh^{-5}A \cdot \sinh A\right.$$
$$-48\cosh^{-5}A \cdot \sinh^3 A + 32\cosh^{-3}A \cdot \sinh A\right)$$
$$= 48k^5 \cosh^{-5}A \left(\cosh^2 A \cdot \sinh A - \sinh A - \sinh^3 A\right)$$
$$= 48k^5 \cosh^{-5}A \left(\sinh A(-1 + \cosh^2 A) - \sinh^3 A\right)$$
$$= 48k^5 \cosh^{-5}A \left(\sinh A(\sinh^2 A) - \sinh^3 A\right) = 0.$$

23. From $f(x, y) = \arctan(y/x)$ we get

$$f_x = \frac{1}{1 + (y/x)^2} \frac{-y}{x^2} = \frac{-y}{x^2 + y^2},$$

and, similarly, $f_y = x/(x^2 + y^2)$. Differentiating again, we get $f_{yy} = -2xy/(x^2 + y^2)^2$ and

$$f_{xx} = -y(-1)(x^2 + y^2)^{-2}(2x) = \frac{2xy}{(x^2 + y^2)^2}.$$

Thus, $f_{xx} + f_{yy} = 0$.

25. From $V = -GMm(x^2 + y^2 + z^2)^{-1/2}$ we get, by the chain rule, $V_x = GMmx(x^2 + y^2 + z^2)^{-3/2}$. Proceed using the product rule and the chain rule:

$$V_{xx} = \frac{GMm(x^2 + y^2 + z^2)^{3/2} - GMmx \frac{3}{2}(x^2 + y^2 + z^2)^{1/2} 2x}{(x^2 + y^2 + z^2)^3}$$
$$= \frac{GMm(x^2 + y^2 + z^2)^{1/2}(x^2 + y^2 + z^2 - 3x^2)}{(x^2 + y^2 + z^2)^3} = \frac{GMm(y^2 + z^2 - 2x^2)}{(x^2 + y^2 + z^2)^{5/2}}.$$

By symmetry,

$$V_{yy} = \frac{GMm(x^2 + z^2 - 2y^2)}{(x^2 + y^2 + z^2)^{5/2}} \quad \text{and} \quad V_{zz} = \frac{GMm(x^2 + y^2 - 2z^2)}{(x^2 + y^2 + z^2)^{5/2}}.$$

Adding up V_{xx}, V_{yy} and V_{zz} we get zero. Notice that the computation works whenever $(x, y, z) \neq (0, 0, 0)$.

27. (a) The initial temperature at left end is $T(0, 0) = 1$; at the right end, $T(\pi, 0) = 1$. When $t = 0$, $T(x, 0) = 1 + \sin x$. Thus, the warmest point is the midpoint of the rod (i.e., when $x = \pi/2$), where $T(\pi/2, 0) = 2$.

(b) From $T_t = -e^{-t}\sin x$, $T_x = e^{-t}\cos x$ and $T_{xx} = -e^{-t}\sin x$ we get $T_t = T_{xx}$.

(c) See figure below. As $t \to \infty$, $T(x,t) \to 1$ for all x.

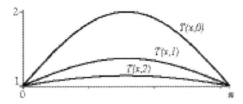

Figure 27

4.2. Taylor's Formula

1. If $h < 0$, then $\int_{x_0}^{x_0+h} |f(t)|\, dt < 0$ and $|\int_{x_0}^{x_0+h} f(t)\, dt| \geq 0$, and so the formula

$$\left| \int_{x_0}^{x_0+h} f(t)\, dt \right| \leq \int_{x_0}^{x_0+h} |f(t)|\, dt$$

no longer works.

The mentioned step is correct because

$$\left| \int_a^b f(t)\, dt \right| = \left| -\int_b^a f(t)\, dt \right| = \left| \int_b^a f(t)\, dt \right|$$

for all a, b.

If $h < 0$, the estimate (4.10) is

$$|R_1(x_0, h)| = \left| \int_{x_0}^{x_0+h} (x_0 + h - t) f''(t)\, dt \right| = \left| \int_{x_0+h}^{x_0} (x_0 + h - t) f''(t)\, dt \right|$$

$$\leq \int_{x_0+h}^{x_0} |x_0 + h - t|\, |f''(t)|\, dt \leq \int_{x_0+h}^{x_0} |h|\, M\, dt = |h|\, M\, (-h) = |h|^2 M,$$

since $h < 0$ and thus $|h| = -h$.

3. We start with

$$f(x_0 + h) = f(x_0) + f'(x_0)h + \frac{f''(x_0)}{2}h^2 + \frac{1}{2}\int_{x_0}^{x_0+h} (x_0 + h - t)^2 f'''(t)\, dt.$$

Applying integration by parts, with $dv = (x_0 + h - t)^2\, dt$ and $u = f'''(t)$, we get ($du = f^{(4)}(t)dt$ and $v = -(x_0 + h - t)^3/3$)

$$\int_{x_0}^{x_0+h} (x_0 + h - t)^2 f'''(t)\, dt = -\frac{(x_0 + h - t)^3}{3} f'''(t) \Big|_{x_0}^{x_0+h} + \int_{x_0}^{x_0+h} \frac{(x_0 + h - t)^3}{3} f^{(4)}\, dt$$

$$= \frac{h^3}{3} f'''(x_0) + \frac{1}{3}\int_{x_0}^{x_0+h} (x_0 + h - t)^3 f^{(4)}(t)\, dt.$$

Thus,

$$f(x_0 + h) = f(x_0) + f'(x_0)h + \frac{f''(x_0)}{2}h^2 + \frac{f'''(x_0)}{3!}h^3 + \frac{1}{3!}\int_{x_0}^{x_0+h}(x_0 + h - t)^3 f^{(4)}(t)\,dt.$$

The remainder

$$R_3(x_0, h) = \frac{1}{3!}\int_{x_0}^{x_0+h}(x_0 + h - t)^3 f^{(4)}(t)\,dt$$

can be estimated (imitating Exercise 2) as

$$|R_3(x_0, h)| \le \frac{1}{3!}\int_{x_0}^{x_0+h}|x_0 + h - t|^3\,|f^{(4)}|\,dt \le \frac{M}{3!}\int_{x_0}^{x_0+h}|x_0 + h - t|^3\,dt,$$

where $|f^{(4)}|(t) \le M$ for all t in $[x_0, x_0+h]$ (this is true as long as f is C^4). Since $|x_0+h-t| \le |h|$ for t in $[x_0, x_0 + h]$, we proceed as follows:

$$|R_3(x_0, h)| \le \frac{M}{3!}|h|^3\int_{x_0}^{x_0+h} dt \le \frac{M|h|^4}{3!}.$$

Clearly,

$$0 \le \lim_{h\to 0}\frac{|R_3(x_0, h)|}{|h|^3} \le \lim_{h\to 0}\frac{M|h|}{3!} = 0.$$

5. From $f(x) = \sqrt{x}$, we get $f(3) = \sqrt{3}$; $f'(x) = \frac{1}{2\sqrt{x}}$, thus $f'(3) = \frac{1}{2\sqrt{3}} = \sqrt{3}/6$; $f''(x) = -\frac{1}{4x^{3/2}}$, thus $f''(3) = -\frac{1}{4(3)\sqrt{3}} = -\sqrt{3}/36$. We will also need $f'''(x) = \frac{3}{8x^{5/2}}$. It follows that

$$T_2(3, h) = \sqrt{3} + \frac{\sqrt{3}}{6}h - \frac{\sqrt{3}}{72}h^2,$$

where $R_2(3, h) = \frac{1}{2}\int_3^{3+h}(3 + h - t)^2 f'''(t)\,dt$. (In (perhaps) more familiar form, $T_2(3, h)$ can be written as

$$T_2(x) = \sqrt{3} + \frac{\sqrt{3}}{6}(x - 3) - \frac{\sqrt{3}}{72}(x - 3)^2,$$

and $R_2(3, h)$ as $R_2(x) = \frac{1}{2}\int_3^x(x - t)^2 f'''(t)\,dt$.)

Given $2 \le x \le 4$, it follows that $x^{5/2} \ge 2^{5/2}$, and so $|f'''(t)| \le \frac{3}{8\cdot 2^{5/2}} = \frac{3}{32\sqrt{2}}$. Thus,

$$|R_2(3, h)| \le \frac{M|h|^3}{2} \le \frac{1}{2}\frac{3}{32\sqrt{2}}1^3 = \frac{3}{64\sqrt{2}},$$

since $|h| \le 1$.

7. From $f(x) = \sin x$, we get $f(\pi/4) = \sqrt{2}/2$; $f'(x) = \cos x$, thus $f'(\pi/4) = \sqrt{2}/2$; $f''(x) = -\sin x$, thus $f''(\pi/4) = -\sqrt{2}/2$. We will also need $f'''(x) = -\cos x$. The second order Taylor formula can be written as

$$T_2(\pi/4, h) = \frac{\sqrt{2}}{2} + \frac{\sqrt{2}}{2}h - \frac{\sqrt{2}}{4}h^2,$$

where $R_2(\pi/4, h) = \frac{1}{2}\int_{\pi/4}^{\pi/4+h}(\frac{\pi}{4} + h - t)^2(-\cos t)\,dt$. Or,

$$T_2(x) = \frac{\sqrt{2}}{2} + \frac{\sqrt{2}}{2}\left(x - \tfrac{\pi}{4}\right) - \frac{\sqrt{2}}{4}\left(x - \tfrac{\pi}{4}\right)^2,$$

where $R_2(x) = \frac{1}{2}\int_{\pi/4}^{x}(x-t)^2(-\cos t)\,dt$. (Keep in mind that $x = x_0 + h$, i.e., $x = \frac{\pi}{4} + h$ and $h = x - \frac{\pi}{4}$.)

9. From $f(x) = \ln x$, we get $f(4) = \ln 4$; $f'(x) = 1/x$, thus $f'(4) = 1/4$; $f''(x) = -1/x^2$, thus $f''(4) = -1/16$. We will also need $f'''(x) = 2/x^3$. The second order Taylor formula can be written as

$$T_2(4,h) = \ln 4 + \frac{1}{4}h - \frac{1}{32}h^2,$$

where $R_2(4,h) = \frac{1}{2}\int_4^{4+h}(4+h-t)^2 2t^{-3}\,dt$. Or (keep in mind that $x = 4+h$ and $h = x - 4$),

$$T_2(x) = \ln 4 + \frac{1}{4}(x-4) - \frac{1}{32}(x-4)^2,$$

where $R_2(x) = \int_4^x (x-t)^2 t^{-3}\,dt$.

11. Using terms up to and including x^2 in the MacLaurin series expansions for $\sin x$ and $\cos x$, we get $T_2(x) = x + 1 - x^2/2 = 1 + x - x^2/2$. Similarly, excluding all terms aboove x^3, we get $T_3(x) = x - x^3/6 + 1 - x^2/2 = 1 + x - x^2/2 - x^3/6$. (Of course, we could have computed T_2 and T_3 in the usual way, by calculating derivatives.)

The approximations are $T_2(0.1) = 1.0950000$ and $T_3(0.1) = 1.0948333$. Exact value (as a matter of fact, just a better approximation) is $\sin 0.1 + \cos 0.1 = 1.0948376$.

13. From $\partial F/\partial y = 3y^2 - 4$ we get that $(\partial F/\partial y)(0,2) = 8 \neq 0$. Using the Implicit Function Theorem, we conclude that $F(x,y) = y^3 - 4y + x^2 = 0$ defines y (uniquely and) implicitly as a function $y = g(x)$, near $(0,2)$.

Computing the derivative of $F(x,y) = y^3 - 4y + x^2 = 0$ with respect to x, we get

$$3y^2 y' - 4y' + 2x = 0,$$

i.e., (substitute $x = 0$, $y = 2$) $8y' = 0$. So, $y'|_{(0,2)} = g'(0) = 0$. Note: we could have obtained the same answer from the Implicit Function Theorem: since $g'(x) = -(\partial F/\partial x)/(\partial F/\partial y) = -2x/(3y^2 - 4)$, we compute $g'(0) = 0$.

Starting from $3y^2 y' - 4y' + 2x = 0$, we compute another derivative:

$$6y(y')^2 + 3y^2 y'' - 4y'' + 2 = 0,$$

i.e., $y'' = (-6y(y')^2 - 2)/(3y^2 - 4)$, and so $g''(0) = y''|_{(0,2)} = -2/8 = -1/4$.

Thus, $T_2(x) = 2 - x^2/8$.

15. Let $f(x) = xe^{-x}$. We will approximate $f(0.2)$ using the the second-order Taylor polynomial T_2 at 0. From $f(x) = xe^{-x}$, we get $f(0) = 0$; $f'(x) = (1-x)e^{-x}$, thus $f'(0) = 1$; $f''(x) = (x-2)e^{-x}$, thus $f''(0) = -2$. We will also need $f'''(x) = (3-x)e^{-x}$. The second order Taylor polynomial is $T_2(0,h) = h - h^2$, and thus $T_2(0,0.2) = 0.16$. The error term is given by

$$R_2(0,h) = \frac{1}{2}\int_0^h (h-t)^2 f'''(t)\,dt = \frac{1}{2}\int_0^h (h-t)^2(3-t)e^{-t}\,dt.$$

Because we are interested in $f(0.2)$, we can assume that $0 \le h \le 1$. Since $0 \le t \le h$, we conclude that $(h-t)^2 \le h^2$. Combining with the inequalities $|3-t| \le 3$ and $e^{-t} \le 1$, we get

$$|R_2(0,h)| \le \frac{1}{2} \int_0^h h^2(3)(1)\, dt = \frac{3}{2} h^3.$$

When $h = 0.2$, we get $|R_2(0, 0.2)| \le 3(0.2)^3/2 = 0.012$.

Indeed, the exact value $0.2e^{-0.2} = 0.16374615$ differs from our estimate $T_2(0, 0.2) = 0.16$ by 0.00374615, which is smaller than 0.012.

17. Using Theorem 4.5, we write

$$T_2(\mathbf{x}_0, \mathbf{h}) = f(\mathbf{x}_0) + \nabla f(\mathbf{x}_0) \cdot \mathbf{h} + \tfrac{1}{2}\mathbf{h}^t Hf(\mathbf{x}_0)\,\mathbf{h},$$

where $\mathbf{x}_0 = (x_0, y_0, z_0)$, $\mathbf{h} = (h_1, h_2, h_3)$, and

$$Hf(\mathbf{x}_0) = \begin{bmatrix} f_{xx}(\mathbf{x}_0) & f_{xy}(\mathbf{x}_0) & f_{xz}(\mathbf{x}_0) \\ f_{yx}(\mathbf{x}_0) & f_{yy}(\mathbf{x}_0) & f_{yz}(\mathbf{x}_0) \\ f_{zx}(\mathbf{x}_0) & f_{zy}(\mathbf{x}_0) & f_{zz}(\mathbf{x}_0) \end{bmatrix}.$$

Thus,

$$T_2(\mathbf{x}_0, \mathbf{h}) = f(\mathbf{x}_0) + f_x(\mathbf{x}_0)h_1 + f_y(\mathbf{x}_0)h_2 + f_z(\mathbf{x}_0)h_3 + \frac{1}{2}\Big(f_{xx}(\mathbf{x}_0)h_1^2$$
$$+ 2f_{xy}(\mathbf{x}_0)h_1h_2 + 2f_{xz}(\mathbf{x}_0)h_1h_3 + f_{yy}(\mathbf{x}_0)h_2^2 + 2f_{yz}(\mathbf{x}_0)h_2h_3 + f_{zz}(\mathbf{x}_0)h_3^2 \Big).$$

If $f = f(x_1, x_2, \ldots, x_m)$ and $\mathbf{h} = (h_1, h_2, \ldots, h_m)$, then

$\mathbf{h}^t Hf(\mathbf{x}_0)\,\mathbf{h}$

$$= [h_1 \quad h_2 \quad \ldots \quad h_m] \begin{bmatrix} f_{x_1 x_1}(\mathbf{x}_0) & f_{x_1 x_2}(\mathbf{x}_0) & \ldots & f_{x_1 x_m}(\mathbf{x}_0) \\ f_{x_2 x_1}(\mathbf{x}_0) & f_{x_2 x_2}(\mathbf{x}_0) & \ldots & f_{x_2 x_m}(\mathbf{x}_0) \\ \vdots & \vdots & & \vdots \\ f_{x_m x_1}(\mathbf{x}_0) & f_{x_m x_2}(\mathbf{x}_0) & \ldots & f_{x_m x_m}(\mathbf{x}_0) \end{bmatrix} \begin{bmatrix} h_1 \\ h_2 \\ \vdots \\ h_m \end{bmatrix}$$

$$= [f_{x_1 x_1}h_1 + f_{x_2 x_1}h_2 + \cdots + f_{x_m x_1}h_m \quad f_{x_1 x_2}h_1 + f_{x_2 x_2}h_2 + \cdots + f_{x_m x_2}h_m$$

$$\cdots \quad f_{x_1 x_m}h_1 + f_{x_2 x_m}h_2 + \cdots + f_{x_m x_m}h_m] \begin{bmatrix} h_1 \\ h_2 \\ \vdots \\ h_m \end{bmatrix}$$

$$= f_{x_1 x_1}h_1^2 + f_{x_2 x_1}h_1 h_2 + \cdots + f_{x_m x_1}h_1 h_m + f_{x_1 x_2}h_1 h_2 + f_{x_2 x_2}h_2^2 + \cdots + f_{x_m x_2}h_2 h_m$$
$$+ \cdots + f_{x_1 x_m}h_1 h_m + f_{x_2 x_m}h_2 h_m + \cdots + f_{x_m x_m}h_m^2$$
$$= \sum_{i=1}^m \sum_{j=1}^m f_{x_i x_j}h_i h_j,$$

where all partial derivatives are computed at $\mathbf{x}_0$. Thus,

$$T_2(\mathbf{x}_0, \mathbf{h}) = f(\mathbf{x}_0) + \sum_{i=1}^m \frac{\partial f}{\partial x_i}(\mathbf{x}_0)\,h_i + \frac{1}{2}\left(\sum_{i=1}^m \sum_{j=1}^m \frac{\partial^2 f}{\partial x_i \partial x_j}(\mathbf{x}_0)\,h_i h_j \right).$$

19. (a) Let $\mathbf{x}_0 = (x_0, y_0)$ and $\mathbf{h} = (h_1, h_2)$. Keeping $\mathbf{x}_0$ and $\mathbf{h}$ fixed, we define the function

$$F(t) = f(\mathbf{x}_0 + t\mathbf{h}) = f(x_0 + th_1, y_0 + th_2).$$

Note that $F(0) = f(\mathbf{x}_0) = f(x_0, y_0)$ and $F(1) = f(\mathbf{x}_0 + \mathbf{h}) = f(x_0 + h_1, y_0 + h_2)$. F is a function of one variable, so we apply (4.11) to get $F(t_0 + h) = F(t_0) + F'(t_0)h + R_1(t_0, h)$, where

$$R_1(t_0, h) = \int_{t_0}^{t_0+h} (t_0 + h - t)F''(t)\, dt.$$

In particular, when $t_0 = 0$ and $h = 1$, we get $F(1) = F(0) + F'(0) + R_1(0, 1)$, and

$$R_1(0, 1) = \int_0^1 (1 - t)F''(t)\, dt.$$

From

$$F''(t) = f_{xx}(\mathbf{x}_0 + t\mathbf{h})\, h_1^2 + f_{xy}(\mathbf{x}_0 + t\mathbf{h})\, h_1 h_2 + f_{yx}(\mathbf{x}_0 + t\mathbf{h})\, h_1 h_2 + f_{yy}(\mathbf{x}_0 + t\mathbf{h})\, h_2^2$$

(see calculation following (4.18)) we obtain

$$R_1(0, 1) = \int_0^1 (1 - t)(f_{xx}(\mathbf{x}_0 + t\mathbf{h})h_1^2 + 2f_{xy}(\mathbf{x}_0 + t\mathbf{h})h_1 h_2 + f_{yy}(\mathbf{x}_0 + t\mathbf{h})h_2^2)\, dt.$$

Since the integral involves $\mathbf{x}_0$ and $\mathbf{h}$, we use the notation $R(\mathbf{x}_0, \mathbf{h})$ instead of $R_1(0, 1)$.

(b) Applying the Mean Value Theorem with $g(t) = f_{xx}(\mathbf{x}_0 + t\mathbf{h})$ and $h(t) = 1 - t$, we conclude that there exists $\mathbf{c}_{11}$ on the line between $\mathbf{x}_0 + 0 \cdot \mathbf{h} = \mathbf{x}_0$ and $\mathbf{x}_0 + 1 \cdot \mathbf{h} = \mathbf{x}_0 + \mathbf{h}$ such that

$$\int_0^1 (1 - t)f_{xx}(\mathbf{x}_0 + t\mathbf{h})h_1^2\, dt = f_{xx}(\mathbf{c}_{11})h_1^2 \int_0^1 (1 - t)\, dt = \tfrac{1}{2}f_{xx}(\mathbf{c}_{11})h_1^2.$$

Similarly, there exist $\mathbf{c}_{12}$ and $\mathbf{c}_{22}$ on the line joining $\mathbf{x}_0$ and $\mathbf{x}_0 + \mathbf{h}$ such that

$$\int_0^1 (1 - t)f_{xy}(\mathbf{x}_0 + t\mathbf{h})h_1 h_2\, dt = \tfrac{1}{2}f_{xy}(\mathbf{c}_{12})h_1 h_2$$

and

$$\int_0^1 (1 - t)f_{yy}(\mathbf{x}_0 + t\mathbf{h})h_2^2\, dt = \tfrac{1}{2}f_{yy}(\mathbf{c}_{22})h_2^2.$$

Thus,

$$R_1(\mathbf{x}_0, \mathbf{h}) = R_1(0, 1) = \tfrac{1}{2}\Big(f_{xx}(\mathbf{c}_{11})h_1^2 + 2f_{xy}(\mathbf{c}_{12})h_1 h_2 + f_{yy}(\mathbf{c}_{22})h_2^2\Big).$$

(c) As above,

$$R_1(\mathbf{x}_0, \mathbf{h}) = \frac{1}{2}\sum_{i=1}^{m}\sum_{j=1}^{m} \frac{\partial^2 f}{\partial x_i \partial x_j}(\mathbf{c}_{ij})\, h_i h_j,$$

where $\mathbf{h} = (h_1, h_2, \ldots, h_m)$ and $\mathbf{c}_{ij}$, $i, j = 1, \ldots, m$, lie on the line joining $\mathbf{x}_0$ and $\mathbf{x}_0 + \mathbf{h}$.

21. As in the derivation of the second-order Taylor formula, we define

$$F(t) = f(\mathbf{x}_0 + t\mathbf{h}) = f(x_0 + th_1, y_0 + th_2).$$

Now Theorem 4.2 gives

$$F(t_0 + h) = F(t_0) + F'(t_0)h + \frac{F''(t_0)}{2!}h^2 + \frac{F'''(t_0)}{3!}h^3 + R_3(t_0, h),$$

where

$$R_3(t_0, h) = \frac{1}{3!} \int_{t_0}^{t_0+h} (t_0 + h - t)^3 F^{(4)}(t)\, dt.$$

When $t_0 = 0$ and $h = 1$, we get

$$F(1) = F(0) + F'(0)h + \frac{F''(0)}{2!}h^2 + \frac{F'''(0)}{3!}h^3 + R_3(0, 1),$$

(keep in mind that $F(0) = f(x_0, y_0)$ and $F(1) = f(x_0 + h_1, y_0 + h_2)$) and

$$R_3(t_0, h) = R_3(0, 1) = \frac{1}{3!} \int_0^1 (1 - t)^3 F^{(4)}(t)\, dt.$$

In the text we already computed $F'(t)$, $F''(t)$ and $F'(0)$, $F''(0)$, see text immediately following (4.18). Staring from $F''(t)$, we compute

$$F'''(t) = f_{xxx} h_1^3 + f_{xxy} h_1^2 h_2 + f_{xyx} h_1^2 h_2 + f_{xyy} h_1 h_2^2 + f_{yxx} h_1^2 h_2 + f_{yxy} h_1 h_2^2$$
$$+ f_{yyx} h_1 h_2^2 + f_{yyy} h_2^3,$$

(all derivatives are computed at $\mathbf{x}_0 + t\mathbf{h} = (x_0 + th_1, y_0 + th_2)$) and thus

$$F'''(0) = f_{xxx}(x_0, y_0) h_1^3 + 3f_{xxy}(x_0, y_0) h_1^2 h_2 + 3f_{xyy}(x_0, y_0) h_1 h_2^2 + f_{yyy}(x_0, y_0) h_2^3.$$

Substituting $F'(0)$, $F''(0)$ and $F'''(0)$, we get

$$f(x_0 + h_1, y_0 + h_2) = f(x_0, y_0) + f_x(x_0, y_0)h_1 + f_y(x_0, y_0)h_2$$
$$+ \frac{1}{2!} \left(f_{xx}(x_0, y_0)h_1^2 + 2f_{xy}(x_0, y_0)h_1 h_2 + f_{yy}(x_0, y_0)h_2^2 \right)$$
$$+ \frac{1}{3!} \left(f_{xxx}(x_0, y_0)h_1^3 + 3f_{xxy}(x_0, y_0)h_1^2 h_2 + 3f_{xyy}(x_0, y_0)h_1 h_2^2 \right.$$
$$\left. + f_{yyy}(x_0, y_0)h_2^3 \right) + R_3(0, 1),$$

where $R_3(0, 1) = \frac{1}{3!} \int_0^1 (1 - t)^3 G(t)\, dt$, and $G(t)$ involves fourth derivatives of f. In particular, as in Exercise 16, we obtain

$$G(t) = f_{xxxx}(x_0 + th_1, y_0 + th_2)h_1^4 + 4f_{xxxy}(x_0 + th_1, y_0 + th_2)h_1^3 h_2$$
$$+ 6f_{xxyy}(x_0 + th_1, y_0 + th_2)h_1^2 h_2^2 + 4f_{xyyy}(x_0 + th_1, y_0 + th_2)h_1 h_2^3$$
$$+ f_{yyyy}(x_0 + th_1, y_0 + th_2)h_2^4.$$

23. From $f(x, y) = x^2 + y^2 - 2xy + 1$, we get $f(1, 1) = 1$. First partial derviatives are $f_x = 2x - 2y$, $f_x(1, 1) = 0$, and $f_y = 2y - 2x$, $f_y(1, 1) = 0$.

Second partial derivatives are $f_{xx} = 2$, $f_{xy} = -2$ and $f_{yy} = 2$. Thus,

$$T_2(x, y) = 1 + 0(x - 1) + 0(y - 1) + \tfrac{1}{2}\left(2(x - 1)^2 - 4(x - 1)(y - 1) + 2(y - 1)^2 \right)$$
$$= 1 + (x - 1)^2 - 2(x - 1)(y - 1) + (y - 1)^2.$$

Note: there was no need to do any calculations!

25. From $f(x, y) = \sin x + \sin 2y$ we get $f(0, \pi/2) = 0$. First partial derviatives are $f_x = \cos x$, $f_x(0, \pi/2) = 1$, and $f_y = 2\cos 2y$, $f_y(0, \pi/2) = -2$.

Second partials are $f_{xx} = -\sin x$, $f_{xx}(0, \pi/2) = 0$, $f_{xy} = 0$, and $f_{yy} = -4\sin 2y$, $f_{yy}(0, \pi/2) = 0$. Thus,

$$T_2(x, y) = 0 + 1(x - 0) - 2(y - \tfrac{\pi}{2}) + \tfrac{1}{2}\left(0(x - 0)^2 + 0(x - 0)(y - \tfrac{\pi}{2}) + 0(y - \tfrac{\pi}{2})^2\right)$$
$$= x - 2y + \pi.$$

27. From $f(x, y) = x^{-1}y^{-1}$, we get $f(1, 2) = 1/2$. First partial derviatives are $f_x = -x^{-2}y^{-1}$, $f_x(1, 2) = -1/2$, and $f_y = -x^{-1}y^{-2}$, $f_y(1, 2) = -1/4$.

Second partials are $f_{xx} = 2x^{-3}y^{-1}$, $f_{xx}(1, 2) = 1$, $f_{xy} = x^{-2}y^{-2}$, $f_{xy}(1, 2) = 1/4$, and $f_{yy} = 2x^{-1}y^{-3}$, $f_{yy}(1, 2) = 1/4$. Thus,

$$T_2(x, y) = \tfrac{1}{2} - \tfrac{1}{2}(x - 1) - \tfrac{1}{4}(y - 2) + \tfrac{1}{2}(x - 1)^2 + \tfrac{1}{4}(x - 1)(y - 2) + \tfrac{1}{8}(y - 2)^2.$$

29. From $f(x, y) = \sqrt{x + 4y - 1}$ we get $f(5, 3) = 4$. First partial derviatives are $f_x = (x + 4y - 1)^{-1/2}/2$, $f_x(5, 3) = 1/8$, and $f_y = 2(x + 4y - 1)^{-1/2}$, $f_y(5, 3) = 1/2$.

Second partials are $f_{xx} = -(x + 4y - 1)^{-3/2}/4$, $f_{xx}(5, 3) = -1/256$, $f_{xy} = -(x + 4y - 1)^{-3/2}$, $f_{xy}(5, 3) = -1/64$, and $f_{yy} = -4(x + 4y - 1)^{-3/2}$, $f_{yy}(5, 3) = -1/16$. First-order Taylor polynomial is

$$T_1(x, y) = 4 + \tfrac{1}{8}(x - 5) + \tfrac{1}{2}(y - 3) = \tfrac{15}{8} + \tfrac{1}{8}x + \tfrac{1}{2}y.$$

Second-order Taylor polynomial is

$$T_2(x, y) = 4 + \tfrac{1}{8}(x - 5) + \tfrac{1}{2}(y - 3) - \tfrac{1}{512}(x - 5)^2 - \tfrac{1}{64}(x - 5)(y - 3) - \tfrac{1}{32}(y - 3)^2$$
$$= \tfrac{671}{512} + \tfrac{49}{256}x + \tfrac{49}{64}y - \tfrac{1}{512}x^2 - \tfrac{1}{64}xy - \tfrac{1}{32}y^2.$$

The values are $T_1(4.9, 3.1) = 4.0375000$, $T_2(4.9, 3.1) = 4.0373242$ and the exact value (actually, just a better approximation) is $f(4.9, 3.1) = 4.0373258$.

31. From $f(x, y) = y\sin x$ we get $f(0, 1) = 0$. First partial derviatives are $f_x = y\cos x$, $f_x(0, 1) = 1$, and $f_y = \sin x$, $f_y(0, 1) = 0$.

Second partials are $f_{xx} = -y\sin x$, $f_{xx}(0, 1) = 0$, $f_{xy} = \cos x$, $f_{xy}(0, 1) = 1$, and $f_{yy} = 0$. Second-order Taylor polynomial is

$$T_2(x, y) = x + \tfrac{1}{2}2x(y - 1) = xy.$$

From $xy = c$ we get $y = c/x$. If $c = 0$, then $xy = 0$ implies that $x = 0$ or $y = 0$. Thus, the contour diagram of f, approximated by T_2 near $(0, 1)$, consists of hyperbolas $y = c/x$ if $c \neq 0$, and of the two coordinate axes if $c = 0$.

33. Let $f(x, y) = e^{x^2 - y^2}$. We approximate $e^{0.03^2 - 0.95^2}$ using the value $T_2(0.03, 0.95)$, where T_2 is the second-order Taylor polynomial of f at $(0, 1)$.

From $f(x, y) = e^{x^2 - y^2}$ we get $f(0, 1) = e^{-1}$. First partial derviatives are $f_x = 2xe^{x^2 - y^2}$, $f_x(0, 1) = 0$, and $f_y = -2ye^{x^2 - y^2}$, $f_y(0, 1) = -2e^{-1}$. Second partials are $f_{xx} = 2e^{x^2 - y^2} +$

$4x^2 e^{x^2-y^2}$, $f_{xx}(0,1) = 2e^{-1}$, $f_{xy} = -4xye^{x^2-y^2}$, $f_{xy}(0,1) = 0$, and $f_{yy} = -2e^{x^2-y^2} + 4y^2 e^{x^2-y^2}$, $f_{yy}(0,1) = 2e^{-1}$. Second-order Taylor polynomial is

$$T_2(x,y) = e^{-1} - 2e^{-1}(y-1) + e^{-1}x^2 + e^{-1}(y-1)^2 = 4e^{-1} - 4e^{-1}y + e^{-1}x^2 + e^{-1}y^2.$$

Thus, $T_2(0.03, 0.95) = 0.405918$ is an approximation of $f(0.03, 0.95) = e^{0.03^2 - 0.95^2}$. Exact value (i.e., just a better approximation, obtained using a calculator, or a computer) is $f(0.03, 0.95) = 0.40591967$.

35. Let $f(x,y) = x \arctan y$. We approximate $3.98 \arctan 0.02$ using the value $T_2(3.98, 0.02)$, where T_2 is the second-order Taylor polynomial of f at $(4, 0)$.

From $f(x,y) = x \arctan y$ we get $f(4,0) = 0$. First partial derviatives are $f_x = \arctan y$, $f_x(4,0) = 0$, and $f_y = x/(1+y^2)$, $f_y(4,0) = 4$. Second partials are $f_{xx} = 0$, $f_{xx}(4,0) = 0$, $f_{xy} = 1/(1+y^2)$, $f_{xy}(4,0) = 1$, and $f_{yy} = -2xy/(1+y^2)^2$, $f_{yy}(4,0) = 0$. Second-order Taylor polynomial is $T_2(x,y) = 4y + (x-4)y = xy$. Thus, $T_2(3.98, 0.02) = 0.0796$ is an approximation of $f(3.98, 0.02) = 3.98 \arctan 0.02$. Exact value is $f(3.98, 0.02) = 0.079589$.

4.3. Extreme Values of Real-Valued Functions

1. Differentiating $f(-x) = f(x)$ we get $f'(-x)(-1) = f'(x)$. Substituting $x = 0$, we get $-f'(0) = f'(0)$, i.e., $f'(0) = 0$.

Computing the partial derivative of $f(-x, -y) = f(x,y)$ with respect to x, we get $-f_x(-x,-y) = f_x(x,y)$. Thus, $-f_x(0,0) = f_x(0,0)$, i.e., $f_x(0,0) = 0$. In the same way we show that $f_y(0,0) = 0$.

3. $\nabla f = \mathbf{0}$ at the point P shown in figure below. Consider a horizontal line through P: ∇f points toward P, i.e. the greatest increase in f is in direction toward P. Along the vertical line through P, ∇f points away from P, i.e., the greatest decrease is in direction toward P. Thus - considering horizontal direction, P is a local maximum, and considering vertical direction, P is a local minimum. We conclude that P is a saddle point.

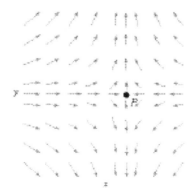

Figure 3

5. Clearly, $\nabla f \neq \mathbf{0}$ at all points, so f has no critical points.

7. Critical points (x, y) of the given function are the solutions of the system $f_x = 2x + y^2 = 0$ and $f_y = 2y + 2xy = 0$. From the second equation, we get $2y(1+x) = 0$; thus $y = 0$ or $x = -1$. Substituting $y = 0$ into $f_x = 2x + y^2 = 0$, we get $2x = 0$ and so $x = 0$. Substituting $x = -1$ into $2x + y^2 = 0$, we get $y^2 = 2$ and so $y = \pm\sqrt{2}$. Thus, there are three critical points: $(0, 0)$, $(-1, \sqrt{2})$ and $(-1, -\sqrt{2})$. The corresponding values of f are $f(0, 0) = 0$, $f(-1, \pm\sqrt{2}) = 1$.

We continue by applying the Second Derivatives Test. From $f_{xx} = 2$, $f_{yy} = 2 + 2x$ and $f_{xy} = f_{yx} = 2y$, we get

$$D(x, y) = 2(2 + 2x) - (2y)^2 = 4 + 4x - 4y^2 = 4(1 + x - y^2).$$

From $D(0, 0) = 4 > 0$ and $f_{xx}(0, 0) = 2 > 0$, we conclude that $f(0, 0) = 0$ is a local minimum. From $D(-1, \pm\sqrt{2}) = 4(-2) = -8 < 0$, it follows that $(-1, \pm\sqrt{2})$ are saddle points.

9. To find critical points, we consider the system (rewrite f as $f(x, y) = xy + \frac{1}{x} + \frac{1}{y}$)

$$f_x = y - 1/x^2 = 0 \qquad \text{and} \qquad f_y = x - 1/y^2 = 0,$$

where $x \neq 0$ and $y \neq 0$. Simplifying, we get $y = 1/x^2$ and $x = 1/y^2$. Substituting $y = 1/x^2$ into the second equation, we get $x = x^4$; i.e., $x - x^4 = x(1 - x^3) = 0$. Thus, $x = 1$ (since $x \neq 0$), and, consequently, $y = 1$. It follows that $(1, 1)$ is the only critical point. From $f_{xx} = 2/x^3$, $f_{yy} = 2/y^3$ and $f_{xy} = f_{yx} = 1$, we get

$$D(x, y) = f_{xx}f_{yy} - (f_{xy})^2 = \frac{4}{x^3 y^3} - 1.$$

Since $D(1, 1) = 3 > 0$ and $f_{xx}(1, 1) = 2 > 0$, we conclude that $f(1, 1) = 3$ is a local minimum.

11. To find critical points, we solve the system

$$f_x = ye^{-x^2-y^2} + xye^{-x^2-y^2}(-2x) = ye^{-x^2-y^2}(1 - 2x^2) = 0$$

and (compute f_y by the product rule and the chain rule as above, or use the fact that f is symmetric in x and y and just interchange x and y)

$$f_y = xe^{-x^2-y^2}(1 - 2y^2) = 0.$$

From the equation $f_x = 0$, it follows that $y = 0$ or $1 - 2x^2 = 0$; i.e., $x = \pm 1/\sqrt{2}$. Substituting $y = 0$ into $f_y = 0$, we get $xe^{-x^2} = 0$ and so $x = 0$. Thus, $(0, 0)$ is a critical point. Substituting $x = 1/\sqrt{2}$ into $f_y = 0$, we get that $1 - 2y^2 = 0$ and $y = \pm 1/\sqrt{2}$. Similarly, $x = -1/\sqrt{2}$ implies that $y = \pm 1/\sqrt{2}$. Consequently, there are four more critical points, $(1/\sqrt{2}, 1/\sqrt{2})$, $(1/\sqrt{2}, -1/\sqrt{2})$, $(-1/\sqrt{2}, 1/\sqrt{2})$ and $(-1/\sqrt{2}, -1/\sqrt{2})$.

From

$$f_{xx} = ye^{-x^2-y^2}(-2x)(1 - 2x^2) + ye^{-x^2-y^2}(-4x) = -2xye^{-x^2-y^2}(3 - 2x^2),$$

(interchange x and y)

$$f_{yy} = -2xye^{-x^2-y^2}(3 - 2y^2)$$

and

$$f_{xy} = f_{yx} = e^{-x^2-y^2}(1-2x^2) + ye^{-x^2-y^2}(-2y)(1-2x^2) = (1-2x^2)e^{-x^2-y^2}(1-2y^2),$$

we get

$$D(x,y) = 4x^2y^2(3-2x^2)(3-2y^2)e^{-2x^2-2y^2} - (1-2x^2)^2(1-2y^2)^2e^{-2x^2-2y^2}.$$

From $D(0,0) = -1 < 0$, it follows that $(0,0)$ is a saddle point.

For the remaining four points, $D(\pm 1/\sqrt{2}, \pm 1/\sqrt{2}) = 4e^{-2} > 0$. Since $f_{xx}(1/\sqrt{2}, 1/\sqrt{2}) = -2e^{-1} < 0$ and $f_{xx}(-1/\sqrt{2}, -1/\sqrt{2}) = -2e^{-1} < 0$, it follows that $f(1/\sqrt{2}, 1/\sqrt{2}) = e^{-1}/2$ and $f(-1/\sqrt{2}, -1/\sqrt{2}) = e^{-1}/2$ are local maxima. Similarly, $f_{xx}(-1/\sqrt{2}, 1/\sqrt{2}) = 2e^{-1} > 0$ and $f_{xx}(1/\sqrt{2}, -1/\sqrt{2}) = 2e^{-1} > 0$ imply that the values $f(-1/\sqrt{2}, 1/\sqrt{2}) = -e^{-1}/2$ and $f(1/\sqrt{2}, -1/\sqrt{2}) = -e^{-1}/2$ are local minima.

13. To find critical points, we consider the system $f_x = \cos y = 0$ and $f_y = -x \sin y = 0$. From the first equation, we get $y = \pi/2 + k\pi$, where k is an integer. Since $\sin y = \sin(\pi/2 + k\pi) = \pm 1$, the second equation forces $x = 0$. Thus, the critical points are $(0, \pi/2 + k\pi)$, where k is an integer. From $f_{xx} = 0$, $f_{yy} = -x \cos y$ and $f_{xy} = f_{yx} = -\sin y$, it follows that

$$D(x,y) = f_{xx}f_{yy} - f_{xy}^2 = -\sin^2 y.$$

Since $D(0, \pi/2 + k\pi) = -1 < 0$, we conclude that all critical points are saddle points.

15. To find critical points, we consider the system $f_x = \sin(x+y) + x\cos(x+y) = 0$ and $f_y = x\cos(x+y) = 0$. From $f_y = 0$, it follows that $x = 0$ or $\cos(x+y) = 0$; in the latter case, $x + y = \pi/2 + \ell\pi$, where ℓ is an integer. Substituting $x = 0$ into the equation $f_x = 0$, we get $\sin y = 0$ and $y = k\pi$ (k is an integer). Thus, the points $(0, k\pi)$, where k is an integer, are critical points of f.

Now consider the case $x + y = \pi/2 + \ell\pi$. Since $\cos(x+y) = \cos(\pi/2 + \ell\pi) = 0$ and $\sin(x+y) = \cos(\pi/2 + \ell\pi) = \pm 1$, the equation $f_x = 0$ has no solutions. Thus, $(0, k\pi)$, where k is an integer, are the only critical points of f.

From $f_{xx} = \cos(x+y) + \cos(x+y) - x\sin(x+y)$, $f_{yy} = -x\sin(x+y)$ and $f_{xy} = f_{yx} = \cos(x+y) - x\sin(x+y)$, it follows that

$$D(x,y) = (2\cos(x+y) - x\sin(x+y))(-x\sin(x+y)) - (\cos(x+y) - x\sin(x+y))^2.$$

Since $D(0, k\pi) = -(\cos(k\pi))^2 = -1 < 0$, we conclude that all critical points $(0, k\pi)$, where k is an integer, are saddle points.

17. Recall that the distance from a point (x, y, z) to the point $(2, 0, 3)$ is given by $d(x, y, z) = \sqrt{(x-2)^2 + y^2 + (z-3)^2}$. In our case, a point (x, y, z) belongs to the plane $x - y + z = 4$. Thus, $z = 4 - x + y$, and the distance formula gives

$$d(x,y) = \sqrt{(x-2)^2 + y^2 + (1-x+y)^2}.$$

Note that, in order to minimize the expression $\sqrt{a}$, it suffices to minimize a; thus, we find the minimum of the function

$$f(x,y) = (x-2)^2 + y^2 + (1-x+y)^2 = 2x^2 + 2y^2 - 6x + 2y - 2xy + 5.$$

To find critical points, we consider the system of equations

$$f_x = 4x - 6 - 2y = 0 \qquad \text{and} \qquad f_y = 4y + 2 - 2x = 0.$$

Multiplying the equation for f_y by 2 and adding up the two equations, we get $6y - 2 = 0$ and $y = 1/3$. From $f_y = 0$, it follows that $x = 2y + 1 = 5/3$ and $z = 4 - x + y = 8/3$.

From $f_{xx} = 4$, $f_{yy} = 4$ and $f_{xy} = f_{yx} = -2$, it follows that $D(x,y) = f_{xx}f_{yy} - f_{xy}^2 = 16 - (-2)^2 = 12$. Thus, $f_{xx}(5/3, 1/3) = 4 > 0$ and $D(5/3, 1/3) = 12 > 0$, and so $f(5/3, 1/3) = (5/3 - 2)^2 + (1/3)^3 + (1 - 8/3)^2 = 1/3$ is a local minimum.

Intuitively, it is clear that there is only one point in a plane that is closest to a given point (that does not lie in the plane). Thus, $f(5/3, 1/3)$ is also a global minimum. It follows that the shortest distance is $1/\sqrt{3}$; it is the distance between the point $(2, 0, 3)$ and the point $(5/3, 1/3, 8/3)$ on the given plane.

19. Label the length, width and height of the box by x, y and z respectively (x, y, $z > 0$). It is given that $xyz = V$; we must minimize the surface area $S = 2xy + 2xz + 2yz$.

Substituting $z = V/xy$ into S, we get

$$S(x,y) = 2xy + \frac{2V}{y} + \frac{2V}{x}.$$

To find the critical points of S, we solve the system

$$S_x = 2y - 2V/x^2 = 0; \qquad \text{i.e.,} \qquad y = V/x^2$$

and

$$S_y = 2x - 2V/y^2 = 0; \qquad \text{i.e.,} \qquad x = V/y^2.$$

Combining the two equations, we get

$$x = \frac{V}{y^2} = \frac{V}{(V/x^2)^2} = \frac{x^4}{V}$$

and so

$$x - \frac{x^4}{V} = x\left(1 - \frac{x^3}{V}\right) = 0.$$

Since $x > 0$, it follows that $x^3/V = 1$ and $x = \sqrt[3]{V}$. Thus, $y = V/x^2 = V/(\sqrt[3]{V})^2 = \sqrt[3]{V}$ and $z = V/xy = V/(\sqrt[3]{V}\sqrt[3]{V}) = \sqrt[3]{V}$. This computation implies that the only critical point is $(x = \sqrt[3]{V}, y = \sqrt[3]{V})$.

Next, we show that it is a point where S has a minimum. From $S_{xx}(x,y) = 4V/x^3$, $S_{yy}(x,y) = 4V/y^3$ and $S_{xy}(x,y) = S_{yx} = 2$, it follows that

$$D(x,y) = S_{xx}(x,y)S_{yy}(x,y) - (S_{xy}(x,y))^2 = \frac{16V^2}{x^3y^3} - 4.$$

Since $D(\sqrt[3]{V}, \sqrt[3]{V}) = 16V^2/V^2 - 4 = 12 > 0$ and $S_{xx}(\sqrt[3]{V}, \sqrt[3]{V}) = 4V/V = 4 > 0$, we conclude that $S(\sqrt[3]{V}, \sqrt[3]{V}) = 6\sqrt[3]{V^2}$ is a minimum. In words, the optimal rectangular box (in the context of this exercise) is a cube of side $\sqrt[3]{V}$.

21. Place the box so that its sides are parallel to the coordinate planes and its center is at the origin. Label its length, width and height by x, y and z respectively ($x, y, z > 0$). Since the box is inscribed into a sphere (whose center must be at the origin, by construction), the eight points $(\pm x/2, \pm y/2, \pm z/2)$ must lie on it. Thus,

$$\left(\pm\frac{x}{2}\right)^2 + \left(\pm\frac{y}{2}\right)^2 + \left(\pm\frac{z}{2}\right)^2 = R^2;$$

i.e., $x^2 + y^2 + z^2 = 4R^2$. We are asked to find the maximum value of the volume $V = xyz$. From $x^2 + y^2 + z^2 = 4R^2$, it follows that $z = \sqrt{4R^2 - x^2 - y^2}$ (since $z > 0$), and so,

$$V(x, y) = xy\sqrt{4R^2 - x^2 - y^2}.$$

To find critical points of V, we must solve the system

$$V_x = y\sqrt{4R^2 - x^2 - y^2} + \frac{xy}{2\sqrt{4R^2 - x^2 - y^2}}(-2x) = \frac{4R^2y - 2x^2y - y^3}{\sqrt{4R^2 - x^2 - y^2}} = 0$$

and (switch x and y)

$$V_y = \frac{4R^2x - 2y^2x - x^3}{\sqrt{4R^2 - x^2 - y^2}} = 0.$$

It follows that $y(4R^2 - 2x^2 - y^2) = 0$ and $x(4R^2 - 2y^2 - x^2) = 0$; since $x > 0$ and $y > 0$, we get $4R^2 - 2x^2 - y^2 = 0$ and $4R^2 - 2y^2 - x^2 = 0$. Combining the two equations, we get $4R^2 - 2x^2 - y^2 = 4R^2 - 2y^2 - x^2$; i.e., $x^2 = y^2$ and (since x and y must be positive) $x = y$. Therefore, from $4R^2 - 2x^2 - y^2 = 0$, we get $4R^2 - 3x^2 = 0$ and $x = y = 2R/\sqrt{3}$. Thus, there is only one critical point, $(2R/\sqrt{3}, 2R/\sqrt{3})$. From $x^2 + y^2 + z^2 = 4R^2$, we get $z = 2R/\sqrt{3}$; thus, the box in question is a cube. We claim that

$$V(2R/\sqrt{3}, 2R/\sqrt{3}) = \left(\frac{2R}{\sqrt{3}}\right)^3 = \frac{8R^3}{3^{3/2}}$$

is a maximum. An alternative to a somewhat messy use of the Second Derivatives Test is to reason intuitively as follows. There must be a box of maximum volume since the volume of a box that is inscribed into a sphere is bounded by the volume of that sphere. The critical point we found cannot be a minimum since there is no minimum — we can inscribe a rectangular box with volume that is as small as we want (by choosing small height, for example).

23. Since $g(0, 0) = 0$, the contour curve that goes through the origin is given by $x^4 - y^4 = 0$; i.e., $y^4 = x^4$ and $y = \pm x$. Thus, the contour curve through the origin consists of a pair of perpendicular lines $y = \pm x$. When $c \neq 0$, the level curve $x^4 - y^4 = c$ is a hyperbola. The asymptotes $y = \pm x$ divide the xy-plane into four regions. In two of these (that contain the x-axis) the value of c (thus, the value of f) is positive, and in the remaining two (those regions contain the y-axis) the value of c is negative. Thus, $(0, 0)$ can be neither minimum nor maximum. So, it is a saddle point.

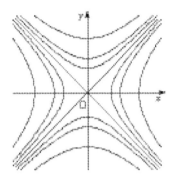

Figure 23

25. The solid cut off by the plane is a tetrahedron with volume $V = abc/6$. The plane passes through $(1,1,1)$, and thus $1/a + 1/b + 1/c = 1$, $a, b, c > 0$. It follows that

$$\frac{1}{c} = 1 - \frac{1}{a} - \frac{1}{b} = \frac{ab - a - b}{ab}$$

and $c = ab/(ab - a - b)$, so that

$$V(a,b) = \frac{1}{6}\frac{a^2b^2}{ab - a - b}.$$

The critical points are computed from

$$V_a = \frac{1}{6}\frac{2ab^2(ab - a - b) - a^2b^2(b - 1)}{(ab - a - b)^2} = \frac{a^2b^3 - a^2b^2 - 2ab^3}{6(ab - a - b)^2} = 0$$

and (interchange a and b)

$$V_b = \frac{a^3b^2 - a^2b^2 - 2a^3b}{6(ab - a - b)^2} = 0.$$

Simplifying, we get $ab^2(ab - a - 2b) = 0$ and $a^2b(ab - b - 2a) = 0$; discarding the cases $a = 0$ and $b = 0$, we end up with the system $ab - a - 2b = 0$ and $ab - b - 2a = 0$. Eliminating ab from both equations, we get $a + 2b = b + 2a$; i.e., $a = b$. Substituting $a = b$ into the equation $ab - a - 2b = 0$, we get $b^2 - b - 2b = 0$ and $b(b - 3) = 0$; thus (since $b > 0$), $b = 3$. It follows that $a = 3$ and $c = ab/(ab - a - b) = 9/3 = 3$. So, $(3,3)$ is the only critical point.

It is intuitively clear that $V(3,3) = 27/6 = 9/2$ is a minimum (from the formula for V it follows that $V \to \infty$ as $a \to \infty$ and $b \to \infty$; thus, V cannot have a maximum). For the record, we do the Second Derivatives Test (we used Maple). From

$$V_{aa} = \frac{1}{3}\frac{b^4}{(ab - a - b)^3},$$

$$V_{bb} = \frac{1}{3}\frac{a^4}{(ab - a - b)^3}$$

and

$$V_{ab} = V_{ba} = \frac{1}{6}\frac{ab(a^2b^2 - 3a^2b - 3ab^2 + 2a^2 + 6ab + 2b^2)}{(ab - a - b)^3},$$

we get

$$D = \frac{-a^2b^2(a^3b^3 - 5a^3b^2 + 8a^3b - 4a^3 - 5a^2b^3 + 20a^2b^2 - 20a^2b + 8ab^3 - 20ab^2 - 4b^3)}{(ab - a - b)^5}.$$

Since $D(3,3) = 3/4 > 0$ and $V_{xx}(3,3) = 1 > 0$, we conclude that $(3,3)$ is a local minimum point.

27. From $f_x = 2x/(x^2 + y + 1) = 0$ and $f_y = 1/(x^2 + y + 1) = 0$, it follows that f has no critical points (since $f_y = 0$ has no solutions). The boundary of D consists of three line segments. Along the line segment $\mathbf{c}_1$ from $(0,0)$ to $(1,0)$, the x-values change from 0 to 1, whereas the y-values are fixed at 0; the restriction of f to $\mathbf{c}_1$ is

$$g_1(x) = f(x,0) = \ln(x^2 + 1), \qquad x \in [0,1].$$

Since $\ln(x^2 + 1)$ is a composition of two increasing functions, it is increasing; consequently, its extreme values are attained at the endpoints. It follows that the candidates for absolute extrema are $g_1(0) = f(0,0) = \ln 1 = 0$ and $g_1(1) = f(1,0) = \ln 2$.

Along the line segment from $(1,0)$ to $(1,1)$ (call it $\mathbf{c}_2$), x is equal to 1 and y changes from 0 to 1; the restriction of f to $\mathbf{c}_2$ is

$$g_2(y) = f(1,y) = \ln(y + 2), \qquad 0 \leq y \leq 1.$$

Again, g_2 is an increasing function, so the only candidates for extreme values are the values at endpoints, $g_2(0) = f(1,0) = \ln 2$ and $g_2(1) = f(1,1) = \ln 3$.

Let $\mathbf{c}_3$ be the line segment joining $(0,0)$ and $(1,1)$; it can be described as $y = x$, $0 \leq x \leq 1$. The restriction of f to $\mathbf{c}_3$ is

$$g_3(x) = f(x,x) = \ln(x^2 + x + 1), \qquad 0 \leq x \leq 1.$$

We could argue as above (g_3 is increasing for $x \geq 0$) to show that there g_3 has no critical points on $[0,1]$. Alternatively, from $g_3'(x) = (2x+1)/(x^2+x+1) = 0$, it follows that $x = -1/2$. The values of g_3 at the endpoints are $g_3(0) = f(0,0) = \ln 1 = 0$ and $g_3(1) = f(1,1) = \ln 3$.

It follows that $f(0,0) = 0$ is the absolute minimum and $f(1,1) = \ln 3$ is the absolute maximum of f on D.

29. From $f_x = y\cos(xy) = 0$, we conclude that either $y = 0$ or $xy = \pi/2$. The derivative of f with respect to y is $f_y = x\cos(xy)$. If $y = 0$, then $f_y = 0$ implies that $x = 0$. Thus, $(0,0)$ is a critical point; however, it does not belong to D, so we ignore it. If $xy = \pi/2$, then $f_y = x\cos(\pi/2) = 0$ implies that x can be any real number. Thus, every point (x,y) such that $xy = \pi/2$ and (x,y) belongs to the interior of D is a critical point. Corresponding value of f is $f(x,y) = \sin(\pi/2) = 1$.

The line segment from $(0,\pi/2)$ to $(\pi/2,\pi/2)$ is given by $y = \pi/2$, $0 \leq x \leq \pi/2$. The corresponding restriction of f is $g_1(x) = f(x,\pi/2) = \sin(\pi x/2)$, $0 \leq x \leq \pi/2$. From $g_1'(x) = \frac{\pi}{2}\cos(\pi x/2) = 0$, it follows that $x = 1$. So the candidates for extreme values are $g_1(1) = f(1,\pi/2) = \sin(\pi/2) = 1$ (critical point of g_1) and $g_1(0) = f(0,\pi/2) = \sin 0 = 0$ and $g_1(\pi/2) = f(\pi/2,\pi/2) = \sin(\pi^2/4)$ (endpoints).

The line segment from $(\pi/2,\pi/2)$ to $(\pi/2,\pi)$ is given by $x = \pi/2$, $\pi/2 \leq y \leq \pi$. The corresponding restriction of f is $g_2(y) = f(\pi/2,y) = \sin(\pi y/2)$, $\pi/2 \leq y \leq \pi$. From

$g_2'(y) = \frac{\pi}{2}\cos(\pi y/2) = 0$, it follows that $y = 3$ (keep in mind the range of y!). This time, the candidates for extreme values are $g_2(3) = f(\pi/2, 3) = \sin(3\pi/2) = -1$ (critical point of g_2) and $g_2(\pi/2) = f(\pi/2, \pi/2) = \sin(\pi^2/4)$ and $g_2(\pi) = f(\pi/2, \pi) = \sin(\pi^2/2)$ (endpoints).

The line segment from $(0, \pi)$ to $(\pi/2, \pi)$ is given by $y = \pi$, $0 \le x \le \pi/2$. The restriction of f is $g_3(x) = f(x, \pi) = \sin(\pi x)$, $0 \le x \le \pi/2$. From $g_3'(y) = \pi\cos(\pi x) = 0$, it follows that $x = 1/2$ and $x = 3/2$. The candidates for extreme values are $g_3(1/2) = f(1/2, \pi) = \sin(\pi/2) = 1$ and $g_3(3/2) = f(3/2, \pi) = \sin(3\pi/2) = -1$ (critical points of g_3) and $g_3(0) = f(0, \pi) = \sin 0 = 0$ and $g_3(\pi/2) = f(\pi/2, \pi) = \sin(\pi^2/2)$ (endpoints).

Finally, the line segment from $(0, \pi/2)$ to $(0, \pi)$ is given by $x = 0$, $\pi/2 \le y \le \pi$. The restriction of f is $g_4(x) = f(0, y) = \sin 0 = 0$, where $\pi/2 \le y \le \pi$. g_4 is a constant function, and hence has no critical points. The values at the endpoints are $g_4(0) = f(0, \pi/2) = \sin 0 = 0$ and $g_4(\pi) = f(0, \pi) = \sin 0 = 0$.

It follows that the absolute maximum is 1; it occurs at all points (x, y) inside D such that $xy = \pi/2$, and also at $(1, \pi/2)$ and $(1/2, \pi)$. The absolute minimum is -1; it occurs at $(3/2, \pi)$ and at $(\pi/2, 3)$.

31. From $p_x = 2xe^{x^2+y^2+y} = 0$ and $p_y = (2y + 1)e^{x^2+y^2+y} = 0$, it follows that $(0, -1/2)$ is the only critical point of p (and it belongs to the interior of D). Thus, $p(0, -1/2) = e^{-1/4} + 1$ is a candidate for absolute extreme value.

Parametrize the boundary circle of D by $\mathbf{c}(t) = (\cos t, \sin t)$, $t \in [0, 2\pi]$. The restriction of p to $\mathbf{c}$ is

$$g(t) = p(\mathbf{c}(t)) = p(\cos t, \sin t) = e^{1+\sin t} + 1, \qquad t \in [0, 2\pi].$$

From $g'(t) = \cos t \, e^{1+\sin t} = 0$, we get the critical points, $t = \pi/2$ and $t = 3\pi/2$, that belong to $[0, 2\pi]$. Consequently, the list of candidates includes $p(\mathbf{c}(\pi/2)) = p(0, 1) = e^2 + 1$ and $p(\mathbf{c}(3\pi/2)) = p(0, -1) = 2$ (critical points) and $p(\mathbf{c}(0)) = p(\mathbf{c}(2\pi)) = p(1, 0) = e + 1$ (endpoints).

Thus, $p(0, 1) = e^2 + 1$ is the absolute maximum and $p(0, -1/2) = e^{-1/4} + 1$ is the absolute minimum of p on the given disk.

4.4. Optimization with Constraints and Lagrange Multipliers

1. Level curves of f are circles centered at the origin. The constraint curve is a line that goes through the origin; it intersects all level curves of f orthogonally. In other words, ∇f is parallel to the constraint curve at all points on it, so f cannot have an extreme value subject to $y = x$.

3. First, we locate all points where gradient of f appears to be perpendicular to the constraint curve. Next, based on analyzing level curves of f, we determine whether a selected point is a minimum or a maximum. Since we have ∇f, we can visualize level curves of f (as we know, they need to be perpendicular to ∇f).

Let P be a point where ∇f is perpendicular to the constraint curve, and label by c_P the level curve of f that goes through P and whose value is c_P. If a nearby level curve of value c looks like the curve c (shown in figure (a) below), then $c > c_P$ (since f increases in the direction of ∇f). Thus, the values of f at the points where the level curve c crosses the constraint curve are larger than $f(P) = c_P$. So, $f(P)$ is minimum subject to the given constraint.

If a nearby level curve of value c looks like c (shown in figure (b) below), then $c < c_P$ (since f decreases in the direction opposite of ∇f). Thus, the values of f at the points where the level curve c crosses the constraint curve are smaller than $f(P) = c_P$. So, $f(P)$ is maximum subject to the given constraint.

Solution to the exercise: There are four points where gradient of f is perpendicular to the constraint curve (see figure below). Subject to the given constraint, P_1 is minimum (case (a)), P_2 is maximum (case (b)), P_3 is minimum (case (a)) and P_4 is maximum (case (b)).

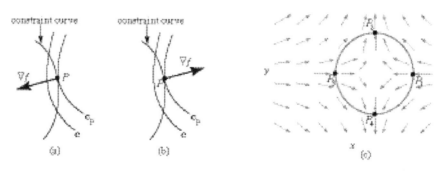

Figure 3

5. We use the discussion given in Exercise 3. There are two points where gradient of f is perpendicular to the constraint curve (see figure below). Subject to the given constraint, P_1 is maximum (case (b)) and P_2 is minimum (case (a)).

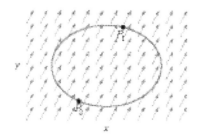

Figure 5

7. (a) Extreme values occur where the level curves are tangent to the constraint curve. In our case, this happens at two points, $(-1, 0)$ and $(3, 0)$. Looking at the diagram, we see that $f(3, 0) = 9$ is maximum, and $f(-1, 0) = 1$ is minimum subject to the given constraint.

(b) Let $g(x, y) = (x - 1)^2 + y^2$, and $f(x, y) = x^2 + y^2$. The constraint is a circle, which is a closed and bounded set. From $\nabla f = \lambda \nabla g$ we get $(2x, 2y) = \lambda(2(x-1), 2y)$; i.e., $2x = (2x - 2)\lambda$ and $2y = 2y\lambda$. It follows that either $y = 0$ or $\lambda = 1$. Substituting $y = 0$ into the constraint, we get $(x - 1)^2 = 4$, i.e., $x = -1$ or $x = 3$. Substituting $\lambda = 1$ into $2x = (2x - 2)\lambda$, we get that $0 = -2$, so there are no solutions. Looking at the diagram, we see that $f(3, 0) = 9$ is maximum, and $f(-1, 0) = 1$ is minimum subject to the given constraint.

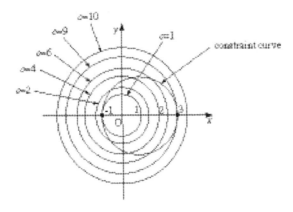

Figure 7

9. Minimize $f(x, y, z) = 2xy + 2xz + 2yz$ subject to the constraint $g(x, y, z) = xyz = 100$. (Keep in mind that $x, y, z > 0$.)

Since $\nabla f = (2y + 2z, 2x + 2z, 2x + 2y)$ and $\nabla g = (yz, xz, xy)$, the requirement $\nabla f = \lambda \nabla g$ implies that

$$2y + 2z = \lambda yz, \qquad 2x + 2z = \lambda xz \qquad \text{and} \qquad 2x + 2y = \lambda xy.$$

Multiplying the first equation by x, the second equation by $-y$ and adding them up, we get $2z(x - y) = 0$ and thus $x = y$ (since $z > 0$). Similarly, we get that $y = z$. Substituting $x = y = z$ into the constraint, we get $x^3 = 100$ and so $x = y = z = \sqrt[3]{100}$. Thus, $f(\sqrt[3]{100}, \sqrt[3]{100}, \sqrt[3]{100}) = 6\sqrt[3]{100^2}$ is the minimum value of the surface area (to prove that the above is really a minimum we can argue as in Example 4.26).

11. Let $g(x, y) - x^2 + y^2$. The constraint $g(x, y) = x^2 + y^2 = 4$ is a circle, which is a closed and bounded set in $\mathbb{R}^2$. The function f is continuous, since it is a polynomial in two variables. Thus, by the Extreme Value Theorem, minimum and maximum of f must exist.

From $\nabla f = \lambda \nabla g$, we get $(3y, 3x) = \lambda(2x, 2y)$; i.e., $3y = 2\lambda x$ and $3x = 2\lambda y$. It follows that $2\lambda = 3y/x$ and $2\lambda = 3x/y$, and thus $y^2 = x^2$. Substituting this equation into the constraint, we get $2x^2 = 4$; so, $x = \pm\sqrt{2}$. Consequently, $y = \pm\sqrt{2}$ for each of the two values of x. So, there are four candidates for extreme values, $(\sqrt{2}, \sqrt{2})$, $(-\sqrt{2}, \sqrt{2})$, $(\sqrt{2}, -\sqrt{2})$ and $(-\sqrt{2}, -\sqrt{2})$.

The requirement $\nabla g = \mathbf{0}$ implies that $x = 0$ and $y = 0$; the point $(0, 0)$ does not lie on the constraint curve. Moreover, the constraint curve has no endpoints, and so parts (b) and (c) of the Lagrange multiplier method give no candidates for constrained extrema.

From $f(\sqrt{2}, \sqrt{2}) = 3$, $f(-\sqrt{2}, \sqrt{2}) = -3$, $f(\sqrt{2}, -\sqrt{2}) = -3$ and $f(-\sqrt{2}, -\sqrt{2}) = 3$, we conclude that $f(\sqrt{2}, \sqrt{2}) = f(-\sqrt{2}, -\sqrt{2}) = 3$ is the maximum value of f, and $f(-\sqrt{2}, \sqrt{2}) = f(\sqrt{2}, -\sqrt{2}) = -3$ is the minimum value of f subject to the given constraint.

13. The constraint $g(x, y) = x^2 + y^2 = 1$ is a circle, which is a closed and bounded set in $\mathbb{R}^2$. The function f is continuous, and thus, by the Extreme Value Theorem, minimum and maximum values of f must exist.

From $\nabla f = \lambda \nabla g$, we get $(4x, -2y) = \lambda(2x, 2y)$; i.e., $4x = 2\lambda x$ and $-2y = 2\lambda y$. The first equation implies that $x(2 - \lambda) = 0$; so, either $x = 0$ or $\lambda = 2$. If $x = 0$, then the constraint equation $x^2 + y^2 = 1$ implies $y^2 = 1$ and $y = \pm 1$. If $\lambda = 2$, then the equation $-2y = 2\lambda y$ implies that $y = 0$; from the constraint, we get $x = \pm 1$. Thus, there are four candidates for extreme values, $(1, 0)$, $(-1, 0)$, $(0, 1)$ and $(0, -1)$.

Parts (b) and (c) of the Lagrange multiplier method give no candidates for constrained extrema since $\nabla g = \mathbf{0}$ implies that $x = 0$ and $y = 0$ (and the point $(0, 0)$ does not lie on the constraint curve), and since the constraint curve has no endpoints.

It follows that $f(1, 0) = f(-1, 0) = 2$ is the maximum of f, and $f(0, 1) = f(0, -1) = -1$ is the minimum of f subject to the given constraint.

15. The constraint curve $g(x, y) = x^3 - y^2 = 0$ is the part of the curve $x = y^{2/3}$ between $(1, -1)$ and $(1, 1)$. It is a closed and bounded set in $\mathbb{R}^2$; f is continuous, and so, the Extreme Value Theorem implies that minimum and maximum of f subject to the given constraint must exist.

The equation $\nabla f = \lambda \nabla g$ implies $(1, -2y) = \lambda(3x^2, -2y)$; thus, $1 = 3\lambda x^2$ and $-2y = -2\lambda y$. The second equation implies that $y = 0$ or $\lambda = 1$. If $y = 0$, then the constraint equation implies that $x = 0$. If $\lambda = 1$, then from $1 = 3\lambda x^2$ we get $x = \pm 1/\sqrt{3}$. Substituting $x = -1/\sqrt{3} = -3^{-1/2}$ into the constraint $x^3 - y^2 = 0$, we get the equation $y^2 = -3^{-3/2}$, which has no solutions. Substituting $x = 1/\sqrt{3} = 3^{-1/2}$ into the constraint, we get $y^2 = 3^{-3/2}$ and $y = \pm 3^{-3/4}$. Thus, the candidates for extreme values are $f(0, 0) = 0$, $f(3^{-1/2}, 3^{-3/4}) = 3^{-1/2} - 3^{-3/2} = 2(3^{-3/2}) \approx 0.3849$, and $f(3^{-1/2}, -3^{-3/4}) = 3^{-1/2} - 3^{-3/2} = 2(3^{-3/2}) \approx 0.3849$.

The equation $\nabla g = \mathbf{0}$ implies that $x = y = 0$; since $(0, 0)$ belongs to the constraint curve, the value $f(0, 0) = 0$ must be included in the list of candidates for extreme values (that is part (b) of the Lagrange multiplier method). Finally, by part (c) of the method, we must include the values of f at the endpoints of the constraint curve, $f(-1, 1) = -2$ and $f(1, 1) = 0$.

It follows that $f(-1,1) = -2$ is the minimum, and $f(3^{-1/2}, 3^{-3/4}) = f(3^{-1/2}, -3^{-3/4}) = 2(3^{-3/2}) \approx 0.3849$, is the maximum of f subject to the given constraint.

17. The constraint surface $g(x, y, z) = x^2 + y^2 + 2z^2 = 4$ is an ellipsoid, which is a closed and bounded set in $\mathbb{R}^3$. The function f is continuous, and by the Extreme Value Theorem, minimum and maximum of f subject to the given constraint must exist.

The equation $\nabla f = \lambda \nabla g$ implies that $(1, 2, -4) = \lambda(2x, 2y, 4z)$; thus, $1 = 2\lambda x$, $2 = 2\lambda y$ and $-4 = 4\lambda z$. It follows that $\lambda = 1/2x$, $\lambda = 1/y$ and $\lambda = -1/z$. Combining the first two equations, we get $x = y/2$. From the second and third equations, we get $z = -y$. Substituting $x = y/2$ and $z = -y$ into the constraint equation, we get $(y/2)^2 + y^2 + 2(-y)^2 = 4$; thus, $13y^2/4 = 4$ and $y = \pm 4/\sqrt{13}$. It follows that $x = y/2 = \pm 2/\sqrt{13}$ and $z = -y = \mp 4/\sqrt{13}$; thus, the candidates for extreme values are $f(2/\sqrt{13}, 4/\sqrt{13}, -4/\sqrt{13}) = -6/\sqrt{13}$ and $f(-2/\sqrt{13}, -4/\sqrt{13}, 4/\sqrt{13}) = -26/\sqrt{13}$.

The equation $\nabla g = \mathbf{0}$ implies that $x = y = z = 0$; since the point $(0, 0, 0)$ does not satisfy $x^2 + y^2 + 2z^2 = 4$, the assumption $\nabla g \neq \mathbf{0}$ (for all points on the constraint surface) holds.

It follows that the value $f(-2/\sqrt{13}, -4/\sqrt{13}, 4/\sqrt{13}) = -26/\sqrt{13}$ is the minimum, and the value $f(2/\sqrt{13}, 4/\sqrt{13}, -4/\sqrt{13}) = -6/\sqrt{13}$ is the maximum of f subject to the given constraint.

19. The constraint curve $g(x, y) = x^2 + 2y^2 = 4$ is an ellipse, which is a closed and bounded set in $\mathbb{R}^2$. By the Extreme Value Theorem, since f is continuous, it must have a minimum and a maximum.

Using $\nabla f = \lambda \nabla g$, we get $(y, x + 2) = \lambda(2x, 4y)$; i.e., $y = 2\lambda x$ and $x + 2 = 4\lambda y$. The first equation implies that $2\lambda = y/x$; combining with the second equation, we get $x + 2 = 2y^2/x$. Rewrite this as $2y^2 = x^2 + 2x$ and substitute into the constraint, thus getting $2x^2 + 2x = 4$; i.e., $x^2 + x - 2 = (x + 2)(x - 1) = 0$. It follows that $x = -2$ (in which case, from $2y^2 = x^2 + 2x$, we compute $y = 0$) or $x = 1$ (and thus, $2y^2 = 3$ and $y = \pm\sqrt{3/2}$).

Parts (b) and (c) of the Lagrange multiplier method give no candidates for constrained extrema, since $\nabla g = (2x, 4y) = \mathbf{0}$ implies that $x = 0$ and $y = 0$ (and the point $(0, 0)$ does not lie on the constraint curve), and since the constraint curve has no endpoints.

From $f(-2, 0) = 0$, $f(1, \sqrt{3/2}) = 3\sqrt{3/2}$ and $f(1, -\sqrt{3/2}) = -3\sqrt{3/2}$, it follows that (subject to the given constraint) the maximum of f is $3\sqrt{3/2}$ and the minimum of f is $-3\sqrt{3/2}$.

21. The distance from a point (x, y, z) to the origin is given by $\sqrt{x^2 + y^2 + z^2}$. To simplify computations, we will minimize its square $f(x, y, z) = x^2 + y^2 + z^2$. The constraint is $g(x, y, z) = x^2 + y^2 - z^2 = 4$. The surface $x^2 + y^2 - z^2 = 4$ is a hyperboloid (its level curves are circles $x^2 + y^2 = 4 + c^2$; the intersections with the xz-plane and the yz-plane are hyperbolas $x^2 - z^2 = 4$ and $y^2 - z^2 = 4$); since it extends to infinity, f does not have a maximum. However, there is a minimum distance — we can argue as follows: take a ball (centered at the origin) of radius 10^{10} (actually, we can choose any radius that guarantees that a part of the hyperboloid $x^2 + y^2 - z^2 = 4$ is inside that ball). Consider the part of the hyperboloid (call it D) that is cut off by that ball together with the boundary points (namely, the points on the intersection of the ball and the hyperboloid). The set D is closed and bounded in $\mathbb{R}^3$, so f, being continuous, must have a minimum (and a maximum, but we do not care for it). That minimum is the one we are looking for.

Using $\nabla f = \lambda \nabla g$, we get $(2x, 2y, 2z) = \lambda(2x, 2y, -2z)$; i.e.,

$$x = \lambda x, \qquad y = \lambda y \qquad \text{and} \qquad z = -\lambda z.$$

Note that the condition $x^2 + y^2 - z^2 = 4$ implies that either $x \neq 0$ or $y \neq 0$ (otherwise, if $x = y = 0$, then $-z^2 = 4$, which has no real solutions for z). If $x \neq 0$, then the first equation implies $\lambda = 1$; from the third equation, we then conclude that $z = 0$; in that case, the constraint implies that $x^2 + y^2 = 4$. Similarly, if $y \neq 0$, then the second and third equations imply that $z = 0$; and again, $x^2 + y^2 = 4$. Thus, all points $(x, y, 0)$, where $x^2 + y^2 = 4$ (that is, the intersection of the hyperboloid and the xy-plane), are the candidates for the minimum.

Note that $\nabla g(x, y, z) = (2x, 2y, -2z) \neq \mathbf{0}$ for all points (x, y, z) that satisfy the constraint $x^2 + y^2 - z^2 = 4$.

It follows that the minimum of f is $x^2 + y^2 + 0^2 = 4$, so that the minimum distance is 2, and it occurs at all points on the circle $x^2 + y^2 = 4$ in the xy-plane.

23. Geometrically, it is clear that such a point exists and is unique (it is obtained as the intersection of the given plane and the line through $(0, 1, 1)$ that is perpendicular to it; as a matter of fact, this describes a possible way of finding it).

We are asked to find the minimum of the distance function

$$d(x, y, z) = \sqrt{x^2 + (y-1)^2 + (z-1)^2},$$

subject to the constraint $g(x, y, z) = x + y + 2z = 11$. For simplicity, we minimize the square of d, namely the function $f(x, y, z) = x^2 + (y-1)^2 + (z-1)^2$.

The equation $\nabla f = \lambda \nabla g$ implies that $(2x, 2(y-1), 2(z-1)) = \lambda(1, 1, 2)$; i.e.,

$$2x = \lambda, \qquad 2(y-1) = \lambda \qquad \text{and} \qquad z - 1 = \lambda.$$

Combining the first two equations, we get $x = y - 1$, and so $y = x + 1$. From the first and third equations, we compute $2x = z - 1$; thus, $z = 2x + 1$. Substituting the expressions for y and z into the constraint, we get $x + (x + 1) + 2(2x + 1) = 11$; i.e., $6x = 8$ and $x = 4/3$. It follows that $y = 7/3$ and $z = 11/3$.

Note that $\nabla g(x, y, z) = (1, 1, 2) \neq \mathbf{0}$ at all points (x, y, z).

It follows that the point $(4/3, 7/3, 11/3)$ in the plane $x + y + 2z = 11$ is closest to the point $(0, 1, 1)$. The distance is $\sqrt{(4/3)^2 + (4/3)^2 + (8/3)^2} = \sqrt{32/3}$.

25. The constraints $g_1(x, y, z) = 2y + z = 6$ and $g_2(x, y, z) = x - 2y = 4$ represent two planes in $\mathbb{R}^3$. We are asked to minimize the restriction of the function $f(x, y, z) = x^2 + y^2 + z^2$ to the points on the intersection of these two planes. Geometrically, we have to minimize the square of the distance between the origin and the points on the line that is the intersection of the two given planes. (Clearly, a maximum does not exist.)

From $\nabla f = \lambda_1 \nabla g_1 + \lambda_2 \nabla g_2$, we get $(2x, 2y, 2z) = \lambda_1(0, 2, 1) + \lambda_2(1, -2, 0)$; i.e.,

$$2x = \lambda_2, \qquad 2y = 2\lambda_1 - 2\lambda_2, \qquad 2z = \lambda_1.$$

Substituting the first and third equations into (the second equation) $y = \lambda_1 - \lambda_2$, we get $y = 2z - 2x$. Thus, we have to solve the system

$$2x + y - 2z = 0$$
$$2y + z = 6$$
$$x - 2y = 4.$$

Computing z from the second equation and x from the third equation and substituting into the first equation, we obtain $2(4 + 2y) + y - 2(6 - 2y) = 0$; i.e., $9y - 4 = 0$ and $y = 4/9$. Thus, $x = 4 + 2y = 44/9$ and $z = 6 - 2y = 46/9$. It follows that the minimum of f subject to the given constraints is $f(44/9, 4/9, 46/9) = (44/9)^2 + (4/9)^2 + (46/9)^2 = 4068/81 \approx 50.2222$.

27. The intersection of the planes $g_1(x, y, z) = x + y - z = 4$ and $g_2(x, y, z) = 2z - y = 0$ is a line. We are asked to minimize the function $f(x, y, z) = x^2 + y^2 + z^2$, which is the square of the distance from the origin to the points on that line.

From $\nabla f = \lambda_1 \nabla g_1 + \lambda_2 \nabla g_2$, we get $(2x, 2y, 2z) = \lambda_1(1, 1, -1) + \lambda_2(2, -1, 0)$; i.e.,

$$2x = \lambda_1 + 2\lambda_2, \qquad 2y = \lambda_1 - \lambda_2, \qquad 2z = -\lambda_1.$$

Multiplying the second equation by 2 and adding it to the first equation, we get $2x + 4y = 3\lambda_1$. Combining with the third equation, we obtain $x + 2y + 3z = 0$. Thus, we have to solve the system

$$x + 2y + 3z = 0$$
$$x + y - z = 4$$
$$2z - y = 0.$$

Subtracting the second equation from the first equation, we get $y + 4z = -4$. Adding this to the third equation gives $6z = -4$ and $z = -2/3$. It follows that $y = 2z = -4/3$ and $x = 4 - y + z = 14/3$. Consequently, the minimum of f subject to the given constraints is $f(14/3, -4/3, -2/3) = (14/3)^2 + (-4/3)^2 + (-2/3)^2 = 216/9 = 24$. So, the minimum distance is $\sqrt{24}$; it is the distance between the origin and the point $(14/3, -4/3, -2/3)$ on the given planes.

29. Place the box so that its sides are parallel to the coordinate planes and its center is at the origin. Label its length, width and height by x, y and z respectively ($x, y, z > 0$). Since the box is inscribed into the sphere (whose center must be at the origin, by construction), the eight points $(\pm x/2, \pm y/2, \pm z/2)$ must lie on it. Thus

$$\left(\pm\frac{x}{2}\right)^2 + \left(\pm\frac{y}{2}\right)^2 + \left(\pm\frac{z}{2}\right)^2 = R^2;$$

i.e., $x^2 + y^2 + z^2 = 4R^2$. Thus, we are asked to find the maximum value of the volume $V = xyz$ subject to the constraint $g(x, y, z) = x^2 + y^2 + z^2 = 4R^2$.

Using $\nabla V = \lambda \nabla g$, we get $(yz, xz, xy) = \lambda(2x, 2y, 2z)$; thus,

$$yz = 2\lambda x, \qquad xz = 2\lambda y, \qquad xy = 2\lambda z$$

(keep in mind that $x, y, z > 0$). Dividing the first two equations, we get $yz/xz = 2\lambda x/2\lambda y$; i.e., $x^2 = y^2$. Similarly, dividing the second equation by the third equation, we conclude that $y^2 = z^2$. From the constraint, we obtain $y^2 + y^2 + y^2 = 4R^2$, and $y = 2R/\sqrt{3}$. It follows that $x = y = z = 2R/\sqrt{3}$. Thus, the largest rectangular box is a cube of radius $2R/\sqrt{3}$. Its volume is $(2R/\sqrt{3})^3 = 8R^3/3^{3/2}$.

31. The solid cut off by the plane is a tetrahedron with volume $V = abc/6$. The plane passes through $(1, 1, 1)$, and thus $1/a + 1/b + 1/c = 1$, $a, b, c > 0$. Let $g(a, b, c) = 1/a + 1/b + 1/c$.

From $\nabla V = \lambda \nabla g$, we get $(bc/6, ac/6, ab/6) = \lambda(-1/a^2, -1/b^2, -1/c^2)$; thus,

$$\frac{bc}{6} = -\frac{\lambda}{a^2}, \qquad \frac{ac}{6} = -\frac{\lambda}{b^2}, \qquad \frac{ab}{6} = -\frac{\lambda}{c^2}.$$

Eliminating λ from the first two equations, we get $\lambda = -\frac{1}{6}a^2bc = -\frac{1}{6}ab^2c$, and (since $a, b, c > 0$) $a = b$. Similarly, combining the second and third equations (or using symmetry), we get $b = c$. The constraint equation implies that $1/b + 1/b + 1/b = 1$; i.e., $b = 3$. It follows that $a = b = c = 3$.

The smallest volume is $V = 3^3/6 = 9/2$. (It is intuitively clear that there must be a positive minimum — no matter how the plane is placed, the resulting solid must contain the cube of side 1 whose (diagonally opposite) vertices are $(0, 0, 0)$ and $(1, 1, 1)$ and whose sides are parallel to the coordinate axes. From the formula for V, it follows that $V \to \infty$ as $a \to \infty$ and $b \to \infty$; thus, V cannot have a maximum.)

4.5. Flow Lines

1. Drop t to simplify notation and use $'$ to denote the derivative with respect to t. Let $\mathbf{c} = (x, y)$. Then $\mathbf{F}(\mathbf{c}) = \mathbf{c}'$ implies that $(x, 2y) = (x', y')$, and hence $x' = x$ and $y' = 2y$. The solutions are (both equations represent a model of an exponential growth) $x = C_1 e^t$ and $y = C_2 e^{2t}$, where C_1 and C_2 are constants. Eliminating t we get $y = C_2(e^t)^2 = C_2 C_1^{-2} x^2$, hence the flow lines are parabolas $y = Cx^2$, $C = $ constant.

Similar computation shows that the flow lines of $\mathbf{F}_1(x, y) = (3x, 6y)$ are given by $x = C_1 e^{3t}$ and $y = C_2 e^{6t}$. Eliminating t, we get a family of parabolas $y = Cx^2$.

The flow lines of $\mathbf{F}_2(x, y) = (-2x, -4y)$ are given by $x = C_1 e^{-2t}$ and $y = C_2 e^{-4t}$; i.e., are (again) parabolas $y = Cx^2$.

However, there is a difference. Since $\|\mathbf{c}'\| = \|\mathbf{F}(\mathbf{c})\|$, it follows that the speed of flow lines of $\mathbf{F}$ is $\sqrt{x^2 + 4y^2}$. Similarly, the speed of flow lines of $\mathbf{F}_1$ is

$$\sqrt{9x^2 + 36y^2} = 3\sqrt{x^2 + 4y^2}$$

and the speed of flow lines of $\mathbf{F}_2$ is

$$\sqrt{4x^2 + 16y^2} = 2\sqrt{x^2 + 4y^2}.$$

So the flow lines of $\mathbf{F}$ are the slowest whereas those of $\mathbf{F}_1$ have the largest speed. Flow lines of $\mathbf{F}_1$ "flow" in the same direction as those of $\mathbf{F}$, but the flow lines of $\mathbf{F}_2$ "flow" in the opposite direction.

3. $\mathbf{F}(x, y) = 2\|\mathbf{r}\|^{-1}\mathbf{r}$ is a radial vector field and therefore its flow lines are straight lines emerging from the origin. In particular, the flow line going through the point $(3, 2)$ has the equation $\mathbf{c}(t) = a(3t, 2t)$, $a \neq 0$. From $\mathbf{F}(\mathbf{c}(t)) = \mathbf{c}'(t)$ it follows that

$$\frac{2(3at, 2at)}{\sqrt{9a^2 t^2 + 4a^2 t^2}} = (3a, 2a) \qquad \text{and} \qquad \frac{2at(3, 2)}{\sqrt{13}|a|\,|t|} = a(3, 2).$$

Now $t/|t| = \pm 1$ implies that $|a| = 2/\sqrt{13}$; i.e., $a = \pm 2/\sqrt{13}$ and so

$$\mathbf{c}(t) = \pm(2/\sqrt{13})(3t, 2t) = \pm\left(6t/\sqrt{13}, 4t/\sqrt{13}\right),$$

$t \geq 0$. Since the flow lines have the same direction as the vector field we have to choose the $+$ sign.

5. Since $\mathbf{F}(x, y) = a\mathbf{i} + b\mathbf{j}$ is a constant vector field, its flow line must have the same slope at all points; thus, it must be a line. So, the flow line that goes through the origin is given by $\mathbf{l}(t) = (0, 0) + t(a, b) = (ta, tb)$, $t \in \mathbb{R}$.

We could easily check that $\mathbf{l}(t)$ is the required flow line: $\mathbf{F}(\mathbf{l}(t)) = \mathbf{F}(ta, tb) = a\mathbf{i} + b\mathbf{j}$ is equal to $\mathbf{l}'(t) = (a, b)$.

7. See Figure 7.

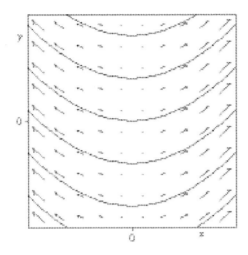

Figure 7

9. See Figure 9.

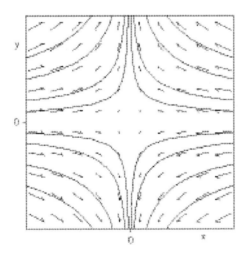

Figure 9

11. Let $\mathbf{F}(x,y,z) = (F_1(x,y,z), F_2(x,y,z), F_3(x,y,z))$. We have to find F_1, F_2 and F_3 so that (write $\mathbf{F}(\mathbf{c}(t)) = \mathbf{c}'(t)$ in components)

$$F_1(t^2, 2t, t) = 2t, \qquad F_2(t^2, 2t, t) = 2 \qquad \text{and} \qquad F_3(t^2, 2t, t) = 1.$$

Take, for example, $F_1(x,y,z) = y$ (then, clearly, $F_1(t^2, 2t, t) = 2t$). Let (keep in mind that we need 2 for the value of F_2) $F_2(x,y,z) = y/z$ and (we need 1 for the value of F_3) $F_3(x,y,z) = 2z/y$. Hence $\mathbf{F}(x,y,z) = (y, y/z, 2z/y)$ is a vector field whose flow line is $\mathbf{c}(t) = (t^2, 2t, t)$.

There are infinitely many answers. For example, $\mathbf{F}(x, y, z) = (2x/z, 2x/z^2, x/z^2)$ or $\mathbf{F}(x, y, z) = (2z, yz/x^2, z^2/x)$ have the same curve $\mathbf{c}(t) = (t^2, 2t, t)$ as their flow line.

13. Notice that the values of both t and $a + b - t$ belong to the interval $[a, b]$; hence $\mathbf{c}$ and $\boldsymbol{\gamma}$ represent the same curve. From $\boldsymbol{\gamma}(a) = \mathbf{c}(b)$ and $\boldsymbol{\gamma}(b) = \mathbf{c}(a)$ it follows that they are of opposite orientations. Assuming that $\mathbf{c}(t)$ is a flow line of $\mathbf{F}$, we get $\mathbf{G}(\boldsymbol{\gamma}(t)) = (-\mathbf{F})(\boldsymbol{\gamma}(t)) = -\mathbf{F}(\mathbf{c}(a+b-t)) = -\mathbf{c}'(a+b-t)$. From $\boldsymbol{\gamma}(t) = \mathbf{c}(a+b-t)$ it follows that $\boldsymbol{\gamma}'(t) = -\mathbf{c}'(a+b-t)$; hence $\mathbf{G}(\boldsymbol{\gamma}(t)) = (-\mathbf{F})(\boldsymbol{\gamma}(t)) = \boldsymbol{\gamma}'(t)$, and so $\boldsymbol{\gamma}$ is a flow line of $-\mathbf{F}$.

This argument implies that the flow lines of $-\mathbf{F}$ are obtained from those of $\mathbf{F}$ by reversing the orientation.

4.6. Divergence and Curl of a Vector Field

1. Since *grad* acts on scalar functions and *grad f* is a vector, the expression *grad (grad f)* has no meaning.

3. *curl* $(\mathbf{F} - \mathbf{G})$ is a vector field; *div* $\mathbf{F}$ is a scalar function; so *grad* (*div* $\mathbf{F}$) is defined, and is a vector field. The expression in question is a vector field, as a cross product of two vector fields.

5. *grad f* is a vector field, and so *curl* (*grad f*) is defined and is a vector field. Since the divergence acts on vector fields, *div* (*curl* (*grad f*)) is defined, and is a scalar function.

7. $\mathbf{F}$ could be any constant field, $\mathbf{F}(\mathbf{x}) = \mathbf{a}$. Or: note that the vector field $\mathbf{F} = f_1(x)\mathbf{i} + f_2(y)\mathbf{j} + f_3(z)\mathbf{k}$, where f_1, f_2 and f_3 are functions of indicated variables, has zero curl. So, any choice of f_1, f_2 and f_3 such that $\partial f_1/\partial x + \partial f_2/\partial y + \partial f_3/\partial z = 0$ will guarantee that *div* $\mathbf{F} = 0$. There are many other possibilities for $\mathbf{F}$.

9. We could look for $\mathbf{F}$ in the form $\mathbf{F} = f_1(x)\mathbf{i} + f_2(y)\mathbf{j} + f_3(z)\mathbf{k}$ (which guarantees zero curl), such that $\partial f_1/\partial x + \partial f_2/\partial y + \partial f_3/\partial z \neq 0$. For instance, let $\mathbf{F} = f_1(x)\mathbf{i}$, where $f_1'(x) \neq 0$.

Alternatively, let $\mathbf{F} = x\mathbf{i} + y\mathbf{j} + z\mathbf{k}$ (i.e., $\mathbf{F}$ is a radial vector field). We could argue geometrically that *div* $\mathbf{F} \neq 0$, or just compute it: *div* $\mathbf{F} = 3$. Clearly, *curl* $\mathbf{F} = 0$.

11. For instance, let $\mathbf{F}(x, y) = f(x)\mathbf{i}$, where f is a positive and increasing function for all x (for instance, $f(x) = e^x$); this guarantees that the total outflux is always positive. Another intuitive example is a radial vector field, $\mathbf{F}(x, y) = x\mathbf{i} + y\mathbf{j}$ (see Example 4.56).

13. By definition,

$$div\,\mathbf{F} = \frac{\partial}{\partial x}\left(y^2 z\right) + \frac{\partial}{\partial y}(-xz) + \frac{\partial}{\partial z}(xyz) = 0 + 0 + xy = xy$$

and

$$curl\,\mathbf{F} = \begin{vmatrix} \mathbf{i} & \mathbf{j} & \mathbf{k} \\ \partial/\partial x & \partial/\partial y & \partial/\partial z \\ y^2 z & -xz & xyz \end{vmatrix} = (xz + x)\,\mathbf{i} - \left(yz - y^2\right)\mathbf{j} + (-z - 2yz)\,\mathbf{k}.$$

15. By definition,

$$div\, \mathbf{F} = \frac{\partial}{\partial x}\big(3(x^2 + y^2 + z^2)\big) + \frac{\partial}{\partial y}\big(x^2 + y^2 + z^2\big) - \frac{\partial}{\partial z}\big(x^2 + y^2 + z^2\big) = 6x + 2y - 2z$$

and

$$curl\, \mathbf{F} = \begin{vmatrix} \mathbf{i} & \mathbf{j} & \mathbf{k} \\ \partial/\partial x & \partial/\partial y & \partial/\partial z \\ 3(x^2 + y^2 + z^2) & x^2 + y^2 + z^2 & -(x^2 + y^2 + z^2) \end{vmatrix}$$
$$= (-2y - 2z)\mathbf{i} - (-2x - 6z))\mathbf{j} + (2x - 6y)\mathbf{k}$$
$$= -2(y + z)\mathbf{i} + 2(x + 3z)\mathbf{j} + 2(x - 3y)\mathbf{k}.$$

17. Let $\mathbf{F}(x, y) = (f(x), 0)$. The mass flowing through the left side of rectangle R (see Figure 4.31) is $M_L = f(x)\Delta t \Delta y$ (particles enter R with speed $f(x)$; if $f(x) > 0$, it is the inflow, and if $f(x) > 0$ is it the outflow). The flow through the right side of R is $M_R = f(x + \Delta x)\Delta t \Delta y$. Thus, the net flow is

$$O(\Delta t) = M_R - M_L = (f(x + \Delta x) - f(x))\Delta t \Delta y$$

(note that $O(\Delta t)$ could be positive, negative, or zero). Thus,

$$\frac{1}{\Delta x \Delta y} \frac{O(\Delta t)}{\Delta t} = \frac{f(x + \Delta x) - f(x)}{\Delta x},$$

i.e.,

$$\frac{1}{\Delta x \Delta y} \frac{O(\Delta t)}{\Delta t} \approx f'(x).$$

Since $\mathbf{F}(x, y) = (f(x), 0)$, it follows that $f'(x) = div\, \mathbf{F}$.

If $\mathbf{F}(x, y) = (0, g(y))$, the the flow through the bottom side of R is $M_B = g(y)\Delta t \Delta x$, and the flow through the top side $M_T = g(y + \Delta y)\Delta t \Delta x$. Again, the net flow is

$$O(\Delta t) = M_T - M_B = (g(y + \Delta y) - g(y))\Delta t \Delta x,$$

and

$$\frac{1}{\Delta x \Delta y} \frac{O(\Delta t)}{\Delta t} = \frac{g(y + \Delta y) - g(y)}{\Delta y} \approx g'(y).$$

Since $\mathbf{F}(x, y) = (0, g(y))$, it follows that $g'(y) = div\, \mathbf{F}$.

19. $\mathbf{r}$ is a radial vector field (its flow lines are lines through the origin). Since there are no rotations within the flow, $curl\, \mathbf{r} = \mathbf{0}$.

21. From

$$curl\, \mathbf{F} = \begin{vmatrix} \mathbf{i} & \mathbf{j} & \mathbf{k} \\ \partial/\partial x & \partial/\partial y & \partial/\partial z \\ \cos y & \sin x & \tan z \end{vmatrix} = (\cos x + \sin y)\mathbf{k} \neq \mathbf{0}$$

it follows that $\mathbf{F}$ is not conservative.

23. The fact that

$$curl\,\mathbf{F} = \begin{vmatrix} \mathbf{i} & \mathbf{j} & \mathbf{k} \\ \partial/\partial x & \partial/\partial y & \partial/\partial z \\ 3x^2 y & x^3 + y^3 & 0 \end{vmatrix} = (3x^2 - 3x^2)\mathbf{k} = \mathbf{0}$$

implies that $\mathbf{F}$ is conservative.

From $\mathbf{F} = -\nabla V$ it follows that $V_x = -3x^2 y$ and $V_y = -x^3 - y^3$. Integrating the first equation with respect to x, we get $V = -x^3 y + C(y)$; substituting this into the expression for V_y, we get $-x^3 + C'(y) = -x^3 - y^3$ and therefore $C'(y) = -y^3$ and $C(y) = -y^4/4 + C$. It follows that all functions of the form $V(x,y) = -x^3 y - y^4/4 + C$ (where C is a constant) are potential functions for $\mathbf{F}$.

25. From

$$curl\,\mathbf{F} = \begin{vmatrix} \mathbf{i} & \mathbf{j} & \mathbf{k} \\ \partial/\partial x & \partial/\partial y & \partial/\partial z \\ -y & -x & -3 \end{vmatrix} = \mathbf{0}$$

it follows that $\mathbf{F}$ is conservative.

The potential function V is determined from $\nabla V = -\mathbf{F} = y\mathbf{i} + x\mathbf{j} + 3\mathbf{k}$; i.e., $V_x = y$, $V_y = x$ and $V_z = 3$. Integrating the first equation with respect to x, we get $V = xy + C(y,z)$; substituting this into the expression for V_y produces a differential equation for $C(y,z)$: $V_y = x + \partial C(y,z)/\partial y = x$, and so $\partial C(y,z)/\partial y = 0$ and $C(y,z) = C(z)$. Consequently, $V = xy + C(z)$. Finally, $V_z = 3$ implies $C'(z) = 3$; i.e., $C(z) = 3z + C$. It follows that $V(x,y,z) = xy + 3z + C$ represents all potential functions of $\mathbf{F}$.

27. By the chain rule,

$$grad\,f = \left(\frac{-2x}{(x^2 + y^2 + z^2)^2}, \frac{-2y}{(x^2 + y^2 + z^2)^2}, \frac{-2z}{(x^2 + y^2 + z^2)^2} \right).$$

The $\mathbf{i}$ component of $curl\,(grad\,f)$ is equal to

$$\frac{\partial}{\partial y}\left(\frac{-2z}{(x^2 + y^2 + z^2)^2} \right) - \frac{\partial}{\partial z}\left(\frac{-2y}{(x^2 + y^2 + z^2)^2} \right)$$
$$= -2z(-2)(x^2 + y^2 + z^2)^{-3}\, 2y - (-2y)(-2)(x^2 + y^2 + z^2)^{-3}\, 2z$$
$$= 8yz(x^2 + y^2 + z^2)^{-3} - 8yz(x^2 + y^2 + z^2)^{-3} = 0.$$

The remaining two components are checked analogously.

29. Since $div\,(curl\,\mathbf{F}) = x^2 + y^2 + z^2 \neq 0$, it follows that there can be no C^2 vector field $\mathbf{F}$ such that $curl\,\mathbf{F} = xy^2\mathbf{i} + yz^2\mathbf{j} + zx^2\mathbf{k}$ (since, for C^2 vector fields, $div\,(curl\,\mathbf{F}) = 0$).

31. By definition,

$$curl\,\mathbf{F} = \begin{vmatrix} \mathbf{i} & \mathbf{j} & \mathbf{k} \\ \partial/\partial x & \partial/\partial y & \partial/\partial z \\ f(x) & g(y) & h(z) \end{vmatrix}$$
$$= \left(\frac{\partial h(z)}{\partial y} - \frac{\partial g(y)}{\partial z} \right)\mathbf{i} - \left(\frac{\partial h(z)}{\partial x} - \frac{\partial f(x)}{\partial z} \right)\mathbf{j} + \left(\frac{\partial g(y)}{\partial x} - \frac{\partial f(x)}{\partial y} \right)\mathbf{k}$$

$$= \mathbf{0}.$$

33. From

$$(x - iy)^3 = x^3 - 3x^2iy + 3xi^2y^2 - i^3y^3 = (x^3 - 3xy^2) + i(-3x^2y + y^3)$$

it follows that the vector field in question is $\mathbf{F} = (x^3 - 3xy^2, -3x^2y + y^3, 0)$. From $div\,\mathbf{F} = 3x^2 - 3y^2 - 3x^2 + 3y^2 = 0$, we conclude that $\mathbf{F}$ is incompressible. The fact that

$$curl\,\mathbf{F} = \begin{vmatrix} \mathbf{i} & \mathbf{j} & \mathbf{k} \\ \partial/\partial x & \partial/\partial y & \partial/\partial z \\ x^3 - 3xy^2 & -3x^2y + y^3 & 0 \end{vmatrix} = (-6xy + 6xy)\mathbf{k} = \mathbf{0}$$

implies that $\mathbf{F}$ is irrotational.

35. Let $\mathbf{F} = grad\,f$. Then $curl\,\mathbf{F} = curl\,(grad\,f) = \mathbf{0}$, and so $\mathbf{F}$ is irrotational. Since $div\,\mathbf{F} = div\,(grad\,f) = \Delta f = 0$ (by assumption), $\mathbf{F}$ is incompressible as well.

37. By definition,

$$div\,\mathbf{F} = 4x + 8y^2z + 3x^3 - 3x - 8y^2z + 2x^3 \neq 0,$$

so $\mathbf{F}$ is not incompressible. However,

$$div\,(xyz^2\mathbf{F}) = div\,\left((2x^3yz^2 + 8x^2y^3z^3)\mathbf{i} + (3x^4y^2z^2 - 3x^2y^2z^2)\mathbf{j} - (4xy^3z^4 + 2x^4yz^3)\mathbf{k}\right)$$

$$= 6x^2yz^2 + 16xy^3z^3 + 6x^4yz^2 - 6x^2yz^2 - 16xy^3z^3 - 6x^4yz^2 = 0;$$

consequently, $\mathbf{G} = xyz^2\mathbf{F}$ is incompressible.

39. Since $f(\|\mathbf{r}\|)\mathbf{r} = f(\sqrt{x^2 + y^2 + z^2})(x\mathbf{i} + y\mathbf{j} + z\mathbf{k})$, it follows that

$$curl\,\mathbf{F} = \begin{vmatrix} \mathbf{i} & \mathbf{j} & \mathbf{k} \\ \partial/\partial x & \partial/\partial y & \partial/\partial z \\ xf(\sqrt{x^2 + y^2 + z^2}) & yf(\sqrt{x^2 + y^2 + z^2}) & zf(\sqrt{x^2 + y^2 + z^2}) \end{vmatrix}.$$

The $\mathbf{i}$ component of $curl\,\mathbf{F}$ is computed by the chain rule to be

$$\frac{\partial}{\partial y}\left(zf(\sqrt{x^2 + y^2 + z^2})\right) - \frac{\partial}{\partial z}\left(yf(\sqrt{x^2 + y^2 + z^2})\right)$$

$$= f'(\sqrt{x^2 + y^2 + z^2})\tfrac{1}{2}(x^2 + y^2 + z^2)^{-1/2}\,2yz$$

$$- f'(\sqrt{x^2 + y^2 + z^2})\tfrac{1}{2}(x^2 + y^2 + z^2)^{-1/2}\,2zy = 0.$$

The remaining two components are computed in the same way.

41. Let $\mathbf{F} = (F_1, F_2, F_3)$. Then $f\mathbf{F} = (fF_1, fF_2, fF_3)$ and (using the product rule)

$$curl\,(f\mathbf{F}) = \begin{vmatrix} \mathbf{i} & \mathbf{j} & \mathbf{k} \\ \partial/\partial x & \partial/\partial y & \partial/\partial z \\ fF_1 & fF_2 & fF_3 \end{vmatrix}$$

$$= \left(\frac{\partial f}{\partial y}F_3 + f\frac{\partial F_3}{\partial y} - \frac{\partial f}{\partial z}F_2 - f\frac{\partial F_2}{\partial z}\right)\mathbf{i} - \left(\frac{\partial f}{\partial x}F_3 + f\frac{\partial F_3}{\partial x} - \frac{\partial f}{\partial z}F_1 - f\frac{\partial F_1}{\partial z}\right)\mathbf{j}$$

$$+ \left(\frac{\partial f}{\partial x}F_2 + f\frac{\partial F_2}{\partial x} - \frac{\partial f}{\partial y}F_1 - f\frac{\partial F_1}{\partial y}\right)\mathbf{k}.$$

On the other hand,

$$f \, curl \, \mathbf{F} + (grad \, f) \times \mathbf{F} = f \begin{vmatrix} \mathbf{i} & \mathbf{j} & \mathbf{k} \\ \partial/\partial x & \partial/\partial y & \partial/\partial z \\ F_1 & F_2 & F_3 \end{vmatrix} + \begin{vmatrix} \mathbf{i} & \mathbf{j} & \mathbf{k} \\ \partial f/\partial x & \partial f/\partial y & \partial f/\partial z \\ F_1 & F_2 & F_3 \end{vmatrix}$$

$$= f \left(\frac{\partial F_3}{\partial y} - \frac{\partial F_2}{\partial z} \right) \mathbf{i} - f \left(\frac{\partial F_3}{\partial x} - \frac{\partial F_1}{\partial z} \right) \mathbf{j} + f \left(\frac{\partial F_2}{\partial x} - \frac{\partial F_1}{\partial y} \right) \mathbf{k}$$

$$+ \left(\frac{\partial f}{\partial y} F_3 - F_2 \frac{\partial f}{\partial z} \right) \mathbf{i} - \left(\frac{\partial f}{\partial x} F_3 - F_1 \frac{\partial f}{\partial z} \right) \mathbf{j} + \left(\frac{\partial f}{\partial x} F_2 - F_1 \frac{\partial f}{\partial y} \right) \mathbf{k}.$$

The assumption needed is that both f and $\mathbf{F}$ are differentiable.

43. $div \, \mathbf{r} = div \, (x\mathbf{i} + y\mathbf{j} + z\mathbf{k}) = 3$.

45. Here is an alternative to a straightforward computation: using Exercise 40 with $f = \|\mathbf{r}\|$ and $\mathbf{F} = \mathbf{r}$ we get

$$div \, (\|\mathbf{r}\|\mathbf{r}) = \|\mathbf{r}\| \, div \, \mathbf{r} + \mathbf{r} \cdot grad \, \|\mathbf{r}\|.$$

Now proceed using Exercises 43 and 44: $div \, (\|\mathbf{r}\|\mathbf{r}) = 3\|\mathbf{r}\| + \mathbf{r} \cdot \mathbf{r}/\|\mathbf{r}\| = 4\|\mathbf{r}\|$, since $\mathbf{r} \cdot \mathbf{r} = \|\mathbf{r}\|^2$.

47. Assuming that f and g are of class C^2 we get

$$div \, (grad \, f \times grad \, g) = div \begin{vmatrix} \mathbf{i} & \mathbf{j} & \mathbf{k} \\ f_x & f_y & f_z \\ g_x & g_y & g_z \end{vmatrix}$$

$$= div \, ((f_y g_z - f_z g_y)\mathbf{i} - (f_x g_z - g_x f_z)\mathbf{j} + (f_x g_y - g_x f_y)\mathbf{k})$$

$$= \frac{\partial}{\partial x}(f_y g_z - f_z g_y) + \frac{\partial}{\partial y}(-(f_x g_z - g_x f_z)) + \frac{\partial}{\partial z}(f_x g_y - g_x f_y)$$

$$= f_{yx} g_z + f_y g_{zx} - f_{zx} g_y - f_z g_{yx} - f_{xy} g_z - f_x g_{zy} + g_{xy} f_z + g_x f_{zy}$$

$$+ f_{xz} g_y + f_x g_{yz} - g_{xz} f_y - g_x f_{yz} = 0.$$

Notice that we used the equality of mixed second partials. This is where the assumption that f and g are of class C^2 is needed.

49. Using Exercise 46, we get

$$div \, (\mathbf{F} \times \mathbf{r}) = \mathbf{r} \cdot curl \, \mathbf{F} - \mathbf{F} \cdot curl \, \mathbf{r} = -\mathbf{F} \cdot curl \, \mathbf{r},$$

since, by assumption, $curl \, \mathbf{F} = \mathbf{0}$. By Exercise 42, $curl \, \mathbf{r} = \mathbf{0}$ and therefore $div \, (\mathbf{F} \times \mathbf{r}) = 0$.

4.7. Implicit Function Theorem

1. Consider the case $F(x, y, g(x, y))$ first. By the chain rule,

$$D_1 F(x, y, g(x, y)) \cdot 1 + D_2 F(x, y, g(x, y)) \cdot 0 + D_3 F(x, y, g(x, y)) \frac{\partial g(x, y)}{\partial x} = 0,$$

and thus $(\partial g/\partial x)(x, y) = -D_1 F(x, y, g(x, y))/D_3 F(x, y, g(x, y))$. Replacing $g(x, y)$ by z, we get

$$\frac{\partial g}{\partial x}(x, y) = -\frac{\frac{\partial F}{\partial x}(x, y, z)}{\frac{\partial F}{\partial z}(x, y, z)}.$$

The formula for $\partial g/\partial y$ is obtained in the same way.

In general, we differentiate the equation $F(x_1, \ldots, x_m, g(x_1, \ldots, x_m)) = 0$ with respect to x_i, $i = 1, \ldots, m$, and use the chain rule:

$$D_1 F \frac{\partial x_1}{\partial x_i} + \cdots + D_i F \frac{\partial x_i}{\partial x_i} + \cdots + D_m F \frac{\partial x_i}{\partial x_m} + D_{m+1} F \frac{\partial g}{\partial x_i} = 0,$$

where $F = F(x_1, \ldots, x_m, g(x_1, \ldots, x_m))$ and $g = g(x_1, \ldots, x_m)$. Since $\partial x_i/\partial x_i = 1$ and $\partial x_j/\partial x_i = 0$ for $j \neq i$, we get $D_i F + (D_{m+1} F)(\partial g/\partial x_i) = 0$, i.e.,

$$\frac{\partial g}{\partial x_i}(x_1, \ldots, x_m) = -\frac{D_i F(x_1, \ldots, x_m, g(x_1, \ldots, x_m))}{D_{m+1} F(x_1, \ldots, x_m, g(x_1, \ldots, x_m))}.$$

Replacing $D_i F$ by $\partial F/\partial x_i$, $D_{m+1} F$ by $\partial F/\partial z$ and $g(x_1, \ldots, x_m)$ by z, we obtain the desired formula.

3. Multiply the first equation in (4.43) by $\sin\theta$, the second by $-\cos\theta$, and add them up, to get $\sin\theta + r(\partial\theta/\partial x) = 0$, i.e., $\partial\theta/\partial x = -\frac{\sin\theta}{r}$.

Computing the partial derivative of (4.42) with respect to y, we get

$$-\frac{\partial r}{\partial y}\cos\theta + r\sin\theta\frac{\partial\theta}{\partial y} = 0$$

$$1 - \frac{\partial r}{\partial y}\sin\theta - r\cos\theta\frac{\partial\theta}{\partial y} = 0. \tag{4.1}$$

Multiplying the first equation by $\cos\theta$, the second equation by $\sin\theta$, and adding them up, we get $\sin\theta - \partial r/\partial y = 0$, i.e., $\partial r/\partial y = \sin\theta$. Similarly, multiplying the first equation by $\sin\theta$, the second equation by $-\cos\theta$, and adding them up, we get $-\cos\theta + r\partial\theta/\partial y = 0$, i.e., $\partial\theta/\partial y = \frac{\cos\theta}{r}$.

5. Differentiating the three equations $F_1 = x - \rho\sin\phi\cos\theta = 0$, $F_2 = y - \rho\sin\phi\sin\theta = 0$ and $F_3 = z - \rho\cos\phi = 0$ with respect to x, we get

$$1 - \frac{\partial\rho}{\partial x}\sin\phi\cos\theta - \rho\cos\phi\cos\theta\frac{\partial\phi}{\partial x} + \rho\sin\phi\sin\theta\frac{\partial\theta}{\partial x} = 0$$

$$-\frac{\partial\rho}{\partial x}\sin\phi\sin\theta - \rho\cos\phi\sin\theta\frac{\partial\phi}{\partial x} - \rho\sin\phi\cos\theta\frac{\partial\theta}{\partial x} = 0$$

$$-\frac{\partial\rho}{\partial x}\cos\phi + \rho\sin\phi\frac{\partial\phi}{\partial x} = 0.$$

Multiplying the first equation by $\cos\theta$, the second equation by $\sin\theta$, and adding them up, we get

$$\cos\theta - \frac{\partial\rho}{\partial x}\sin\phi - \rho\cos\phi\frac{\partial\phi}{\partial x} = 0.$$

Multiplying this equation by $\sin\phi$, and the third equation from above by $\cos\phi$ and adding them up, we get

$$\sin\phi\cos\theta - \frac{\partial\rho}{\partial x}\sin^2\phi - \rho\sin\phi\cos\phi\frac{\partial\phi}{\partial x} - \frac{\partial\rho}{\partial x}\cos^2\phi + \rho\sin\phi\cos\phi\frac{\partial\phi}{\partial x} = 0,$$

i.e., $\partial\rho/\partial x = \sin\phi\cos\theta$. Combining the same equations, we get

$$-\cos\theta\cos\phi + \frac{\partial\rho}{\partial x}\sin\phi\cos\phi + \rho\cos^2\phi\frac{\partial\phi}{\partial x} - \frac{\partial\rho}{\partial x}\sin\phi\cos\phi + \rho\sin^2\phi\frac{\partial\phi}{\partial x} = 0,$$

and thus $\rho(\partial\phi/\partial x) = \cos\theta\cos\phi$, and $\partial\phi/\partial x = \frac{\cos\theta\cos\phi}{\rho}$.

7. From $\partial F/\partial w = \cos(x+y+w) - z\sin(x+y+w)$ it follows that $(\partial F/\partial w)(0,0,1,0) = \cos 0 - \sin 0 = 1 \neq 0$. Thus, the equation $F(x,y,z,w) = \sin(x+y+w) + z\cos(x+y+w) - 1 = 0$ can be solved uniquely, locally, for w as a function of x, y and z near $(x,y,z,w) = (0,0,1,0)$.

Differentiating $F(x,y,z,w) = \sin(x+y+w) + z\cos(x+y+w) - 1 = 0$ with respect to x, we get

$$\cos(x+y+w)\left(1 + \frac{\partial w}{\partial x}\right) - z\sin(x+y+w)\left(1 + \frac{\partial w}{\partial x}\right) = 0,$$

and so $1 + (\partial w/\partial x)(0,0,1) = 0$, i.e., $(\partial w/\partial x)(0,0,1) = -1$. Similarly, differentiating the equation $F(x,y,z,w) = \sin(x+y+w) + z\cos(x+y+w) - 1 = 0$ with respect to z, we get

$$\cos(x+y+w)\frac{\partial w}{\partial z} + \cos(x+y+w) - z\sin(x+y+w)\frac{\partial w}{\partial z} = 0,$$

and so $(\partial w/\partial z)(0,0,1) + 1 = 0$, i.e., $(\partial w/\partial z)(0,0,1) = -1$.

9. Rewrite the system as

$$F_1(u,v,w,x,y,z) = u - x^2 - y^2 = 0$$
$$F_2(u,v,w,x,y,z) = v - y^2 + z^2 = 0$$
$$F_3(u,v,w,x,y,z) = w - xyz = 0.$$

The matrix in (4.39) is equal to

$$\begin{bmatrix} \partial F_1/\partial x & \partial F_1/\partial y & \partial F_1/\partial z \\ \partial F_2/\partial x & \partial F_2/\partial y & \partial F_2/\partial z \\ \partial F_3/\partial x & \partial F_3/\partial y & \partial F_3/\partial z \end{bmatrix} = \begin{bmatrix} -2x & -2y & 0 \\ 0 & -2y & 2z \\ -yz & -xz & -xy \end{bmatrix}.$$

Its determinant is

$$\Delta = -2x(2xy^2 + 2xz^2) + 2y(2yz^2) = 4(-x^2y^2 - x^2z^2 + y^2z^2).$$

Using the Implicit Function theorem, we conclude that the given system can be solved uniquely for x, y, z in terms of u, v, w near any point (x,y,z,u,v,w) whose coordinates satisfy $x^2y^2 + x^2z^2 - y^2z^2 \neq 0$.

11. We compute

$$\Delta = \det\begin{bmatrix} \partial F_1/\partial x & \partial F_1/\partial y \\ \partial F_2/\partial x & \partial F_2/\partial y \end{bmatrix} = \det\begin{bmatrix} a & b \\ c & d \end{bmatrix} = ad - bc.$$

By the Implicit Function Theorem, given system can be solved uniquely for x and y as functions of u and v in a neighbourhood of any point (x, y, u, v), as long as $ad - bc \neq 0$.

To solve for x and y we first multiply $ax + by - u = 0$ by d, $cx + dy - v = 0$ by $-b$ and add, to obtain

$$adx + bdy - du - bcx - bdy + bv = 0,$$

and thus $x = (du - bv)/(ad - bc)$. Similarly, we compute $y = (-cu + av)/(ad - bc)$.

13. Recall that the linear approximation of g at $(3, 0)$ is given by

$$L_{(3,0)}(x, y) = g(3, 0) + \frac{\partial g}{\partial x}(3, 0)(x - 3) + \frac{\partial g}{\partial y}(3, 0)(y - 0).$$

From $F(3, 0, 1) = 0$ we see that $g(3, 0) = 1$. Using the Implicit Function Theorem, we get

$$\frac{\partial g}{\partial x}(3, 0) = -\frac{(\partial F/\partial x)(3, 0, 1)}{(\partial F/\partial z)(3, 0, 1)} = -\frac{-4}{6} = \frac{2}{3}$$

and

$$\frac{\partial g}{\partial y}(3, 0) = -\frac{(\partial F/\partial y)(3, 0, 1)}{(\partial F/\partial z)(3, 0, 1)} = -\frac{-1}{6} = \frac{1}{6}.$$

(Partial derivatives of F are read from $\nabla F(3, 0, 1) = (-4, -1, 6)$.) Thus, the desired linear approximation is $L_{(3,0)}(x, y) = 1 + \frac{2}{3}(x - 3) + \frac{1}{6}y = \frac{2}{3}x + \frac{1}{6}y - 1$.

4.8. Appendix: Some Identities of Vector Calculus

1. By the product rule,

$$
\mathrm{curl}\,(f\mathbf{F}) = \begin{vmatrix} \mathbf{i} & \mathbf{j} & \mathbf{k} \\ \partial/\partial x & \partial/\partial y & \partial/\partial z \\ fF_1 & fF_2 & fF_3 \end{vmatrix}
$$

$$
= \left(\frac{\partial(fF_3)}{\partial y} - \frac{\partial(fF_2)}{\partial z} \right)\mathbf{i} + \left(\frac{\partial(fF_1)}{\partial z} - \frac{\partial(fF_3)}{\partial x} \right)\mathbf{j} + \left(\frac{\partial(fF_2)}{\partial x} - \frac{\partial(fF_1)}{\partial y} \right)\mathbf{k}
$$

$$
= \left(F_3\frac{\partial f}{\partial y} + f\frac{\partial F_3}{\partial y} - F_2\frac{\partial f}{\partial z} - f\frac{\partial F_2}{\partial z} \right)\mathbf{i}
$$

$$
+ \left(F_1\frac{\partial f}{\partial z} + f\frac{\partial F_1}{\partial z} - F_3\frac{\partial f}{\partial x} - f\frac{\partial F_3}{\partial x} \right)\mathbf{j}
$$

$$
+ \left(F_2\frac{\partial f}{\partial x} + f\frac{\partial F_2}{\partial x} - F_1\frac{\partial f}{\partial y} - f\frac{\partial F_1}{\partial y} \right)\mathbf{k}.
$$

The right side is computed to be

$$
f\,\mathrm{curl}\,\mathbf{F} + \mathrm{grad}\,f \times \mathbf{F} = f\begin{vmatrix} \mathbf{i} & \mathbf{j} & \mathbf{k} \\ \partial/\partial x & \partial/\partial y & \partial/\partial z \\ F_1 & F_2 & F_3 \end{vmatrix} + \begin{vmatrix} \mathbf{i} & \mathbf{j} & \mathbf{k} \\ \partial f/\partial x & \partial f/\partial y & \partial f/\partial z \\ F_1 & F_2 & F_3 \end{vmatrix}
$$

$$
= f\left(\left(\frac{\partial F_3}{\partial y} - \frac{\partial F_2}{\partial z} \right)\mathbf{i} + \left(\frac{\partial F_1}{\partial z} - \frac{\partial F_3}{\partial x} \right)\mathbf{j} + \left(\frac{\partial F_2}{\partial x} - \frac{\partial F_1}{\partial y} \right)\mathbf{k} \right)
$$

$$+ \left(F_3 \frac{\partial f}{\partial y} - F_2 \frac{\partial f}{\partial z} \right) \mathbf{i} + \left(F_1 \frac{\partial f}{\partial z} - F_3 \frac{\partial f}{\partial x} \right) \mathbf{j} + \left(F_2 \frac{\partial f}{\partial x} - F_1 \frac{\partial f}{\partial y} \right) \mathbf{k}$$

$$= \left(f \frac{\partial F_3}{\partial y} - f \frac{\partial F_2}{\partial z} + F_3 \frac{\partial f}{\partial y} - F_2 \frac{\partial f}{\partial z} \right) \mathbf{i}$$

$$+ \left(f \frac{\partial F_1}{\partial z} - f \frac{\partial F_3}{\partial x} + F_1 \frac{\partial f}{\partial z} - F_3 \frac{\partial f}{\partial x} \right) \mathbf{j}$$

$$+ \left(f \frac{\partial F_2}{\partial x} - f \frac{\partial F_1}{\partial y} + F_2 \frac{\partial f}{\partial x} - F_1 \frac{\partial f}{\partial y} \right) \mathbf{k}.$$

3. For simplicity, assume that $f, g \colon \mathbb{R}^2 \to \mathbb{R}$ (the general case of functions of m variables, $m \geq 2$, is dealt with analogously). The left side is computed to be

$$\Delta(fg) = div\,(grad\,(fg)) = div\,((f_x g + f g_x)\mathbf{i} + (f_y g + f g_y)\mathbf{j})$$

$$= f_{xx} g + 2 f_x g_x + f g_{xx} + f_{yy} g + 2 f_y g_y + f g_{yy}.$$

The first two terms on the right side are

$$g \Delta f + f \Delta g = g(f_{xx} + f_{yy}) + f(g_{xx} + g_{yy});$$

the remaining term is equal to

$$2\,grad\,f \cdot grad\,g = 2\,(f_x \mathbf{i} + f_y \mathbf{j}) \cdot (g_x \mathbf{i} + g_y \mathbf{j}) = 2(f_x g_x + f_y g_y).$$

It follows that the two sides are equal.

Here is an alternative proof: using product rules for the gradient and the divergence, we get

$$\Delta(fg) = div\,(grad\,(fg)) = div\,(g\,grad\,f + f\,grad\,g)$$

$$= g\,div\,(grad\,f) + grad\,g \cdot grad\,f + f\,div\,(grad\,g) + grad\,f \cdot grad\,g$$

$$= g \Delta f + f \Delta g + 2\,grad\,f \cdot grad\,g.$$

5. One way to prove this identity is by a straightforward computation, using appropriate definitions and the product rule. Here is a shorter alternative: start with a product rule for the divergence

$$div(g\,grad\,f \times f\,grad\,g) = (f\,grad\,g) \cdot curl\,(g\,grad\,f) - (g\,grad\,f) \cdot curl\,(f\,grad\,g),$$

apply a product rule for *curl*

$$= (f\,grad\,g) \cdot [g\,curl\,(grad\,f) + (grad\,g) \times (grad\,f)]$$

$$- (g\,grad\,f) \cdot [f\,curl\,(grad\,g) + (grad\,f) \times (grad\,g)]$$

and use the fact that the curl of the gradient is a zero vector

$$= f\,(grad\,g) \cdot ((grad\,g) \times (grad\,f)) - g\,(grad\,f) \cdot ((grad\,f) \times (grad\,g)) = 0.$$

In the last step we used the fact that the mixed product (of three vectors) with two equal factors is zero.

7. Using the product rule $div\,(f\mathbf{F}) = fdiv\,\mathbf{F} + \mathbf{F} \cdot grad\,f$ with $\mathbf{F} = grad\,g$, we get

$$div\,(f\,grad\,g) = f\,div\,(grad\,g) + (grad\,g) \cdot (grad\,f).$$

Now interchange f and g to get

$$div\,(g\,grad\,f) = g\,div\,(grad\,f) + (grad\,f) \cdot (grad\,g).$$

Subtracting the two equations and using the definition of the Laplace operator, we get

$$div(f\,grad\,g - g\,grad\,f)$$
$$= f\,div\,(grad\,g) + (grad\,g) \cdot (grad\,f) - [g\,div\,(grad\,f) + (grad\,f) \cdot (grad\,g)]$$
$$= f\Delta g - g\Delta f.$$

9. Using the product rule $div\,(f\mathbf{F}) = fdiv\,\mathbf{F} + \mathbf{F} \cdot grad\,f$ with $f = \|\mathbf{r}\|^{-3}$ and $\mathbf{F} = \mathbf{r}$, we get

$$div\,\|\mathbf{r}\|^{-3}\mathbf{r} = \|\mathbf{r}\|^{-3}div\,\mathbf{r} + \mathbf{r} \cdot grad\,(\|\mathbf{r}\|^{-3}) = 3\|\mathbf{r}\|^{-3} + \mathbf{r} \cdot grad\,(\|\mathbf{r}\|^{-3}),$$

since $div\,\mathbf{r} = 3$. From

$$grad\,(\|\mathbf{r}\|^{-3}) = grad\,(x^2 + y^2 + z^2)^{-3/2}$$
$$= \left(\frac{-3x}{(x^2 + y^2 + z^2)^{5/2}}, \frac{-3y}{(x^2 + y^2 + z^2)^{5/2}}, \frac{-3z}{(x^2 + y^2 + z^2)^{5/2}}\right)$$
$$= -3\|\mathbf{r}\|^{-5}\mathbf{r}$$

it follows that

$$div\,\|\mathbf{r}\|^{-3}\mathbf{r} = 3\|\mathbf{r}\|^{-3} + \mathbf{r}(-3)\|\mathbf{r}\|^{-5}\mathbf{r} = 3\|\mathbf{r}\|^{-3} - 3\|\mathbf{r}\|^{-3} = 0.$$

11. Using the chain rule,

$$grad\,f = \left(\frac{\partial f}{\partial x}, \frac{\partial f}{\partial y}\right)$$
$$= \left(D_1g\frac{\partial u}{\partial x} + D_2g\frac{\partial v}{\partial x} + D_3g, D_1g\frac{\partial u}{\partial y} + D_2g\frac{\partial v}{\partial y} + D_4g\right),$$

where D_ig, $i = 1, 2, 3, 4$, are partial derivatives of g with respect to its variables; D_ig are evaluated at $(u(x, y), v(x, y), x, y)$.

13. Since $\|\mathbf{r}\| = \sqrt{x^2 + y^2 + z^2}$, it follows that

$$grad\,(\ln\|\mathbf{r}\|) = \tfrac{1}{2}grad\,(\ln(x^2 + y^2 + z^2))$$
$$= \tfrac{1}{2}\left(\frac{2x}{x^2 + y^2 + z^2}, \frac{2y}{x^2 + y^2 + z^2}, \frac{2z}{x^2 + y^2 + z^2}\right) = \frac{1}{x^2 + y^2 + z^2}(x, y, z) = \frac{\mathbf{r}}{\|\mathbf{r}\|^2}.$$

15. Using the result of Exercise 13, we get

$$\Delta(\ln\|\mathbf{r}\|) = div\,(grad\,(\ln\|\mathbf{r}\|)) = div\,(\|\mathbf{r}\|^{-2}\mathbf{r}).$$

We proceed by a product rule for the divergence:

$$div\,(\|\mathbf{r}\|^{-2}\mathbf{r}) = \|\mathbf{r}\|^{-2}div\,\mathbf{r} + \mathbf{r} \cdot grad\,\|\mathbf{r}\|^{-2} = 3\|\mathbf{r}\|^{-2} + \mathbf{r} \cdot grad\,(\|\mathbf{r}\|^{-2}),$$

since $div \, \mathbf{r} = 3$. From

$$grad \left(\|\mathbf{r}\|^{-2} \right) = grad \left(\frac{1}{x^2 + y^2 + z^2} \right)$$

$$= \left(-\frac{2x}{(x^2 + y^2 + z^2)^2}, -\frac{2y}{(x^2 + y^2 + z^2)^2}, -\frac{2z}{(x^2 + y^2 + z^2)^2} \right)$$

$$= -\frac{2}{(x^2 + y^2 + z^2)^2}(x, y, z) = -2\|\mathbf{r}\|^{-4}\mathbf{r},$$

it follows that

$$div \left(\|\mathbf{r}\|^{-2}\mathbf{r} \right) = 3\|\mathbf{r}\|^{-2} - 2\|\mathbf{r}\|^{-4}\mathbf{r} \cdot \mathbf{r} = \|\mathbf{r}\|^{-2}.$$

17. Since $\mathbf{F}(r, \theta, z) = r^2\mathbf{e}_r + rz\mathbf{e}_\theta + \sin\theta\mathbf{e}_z$ it follows that $F_r = r^2$, $F_\theta = rz$ and $F_z = \sin\theta$. Using the formulas derived in this section, we get

$$div \, \mathbf{F} = \frac{1}{r}\frac{\partial(r^3)}{\partial r} + \frac{1}{r}\frac{\partial(rz)}{\partial\theta} + \frac{\partial(\sin\theta)}{\partial z} = 3r$$

and

$$curl \, \mathbf{F} = \frac{1}{r} \begin{vmatrix} \mathbf{e}_r & r\mathbf{e}_\theta & \mathbf{e}_z \\ \partial/\partial r & \partial/\partial\theta & \partial/\partial z \\ r^2 & r^2 z & \sin\theta \end{vmatrix}$$

$$= \frac{1}{r}[(\cos\theta - r^2)\mathbf{e}_r - (0)r\mathbf{e}_\theta + 2rz\mathbf{e}_z] = \frac{1}{r}(\cos\theta - r^2)\mathbf{e}_r + 2z\mathbf{e}_z.$$

19. The components of $\mathbf{F}$ are $F_\rho = \theta$, $F_\theta = \cos\theta\cos\phi$ and $F_\phi = -\rho$. Therefore,

$$div \, \mathbf{F} = \frac{1}{\rho^2}\frac{\partial}{\partial\rho}\left(\rho^2\theta\right) + \frac{1}{\rho\sin\phi}\frac{\partial}{\partial\theta}\left(\cos\theta\cos\phi\right) + \frac{1}{\rho\sin\phi}\frac{\partial}{\partial\phi}\left(-\rho\sin\phi\right)$$

$$= \frac{2\theta}{\rho} - \frac{\sin\theta\cot\phi}{\rho} - \cot\phi$$

and

$$curl \, \mathbf{F} = \frac{1}{\rho^2\sin\phi} \begin{vmatrix} \mathbf{e}_\rho & \rho\mathbf{e}_\phi & \rho\sin\phi\mathbf{e}_\theta \\ \frac{\partial}{\partial\rho} & \frac{\partial}{\partial\phi} & \frac{\partial}{\partial\theta} \\ \theta & -\rho^2 & \rho\sin\phi\cos\phi\cos\theta \end{vmatrix}$$

$$= \frac{1}{\rho^2\sin\phi}\left[\rho\cos\theta\cos(2\phi)\mathbf{e}_\rho - \rho(\tfrac{1}{2}\cos\theta\sin(2\phi) - 1)\mathbf{e}_\phi + \rho\sin\phi(-2\rho)\mathbf{e}_\theta\right]$$

$$= \frac{\cos\theta\cos(2\phi)}{\rho\sin\phi}\mathbf{e}_\rho - \frac{\tfrac{1}{2}\cos\theta\sin(2\phi) - 1}{\rho\sin\phi}\mathbf{e}_\phi - 2\mathbf{e}_\theta.$$

Chapter Review

True/false Quiz

1. False. The vector $\mathbf{F}(\mathbf{c}(t)) = \mathbf{F}(2\sin t, 2\cos t) = 2\sin t\mathbf{i} + 2\cos t\mathbf{j}$ is not equal to $\mathbf{c}'(t) = 2\cos t\mathbf{i} - 2\sin t\mathbf{j}$.

3. False. Other conditions are needed. For instance, $f(x,y) = x^3 + (y-1)^2$ satisfies $f_x(0,0) = 0$. However, $(0,0)$ is not a critical point, since $f_y(0,0) = -2 \neq 0$.

5. True. The point $(3,1)$ must belong to the constraint curve.

7. True. For all (x,y) near $(0,0)$, $f(0,0) = |0| \le |x| = f(x,y)$.

9. True. The set in question (open ball centred at $(2,3)$ of radius 10) can be enclosed in a ball of radius 100 centered at the origin.

11. False. $D = f_{xx}(1,1)f_{yy}(1,1) - (f_{xy}(1,1))^2$ could be negative (in which case $(1,1)$ is a saddle point).

13. False. From $grad\, f = xe^y \mathbf{i} + e^y \mathbf{j}$ we get that $f_x = xe^y$ and $f_y = e^y$. Thus, $f_{xy} = xe^y$ and $f_{yx} = 0$. We conclude that such f cannot exist, since no C^2 function satisfies $f_{xy} \neq f_{yx}$.

Review Exercises and Problems

1. Denoting $2k\left(x - 4k^2 t\right)$ by A we get $u(x,t) = e^A$,
$$u_t = e^A \cdot 2k(-4k^2) = -8k^3 e^A,$$
and
$$u_x = 2ke^A, \quad u_{xx} = 2ke^A \cdot 2k = 4k^2 e^A \quad \text{and} \quad u_{xxx} = 4k^2 e^A \cdot 2k = 8k^3 e^A.$$
Clearly, $u_t + u_{xxx} = 0$.

3. We use the chain rule to compute the derivative of $f(\mathbf{c}(t))$:
$$\frac{d}{dt} f(\mathbf{c}(t)) = Df(\mathbf{c}(t)) \cdot \mathbf{c}'(t) = grad\, f(\mathbf{c}(t)) \cdot \mathbf{c}'(t)$$
$$= -\mathbf{F}(\mathbf{c}(t)) \cdot \mathbf{c}'(t) = -\mathbf{c}'(t) \cdot \mathbf{c}'(t) = -\|\mathbf{c}'(t)\|^2,$$
since $\mathbf{F}(\mathbf{c}(t)) = \mathbf{c}'(t)$ by the definition of a flow line. It follows that the derivative of $f(\mathbf{c}(t))$ is always negative; i.e., f is a decreasing function.

5. Differentiating $u_x = v_y$ with respect to x gives $u_{xx} = v_{yx}$, and differentiating $u_y = -v_x$ with respect to y gives $u_{yy} = -v_{xy} = -v_{yx}$ (since v_{xy} is continuous, $v_{xy} = v_{yx}$). Adding up the two equations, we get $u_{xx} + u_{yy} = 0$. The Laplace's equation for v is obtained analogously.

7. Rather than computing derivatives (we would need to compute three first and six second partial derivatives) we use the MacLaurin expansions for sin and cos. Thus
$$f(x,y,z) = 1 + \sin(x + y + z^2) + 3\cos(2x - z)$$
$$= 1 + (x + y + z^2) - \tfrac{1}{6}(x + y + z^2)^3 + \cdots$$
$$+ 3\left(1 - \tfrac{1}{2}(2x - z)^2 + \tfrac{1}{4!}(2x - z)^4 + \cdots\right)$$

We are looking for T_2, so we only take the terms up to and including second order (i.e., sum of degrees in all variables should not exceed 2):

$$T_2(x, y, z) = 1 + (x + y + z^2) + 3 - \tfrac{3}{2}(2x - z)^2 = 4 + x + y - 6x^2 - \tfrac{1}{2}z^2 + 6xz.$$

9. Consider $\mathbf{F}(x, y) = (f(x), 0)$, where $f(x)$ is a desreasing function for all x. Then $div\,\mathbf{F} = f'(x) < 0$. To make divergence constant, we need to make $f'(x)$ constant (so, take $f(x)$ to be a linear function with negative slope). For instance, $f(x) = -2x$.

Thus, one field that satisfies given condition is $\mathbf{F}(x, y) = (-2x, 0)$. There are many others, for instance $\mathbf{F}(x, y) = (-2x, -3y)$.

11. We compute $f_x = \sin y = 0$ and $f_y = x \cos y = 0$. If $x \neq 0$, then there are no critical points since there is no y such that $\cos y = 0$ and $\sin y = 0$. Thus, $x = 0$. From $f_x = \sin y = 0$ we conclude that $y = k\pi$, where k is an integer. Thus, critical points of f are $(0, k\pi)$.

Compute $f_{xx} = 0$, $f_{xy} = \cos y$ and $f_{yy} = -x \sin y$, and

$$D(x, y) = (0)(-x \sin y) - (\cos y)^2 = -\cos^2 y.$$

Since $D(0, k\pi) = -1$ for all k, we conclude that all critical points are saddle points.

13. The distance from (x, y, z) to the origin is given by $d(x, y, z) = \sqrt{x^2 + y^2 + z^2}$. To simplify calculations, we consider its square; i.e., we will minimize $f(x, y, z) = x^2 + y^2 + z^2$ subject to $g(x, y, z) = ax + by + cz + d = 0$. (Note that f does not have a maximum subject to this constraint).

The gradient vectors are $\nabla f = (2x, 2y, 2z)$ and $\nabla g = (a, b, c)$. From $\nabla f = \lambda \nabla g$ we get $2x = a\lambda$, $2y = b\lambda$ and $2z = c\lambda$. Assume that $a \neq 0$. Combining $\lambda = 2x/a$ and $\lambda = 2y/b$ we get $bx = ay$, i.e., $y = bx/a$. (If $b = 0$, then $2y = b\lambda$ implies that $y = 0$, so $y = bx/a$ still holds.) Combining $\lambda = 2x/a$ and $\lambda = 2z/c$ we get $z = cx/a$. (If $c = 0$, then $2z = c\lambda$ implies that $z = 0$, so $z = cx/a$ still holds.)

Substituting $y = bx/a$ and $z = cx/a$ into the constraint, we get

$$ax + \frac{b^2}{a}x + \frac{c^2}{a}x + d = 0,$$

i.e., $x = -ad/(a^2 + b^2 + c^2)$. Thus,

$$y = \frac{b}{a} \frac{-ad}{a^2 + b^2 + c^2} = \frac{-bd}{a^2 + b^2 + c^2} \quad \text{and} \quad z = \frac{c}{a} \frac{-ad}{a^2 + b^2 + c^2} = \frac{-cd}{a^2 + b^2 + c^2}.$$

It follows that the point $(-ad/(a^2 + b^2 + c^2), -bd/(a^2 + b^2 + c^2), -cd/(a^2 + b^2 + c^2))$ is the point in the given plane that is closest to the origin. The distance is given by

$$d = \left(\frac{a^2 d^2}{(a^2 + b^2 + c^2)^2} + \frac{b^2 d^2}{(a^2 + b^2 + c^2)^2} + \frac{c^2 d^2}{(a^2 + b^2 + c^2)^2} \right)^{1/2}$$

$$= \left(\frac{d^2}{a^2 + b^2 + c^2} \right)^{1/2} = \frac{|d|}{\sqrt{a^2 + b^2 + c^2}^{1/2}}.$$

If $a = 0$, we repeat the above assuming either $b \neq 0$ or $c \neq 0$.

15. From $f(x, y) = \ln(x - 2y + 1)$ we get $f(2, 1) = 0$. First partial derviatives are $f_x = 1/(x - 2y + 1)$, $f_x(2, 1) = 1$, and $f_y = -2/(x - 2y + 1)$, $f_y(2, 1) = -2$.

Second partials are $f_{xx} = -1/(x - 2y + 1)^2$, $f_{xx}(2, 1) = -1$, $f_{xy} = 2/(x - 2y + 1)^2$, $f_{xy}(2, 1) = 2$, and $f_{yy} = -4/(x - 2y + 1)^2$, $f_{yy}(2, 1) = -4$. Thus,

$$T_2(x, y) = (x - 2) - 2(y - 1) - \tfrac{1}{2}(x - 2)^2 + 2(x - 2)(y - 1) - 2(y - 1)^2$$
$$= x - 2y - \tfrac{1}{2}x^2 + 2xy - 2y^2.$$

5. INTEGRATION ALONG PATHS

5.1. Paths and Parametrizations

1. $\phi(t)$ is bijective: its inverse $\phi^{-1}\colon[0,\ln 2] \to [0,1]$ is $\phi^{-1}(t) = e^t - 1$. The derivative $\phi'(t) = 1/(t+1)$ is defined and continuous on $[0,1]$, and therefore ϕ can be a reparametrization of a path.

3. Since $\phi'(t) = t^{-2/3}/3$, it follows that $\phi'(0)$ is not defined. Consequently, ϕ cannot be a reparametrization of a path.

5. $\phi(t) = e^t$ is bijective: its inverse $\phi^{-1}\colon[1,e] \to [0,1]$ is $\phi^{-1}(t) = \ln t$. The derivative $\phi'(t) = e^t$ is defined and continuous on $[0,1]$, and therefore ϕ can be a reparametrization of a path.

7. ϕ is not bijective, since, for example, $\phi(-1) = \phi(1) = 1$. It is not differentiable either, since $\phi'(0)$ does not exist. Either of these two facts proves that ϕ cannot be a reparametrization of a path.

9. Suppose that $\phi'(t) > 0$ for all $t \in (\alpha,\beta)$. We need to prove that ϕ is one-to-one. Assume not, i.e., assume that there exist $t_1 \neq t_2$ in (α,β) such that $\phi(t_1) = \phi(t_2)$. The Mean Value Theorem implies that there is a number t_0 in (α,β) where

$$\phi'(t_0) = \frac{\phi(t_2) - \phi(t_1)}{t_2 - t_1}.$$

Since $\phi(t_1) = \phi(t_2)$, it follows that $\phi'(t_0) = 0$, which contradicts our assumption.

The proof in the case $\phi'(t) < 0$ is analogous. Note that the reverse implication is not true: for instance, $\phi(t) = t^3$ is one-to-one on $[-1,1]$, but $\phi'(0) = 0$.

11. $\mathbf{c}(t)$ is not one-to-one; for example, $\mathbf{c}(0) = (0,1,4\pi^2)$ and $\mathbf{c}(4\pi) = (0,1,4\pi^2)$. It follows that $\mathbf{c}(t)$ is not a simple curve. Since $\mathbf{c}(-2\pi) = (0,1,16\pi^2)$ and $\mathbf{c}(6\pi) = (0,1,16\pi^2)$, it follows that $\mathbf{c}(t)$ is closed. However, since $\mathbf{c}(0) = \mathbf{c}(4\pi)$, $\mathbf{c}$ is not one-to-one on $[-2\pi,6\pi)$, and so it is not a simple closed curve.

13. From $\|\mathbf{c}(t)\| = \sqrt{t^2\sin^2 t + t^2\cos^2 t} = \sqrt{t^2} = |t| = t$ (since $t \geq 0$), it follows that the distance $\|\mathbf{c}(t) - \mathbf{0}\| = \|\mathbf{c}(t)\|$ between $\mathbf{c}(t)$ and the origin is t. Therefore, no two points $\mathbf{c}(t_1)$ and $\mathbf{c}(t_2)$, with $t_1 \neq t_2$, are at the same distance from the origin, and so $\mathbf{c}$ is simple (that is, it does not intersect itself). Since $\mathbf{c}(0) = (0,0)$ and $\mathbf{c}(2\pi) = (0,2\pi)$, $\mathbf{c}$ is not closed. Consequently, $\mathbf{c}$ cannot be a simple closed curve.

15. Assume that $\mathbf{c}(t_1) = \mathbf{c}(t_2)$. Then

$$\left(t_1 - \frac{1}{t_1}, t_1 + \frac{1}{t_1}\right) = \left(t_2 - \frac{1}{t_2}, t_2 + \frac{1}{t_2}\right)$$

implies

$$t_1 - \frac{1}{t_1} = t_2 - \frac{1}{t_2} \quad \text{and} \quad t_1 + \frac{1}{t_1} = t_2 + \frac{1}{t_2}.$$

Adding the two equations we get $t_1 = t_2$. In other words, $c(t)$ is one-to-one (whenever defined, and, in particular, on $[1, 2]$) and the curve c is simple. Since $c(1) = (0, 2)$ and $c(2) = (3/2, 5/2)$, c is not closed. In that case, it cannot be a simple closed curve.

17. Let $x = t$; then $\sqrt{t^2 + 1}$, and hence

$$c(t) = (t, \sqrt{t^2 + 1}), \qquad t \in [-1, 1]$$

is one possible parametrization of the given curve. It is continuous, and since $c'(t) = (1, t/\sqrt{t^2 + 1})$, it is also differentiable and C^1.

19. (a) Since $c_1'(t) = (1, 0)$, it follows that $c_1(t)$ is C^1, and hence also differentiable and continuous.

(b) The paths $c_{21}(t) = (-t^2, 1)$, $t \in [-1, 0]$, and $c_{22}(t) = (t^2, 1)$, $t \in [0, 1]$, are continuous. Since $c_{21}(0) = (0, 1) = c_{22}(0)$, it follows that $c_2(t)$ is a continuous path. The paths $c_{21}'(t) = (-2t, 0)$, $t \in [-1, 0]$, and $c_{22}'(t) = (2t, 0)$, $t \in [0, 1]$, are both continuous. Since $c_{21}'(0) = (0, 0) = c_{22}'(0)$, it follows that $c_2'(t)$ is a continuous path; therefore, $c_2(t)$ is a C^1 path.

(c) $c_3(t) = (t^{1/3}, 1)$ is continuous; however, it is not differentiable, since $c_3'(t) = (t^{-2/3}/3, 0)$ is not defined at $t = 0$. For the same reason, c_3 is not piecewise C^1 or C^1.

(d) Since $c_4'(t) = (3t^2, 0)$ is continuous, it follows that $c_4(t)$ is a C^1 path.

(e) $c_5(t)$ traverses the given line segment as follows: it goes from $(-1, 1)$ to $(0, 1)$, then jumps to $(1, 1)$ and goes back to $(0, 1)$. So c_5 is not continuous. Hovever, both $(t, 1)$ and $(1 - t, 1)$ are differentiable and have continuous derivatives; so c_5 is piecewise C^1.

21. Drop t for simplicity of notation. Let $\boldsymbol{\gamma} = c(\phi) = (\phi^2, 2 - \phi^2)$. We have to find ϕ so that $\|\boldsymbol{\gamma}'\| = C_1$ (C_1 is a constant). From $\boldsymbol{\gamma}' = (2\phi\phi', -2\phi\phi')$ it follows that

$$\|\boldsymbol{\gamma}'\| = \sqrt{4\phi^2(\phi')^2 + 4\phi^2(\phi')^2} = |\phi\phi'|\sqrt{8} = C_1;$$

i.e., $\phi\phi' = \pm C_1/\sqrt{8}$. Integrating both sides with respect to t, we get

$$\tfrac{1}{2}\phi^2 = \pm \frac{C_1 t}{\sqrt{8}} + D_1,$$

where D_1 is a constant. Therefore,

$$\phi = \pm\sqrt{\pm 2C_1 t/\sqrt{8} + 2D_1}$$

or (simplify the expressions involving constants; let $C = \pm 2C_1/\sqrt{8}$ and $D = 2D_1$) $\phi(t) = \pm\sqrt{Ct + D}$. It follows that the reparametrization of c by $\phi(t) = \pm\sqrt{Ct + D}$, $t \in [1, 3]$, where C and D are chosen so that $\phi(t)$ is defined, will have constant speed.

23. Recall that the speed of the curve $c(t) = (\cos(at), \sin(at))$, $a > 0$, is given by

$$\|c'(t)\| = \sqrt{(-a\sin(at))^2 + (a\cos(at))^2} = a.$$

Therefore, to adjust the speed we have to choose the arguments of sin and cos correctly. Let $c_1(t) = (\cos t, \sin t)$, $t \in [0, 2\pi]$; i.e., the speed of c_1 is 1. Since $c(0) = (1, 0)$, and both components of $c_1(t)$ are positive as t starts increasing from 0, it follows that $c_1(t)$ is oriented counterclockwise. That answers the second part of the question.

To reverse the orientation, let $c_2(t) = (\cos(-t), \sin(-t)) = (\cos t, -\sin t)$, where $t \in [0, 2\pi]$. This time, $c(0) = (1, 0)$, and, as t starts increasing from 0, the y-component is negative; so c_2 is oriented clockwise. To get the correct speed, we use

$$c_2(t) = (\cos(St), -\sin(St)), \qquad t \in [0, 2\pi/S].$$

This is the answer to the first part of the question.

5.2. Path Integrals of Real-Valued Functions

1. Recall that $\int_c f\, ds$ represents the area of the region along c and below the graph of f (assuming that $f \geq 0$). The fact that $f(x, y)$ is linear implies that the region is bounded from above by a straight line.

(a) The region is a trapezoid whose parallel sides are of length 2 and 14 and the distance between them is 4. The area of the trapezoid is $(2 + 14)(4)/2 = 32$. Thus, $\int_c f\, ds = 32$.

(b) The region is a trapezoid whose parallel sides are of length 2 and 14 and the distance between them is $\sqrt{(8-1)^2 + (5-1)^2} = \sqrt{65}$. Thus,

$$\int_c f\, ds = \frac{1}{2}(2 + 14)\sqrt{65} = 8\sqrt{65}.$$

3. The values of f along c are

$$f(c(t)) = 2(e^t + 1) - (e^t - 2) = e^t + 4.$$

From $c'(t) = (e^t, e^t)$ it follows that $\|c'(t)\| = \sqrt{e^{2t} + e^{2t}} = e^t\sqrt{2}$ and so

$$\int_c (2x - y)ds = \int_0^{\ln 2} (e^t + 4)e^t\sqrt{2}\, dt = \sqrt{2}\int_0^{\ln 2} (e^{2t} + 4e^t)\, dt$$

$$= \sqrt{2}\left(\tfrac{1}{2}e^{2t} + 4e^t\right)\Big|_0^{\ln 2} = \frac{11\sqrt{2}}{2} \approx 7.7782.$$

5. The values of f along c are $f(c(t)) = (t^2 + t^2 + t^2)^{-1} = (3t^2)^{-1} = t^{-2}/3$. From $c'(t) = (1, 1, 1)$ it follows that $\|c'(t)\| = \sqrt{3}$. Let $c_b(t) = (t, t, t)$, $1 \leq t \leq b$; then

$$\int_{c_b} (x^2 + y^2 + z^2)^{-1}ds = \int_1^b \frac{t^{-2}}{3}\sqrt{3}\, dt = \frac{-\sqrt{3}}{3}\left(\frac{1}{t}\right)\Big|_1^b = \frac{\sqrt{3}}{3}\left(1 - \frac{1}{b}\right).$$

Therefore,

$$\int_c (x^2 + y^2 + z^2)^{-1}ds = \lim_{b \to \infty} \int_{c_b} (x^2 + y^2 + z^2)^{-1}ds = \frac{\sqrt{3}}{3}.$$

7. The values of f along $\mathbf{c}$ are $f(\mathbf{c}(t)) = f(t^2\mathbf{i} + \ln t\mathbf{j} + 2t\mathbf{k}) = \ln t - 4t^2$. From $\mathbf{c}'(t) = 2t\mathbf{i} + t^{-1}\mathbf{j} + 2\mathbf{k}$ we get

$$\|\mathbf{c}'(t)\| = \sqrt{4t^2 + 4 + t^{-2}} = \sqrt{(2t + t^{-1})^2} = 2t + t^{-1},$$

since $t \geq 0$. It follows that

$$\int_{\mathbf{c}} f\,ds = \int_1^4 (\ln t - 4t^2)(2t + t^{-1})\,dt$$

$$= 2\int_1^4 t\ln t\,dt - 8\int_1^4 t^3\,dt + \int_1^4 \frac{\ln t}{t}\,dt - 4\int_1^4 t\,dt.$$

Using integration by parts with $u = \ln t$ and $dv = t\,dt$, we get ($du = dt/t$ and $v = t^2/2$)

$$\int t\ln t\,dt = \tfrac{1}{2}t^2\ln t - \int \tfrac{1}{2}t\,dt = \tfrac{1}{2}t^2\ln t - \tfrac{1}{4}t^2 + C,$$

and therefore

$$\int_1^4 t\ln t\,dt = \left(\tfrac{1}{2}t^2\ln t - \tfrac{1}{4}t^2\right)\Big|_1^4 = 8\ln 4 - \tfrac{15}{4}.$$

Using the substitution $u = \ln t$, $du = (1/t)dt$, we get

$$\int_1^4 \frac{\ln t}{t}\,dt = \int_0^{\ln 4} u\,du = \tfrac{1}{2}u^2\Big|_0^{\ln 4} = \tfrac{1}{2}(\ln 4)^2.$$

Hence

$$\int_{\mathbf{c}} f\,ds = 2\left(8\ln 4 - \tfrac{15}{4}\right) - 2t^4\Big|_1^4 + \tfrac{1}{2}(\ln 4)^2 - 2t^2\Big|_1^4$$

$$= 16\ln 4 + \tfrac{1}{2}(\ln 4)^2 - 547.5 \approx -524.3584.$$

9. The values of f along $\mathbf{c}$ are $f(2\sin t, 4t, 2\cos t) = (2\sin t)(4t)(2\cos t) = 16t\sin t\cos t = 8t\sin 2t$. From $\mathbf{c}'(t) = (2\cos t, 4, -2\sin t)$ it follows that $\|\mathbf{c}'(t)\| = \sqrt{4\cos^2 t + 16 + 4\sin^2 t} = \sqrt{20}$ and so

$$\int_{\mathbf{c}} xyz\,ds = \int_0^{6\pi} 8t\sin 2t\,\sqrt{20}\,dt = 8\sqrt{20}\int_0^{6\pi} t\sin 2t\,dt.$$

Using integration by parts with $u = t$ and $dv = \sin 2t\,dt$, we get ($du = dt$ and $v = -(\cos 2t)/2$)

$$\int t\sin 2t\,dt = -\tfrac{1}{2}t\cos 2t + \int \tfrac{1}{2}\cos 2t\,dt = -\tfrac{1}{2}t\cos 2t + \tfrac{1}{4}\sin 2t + C,$$

and therefore

$$\int_0^{6\pi} t\sin 2t\,dt = \left(-\tfrac{1}{2}t\cos 2t + \tfrac{1}{4}\sin 2t\right)\Big|_0^{6\pi} = -3\pi.$$

Finally,

$$\int_{\mathbf{c}} xyz\,ds = 8\sqrt{20}(-3\pi) = -48\pi\sqrt{5} \approx -337.1911$$

11. Parametrize $\mathbf{c}$ by

$$\mathbf{c}(t) = (0,0) + t(3,-4) = (3t, -4t), \qquad 0 \leq t \leq 1.$$

Then $\mathbf{c}'(t) = (3, -4)$, $\|\mathbf{c}'(t)\| = 5$ and

$$\int_{\mathbf{c}} e^{x+3y} \, ds = \int_0^1 5e^{-9t} \, dt = -\tfrac{5}{9} e^{-9t} \Big|_0^1 = -\tfrac{5}{9}(e^{-9} - 1) = \tfrac{5}{9}(1 - e^{-9}) \approx 0.5556$$

13. The curve $\mathbf{c}$ consists of two curves, parametrized by

$$\mathbf{c}_1(t) = (4, 2, 0) + t(-4, 0, 0) = (4 - 4t, 2, 0), \quad t \in [0, 1],$$

and

$$\mathbf{c}_2(t) = (0, 2\cos t, 2\sin t), \quad t \in [0, \pi];$$

therefore,

$$\int_{\mathbf{c}} f \, ds = \int_{\mathbf{c}_1} f \, ds + \int_{\mathbf{c}_2} f \, ds.$$

From $f(\mathbf{c}_1(t)) = 4 - 4t - 8 = -4 - 4t$, $\mathbf{c}_1'(t) = (-4, 0, 0)$ and $\|\mathbf{c}_1'(t)\| = 4$ it follows that

$$\int_{\mathbf{c}_1} (x - 4y + z) \, ds = \int_0^1 (-4 - 4t) \, 4 \, dt = -16(t + \tfrac{1}{2}t^2) \Big|_0^1 = -24.$$

From $f(\mathbf{c}_2(t)) = -8\cos t + 2\sin t$, $\mathbf{c}_2'(t) = (0, -2\sin t, 2\cos t)$ and $\|\mathbf{c}_2'(t)\| = 2$ it follows that

$$\int_{\mathbf{c}_2} (x - 4y + z) \, ds = \int_0^\pi 2(-8\cos t + 2\sin t) \, dt = 2(-8\sin t - 2\cos t) \Big|_0^\pi = 2(2 - (-2)) = 8.$$

It follows that $\int_{\mathbf{c}} f \, ds = -24 + 8 = -16$.

15. It is possible. The path integral of a real-valued function is independent of a parametrization of a curve; that is, it is independent of the way the curve is traversed. However, in general, it depends on the curve that is used to join the given points.

Consider the following intuitive example: let $A = (1, 0)$ and $B = (2, 0)$ and $f(x, y) = y^2$. The path integral along the straight line from A to B is zero, since the value of f at each point on that line is zero. On the other hand, the path integral along a curve in the first quadrant joining A and B is positive, since $f(x, y) = y^2 > 0$.

17. We are asked to find the area of the "fence" in the shape of the cylinder $x^2 + y^2 = 4$ whose height is determined by $z = y + 2$. Therefore, area $= \int_{\mathbf{c}} f \, ds$, where $f(x, y) = y + 2$ and $\mathbf{c}$ is the circle $x^2 + y^2 = 4$.

Parametrize $\mathbf{c}$ by $\mathbf{c}(t) = (2\cos t, 2\sin t)$, $t \in [0, 2\pi]$. Then $\mathbf{c}'(t) = (-2\sin t, 2\cos t)$, $\|\mathbf{c}'(t)\| = 2$ and

$$\text{area} = \int_{\mathbf{c}} (y + 2) \, ds = \int_0^{2\pi} 2(2\sin t + 2) \, dt = 4(-\cos t + t) \Big|_0^{2\pi} = 8\pi.$$

19. We are asked to find the area of the "fence" in the shape of the graph of $y = \sin x$, whose height is determined by $z = \sin x \cos x$. Therefore, area $= \int_{\mathbf{c}} f \, ds$, where $f(x, y) = \sin x \cos x$ and $\mathbf{c}$ is the curve $y = \sin x$, $0 \le x \le \pi/2$.

Parametrize $\mathbf{c}$ by $\mathbf{c}(t) = (t, \sin t)$, $t \in [0, \pi/2]$. Then $\mathbf{c}'(t) = (1, \cos t)$, and its length is $\|\mathbf{c}'(t)\| = \sqrt{1 + \cos^2 t}$ and

$$\text{area} = \int_0^{\pi/2} \sin t \cos t \sqrt{1 + \cos^2 t}\, dt.$$

Using the substitution $u = 1 + \cos^2 t$, $du = -2 \sin t \cos t\, dt$, we get

$$\text{area} = -\frac{1}{2} \int_2^1 u^{1/2}\, du = -\frac{1}{3} u^{3/2} \Big|_2^1 = \frac{1}{3}(2^{3/2} - 1) \approx 0.6095.$$

21. The function $f(x, y) = e^{x^2 + y^2}$ is constant along the given circle; i.e., since $x^2 + y^2 = 16$, it follows that $f(x, y) = e^{16}$ for all points on that circle. The path integral $\int_{\mathbf{c}} e^{x^2 + y^2}\, ds$ is the area of the "fence" defined by $x^2 + y^2 = 16$ of height e^{16}. In other words, it is the surface of the cylinder of radius 4 and height e^{16}. Its surface area is $2\pi(4)(e^{16}) = 8\pi e^{16}$. Therefore, $\int_{\mathbf{c}} e^{x^2 + y^2}\, ds = 8\pi e^{16}$.

23. Since $f(x, y) \leq 1$ for all x and y, it is not possible that the average value of f be any number greater than 1.

25. Since $\mathbf{c}$ is a circle of radius 3, its length is $\ell(\mathbf{c}) = 6\pi$. Hence

$$\overline{f} = \frac{1}{\ell(\mathbf{c})} \int_{\mathbf{c}} f\, ds = -\frac{1}{6\pi} \int_{\mathbf{c}} \sqrt{x^2 + z^2}\, ds.$$

From $\mathbf{c}'(t) = -3 \sin t\, \mathbf{j} + 3 \cos t\, \mathbf{k}$ we get $\|\mathbf{c}'(t)\| = 3$ and

$$\overline{f} = -\frac{1}{6\pi} \int_0^{2\pi} 3\sqrt{9 \cos^2 t}\, dt = -\frac{3}{2\pi} \int_0^{2\pi} |\cos t|\, dt.$$

Since

$$\int_0^{2\pi} |\cos t|\, dt = \int_0^{\pi/2} |\cos t|\, dt + \int_{\pi/2}^{3\pi/2} |\cos t|\, dt + \int_{3\pi/2}^{2\pi} |\cos t|\, dt$$

$$= \int_0^{\pi/2} \cos t\, dt - \int_{\pi/2}^{3\pi/2} \cos t\, dt + \int_{3\pi/2}^{2\pi} \cos t\, dt$$

$$= 1 + 2 + 1 = 4,$$

it follows that $\overline{f} = -6/\pi$.

27. As in the construction of the path integral at the beginning of the section, we break up $[a, b]$ into n subintervals and approximate the curve $\mathbf{c}$ with a polygonal path. The mass of the segment $\mathbf{c}_i$ connecting $\mathbf{c}(t_i)$ and $\mathbf{c}(t_{i+1})$ is given by $\ell(\mathbf{c}_i)\rho(\mathbf{c}(t_i))$, where $\rho(\mathbf{c}(t_i))$ is the density at the point $\mathbf{c}(t_i)$ and $\ell(\mathbf{c}_i)$ is the length of the segment $\mathbf{c}_i$. The density here is the linear density, i.e., its units are mass/length. We assume that the density on each of the small segments $\mathbf{c}_i$ is constant. Since $\ell(\mathbf{c}_i)\rho(\mathbf{c}(t_i)) \approx \rho(\mathbf{c}(t_i))\|\mathbf{c}'(t_i)\|\Delta t_i$, it follows that

$$\sum_{i=1}^{n} \rho(\mathbf{c}(t_i))\|\mathbf{c}'(t_i)\|\Delta t_i$$

approximates the mass of the wire. Taking the limit as $n \to \infty$, we obtain that the mass of the wire is given by

$$\int_a^b \rho(\mathbf{c}(t))\|\mathbf{c}'(t)\| \, dt = \int_\mathbf{c} \rho ds.$$

29. Parametrize $\mathbf{c}$ by $\mathbf{c}(\theta) = (r\cos\theta, r\sin\theta)$, $\theta_1 \leq \theta \leq \theta_2$. Then

$$\mathbf{c}'(\theta) = (r'\cos\theta - r\sin\theta, r'\sin\theta + r\cos\theta).$$

Keep in mind that θ is viewed as independent variable, i.e., all derivatives are taken with respect to θ. We compute

$$\|\mathbf{c}'(\theta)\| = \Big((r')^2\cos^2\theta + r^2\sin^2\theta - 2rr'\sin\theta\cos\theta$$
$$+ (r')^2\sin^2\theta + r^2\cos^2\theta + 2rr'\sin\theta\cos\theta\Big)^{1/2} = (r^2 + (r')^2)^{1/2}.$$

By definition,

$$\int_\mathbf{c} f \, ds = \int_{\theta_1}^{\theta_2} f(\mathbf{c}(\theta))\,\|\mathbf{c}'(\theta)\|\, d\theta = \int_{\theta_1}^{\theta_2} f(r\cos\theta, r\sin\theta)\,\sqrt{r^2 + \left(\frac{dr}{d\theta}\right)^2}\, d\theta,$$

keeping in mind that $r' = dr/d\theta$.

5.3. Path Integrals of Vector Fields

1. Recall that

$$\int_\mathbf{c} \mathbf{F} \cdot d\mathbf{s} = \int_a^b \mathbf{F}(\mathbf{c}(t)) \cdot \mathbf{c}'(t)\, dt = \int_a^b \|\mathbf{F}(\mathbf{c}(t))\|\,\|\mathbf{c}'(t)\| \cos\theta(t)\, dt,$$

where $\theta(t)$ is the angle between $\mathbf{F}(\mathbf{c}(t))$ and $\mathbf{c}'(t)$.

(a) Since $\mathbf{F}(\mathbf{c}_1(t))$ and $\mathbf{c}_1'(t)$ are of opposite orientations, $\theta(t) = \pi$ and so $\cos\theta(t) = -1$. Consequently, the integrand $\mathbf{F}(\mathbf{c}_1(t)) \cdot \mathbf{c}_1'(t)$ is negative, and thus $\int_{\mathbf{c}_1} \mathbf{F} \cdot d\mathbf{s} < 0$.

(b) Both $\mathbf{c}_2$ and $\mathbf{c}_3$ are parallel to $\mathbf{F}$, and are of the same orientation as $\mathbf{F}$; i.e., $\theta(t) = 0$ for both. Since $\|\mathbf{F}\|$ is larger in the region where $\mathbf{c}_3$ lies and the two paths have the same length and the same speed, it follows that $\int_{\mathbf{c}_3} \mathbf{F} \cdot d\mathbf{s} > \int_{\mathbf{c}_2} \mathbf{F} \cdot d\mathbf{s}$.

3. (a) The length of each subinterval is $(4 - 1)/n = 3/n$. Thus, dividing the interval $[1, 4]$ into n subintervals, we obtain the following subdivision points:

$$t_1 = 1, \ t_2 = 1 + \frac{3}{n}, \ t_3 = 1 + \frac{3}{n}2, \dots, \ t_{n+1} = 1 + \frac{3}{n}n = 4.$$

The corresponding points on the curve are

$$\mathbf{c}(t_1) = (1, 1), \ \mathbf{c}(t_2) = (1 + 3/n, 1), \mathbf{c}(t_3) = (1 + 6/n, 1), \dots, \ \mathbf{c}(t_{n+1}) = (4, 1),$$

i.e., $\mathbf{c}(t_i) = (1 + 3(i - 1)/n, 1)$, where $i = 1, 2, \dots, n + 1$.

(b) We compute

$$\Delta \mathbf{c}_i = \mathbf{c}(t_{i+1}) - \mathbf{c}(t_i) = \left(1 + \frac{3}{n}i, 1\right) - \left(1 + \frac{3}{n}(i-1), 1\right) = \left(\frac{3}{n}, 0\right) = \frac{3}{n}\mathbf{i},$$

and $\mathbf{F}(\mathbf{c}(t_i)) = \left(1 + \frac{3}{n}(i-1)\right)\mathbf{i} + \mathbf{j}$. Thus,

$$W_n = \sum_{i=1}^{n} \mathbf{F}(\mathbf{c}(t_i)) \cdot \Delta \mathbf{c}_i = \sum_{i=1}^{n} \frac{3}{n}\left(1 + \frac{3}{n}(i-1)\right) = \frac{3}{n}\sum_{i=1}^{n}\left(1 + \frac{3}{n}(i-1)\right)$$
$$= \frac{3}{n}\left(n + \frac{3}{n}\frac{n(n-1)}{2}\right) = 3 + \frac{9}{2n}(n-1).$$

(c) Taking the limit, we get

$$W = \lim_{n\to\infty} W_n = \lim_{n\to\infty}\left(3 + \frac{9}{2n}(n-1)\right) = 3 + \frac{9}{2} = \frac{15}{2}.$$

We check our calculation by computing the integral:

$$W = \int_{\mathbf{c}} \mathbf{F} \cdot d\mathbf{s} = \int_{1}^{4} (t, 1) \cdot (1, 0)\, dt = \int_{1}^{4} t\, dt = \frac{15}{2}.$$

5. Parametrize $\mathbf{c}$ by $\mathbf{c}(t) = t\mathbf{i} + t^2\mathbf{j}$, $t \in [-1, 1]$. Then $\mathbf{c}'(t) = \mathbf{i} + 2t\mathbf{j}$, $\mathbf{F}(\mathbf{c}(t)) = \mathbf{F}(t\mathbf{i} + t^2\mathbf{j}) = t^4\mathbf{i} - t^2\mathbf{j}$, and

$$\int_{\mathbf{c}} (y^2\mathbf{i} - x^2\mathbf{j}) \cdot d\mathbf{s} = \int_{-1}^{1}(t^4\mathbf{i} - t^2\mathbf{j}) \cdot (\mathbf{i} + 2t\mathbf{j})\, dt = \int_{-1}^{1}(t^4 - 2t^3)\, dt = \left(\tfrac{1}{5}t^5 - \tfrac{1}{2}t^4\right)\Big|_{-1}^{1} = \tfrac{2}{5}.$$

7. Parametrize the segment from $(0, 0)$ to $(0, 1)$ by $\mathbf{c}_1(t) = (0, t), 0 \le t \le 1$. Then $\mathbf{c}_1'(t) = (0, 1)$ and

$$\int_{\mathbf{c}_1} (e^{x+y}\mathbf{i} - \mathbf{j}) \cdot d\mathbf{s} = \int_{0}^{1}(e^t, -1) \cdot (0, 1)\, dt = \int_{0}^{1}(-1)\, dt = -1.$$

Parametrize the segment from $(0, 1)$ to $(1, 1)$ by $\mathbf{c}_2(t) = (t, 1), 0 \le t \le 1$. Then $\mathbf{c}_2'(t) = (1, 0)$ and

$$\int_{\mathbf{c}_2} (e^{x+y}\mathbf{i} - \mathbf{j}) \cdot d\mathbf{s} = \int_{0}^{1}(e^{t+1}, -1) \cdot (1, 0)\, dt = \int_{0}^{1}e^{t+1}\, dt = e^{t+1}\Big|_{0}^{1} = e^2 - e.$$

Parametrize the segment from $(1, 1)$ to $(1, 0)$ by $\mathbf{c}_3(t) = (1, 1-t), 0 \le t \le 1$. Then $\mathbf{c}_3'(t) = (0, -1)$ and

$$\int_{\mathbf{c}_3} (e^{x+y}\mathbf{i} - \mathbf{j}) \cdot d\mathbf{s} = \int_{0}^{1}(e^{2-t}, -1) \cdot (0, -1)\, dt = \int_{0}^{1} dt = 1.$$

Finally, parametrize the segment from $(1, 0)$ to $(0, 0)$ by $\mathbf{c}_4(t) = (1-t, 0), 0 \le t \le 1$. Then $\mathbf{c}_4'(t) = (-1, 0)$ and

$$\int_{\mathbf{c}_4} (e^{x+y}\mathbf{i} - \mathbf{j}) \cdot d\mathbf{s} = \int_{0}^{1}(e^{1-t}, -1) \cdot (-1, 0)\, dt = \int_{0}^{1}(-e^{1-t})\, dt = e^{1-t}\Big|_{0}^{1} = 1 - e.$$

It follows that

$$\int_{\mathbf{c}} (e^{x+y}\mathbf{i} - \mathbf{j}) \cdot d\mathbf{s} = \int_{\mathbf{c}_1} (e^{x+y}\mathbf{i} - \mathbf{j}) \cdot d\mathbf{s} + \int_{\mathbf{c}_2} (e^{x+y}\mathbf{i} - \mathbf{j}) \cdot d\mathbf{s} + \int_{\mathbf{c}_3} (e^{x+y}\mathbf{i} - \mathbf{j}) \cdot d\mathbf{s}$$
$$+ \int_{\mathbf{c}_4} (e^{x+y}\mathbf{i} - \mathbf{j}) \cdot d\mathbf{s}$$
$$= -1 + e^2 - e + 1 + 1 - e = e^2 - 2e + 1.$$

9. The value of $\mathbf{F}$ along $\mathbf{c}$ is $\mathbf{F}(\mathbf{c}(t)) = \mathbf{F}(\sin t, \cos t, t^2) = (\sin^2 t, \sin t \cos t, 2t^4)$. Since $\mathbf{c}'(t) = (\cos t, -\sin t, 2t)$, it follows that

$$\int_{\mathbf{c}} (x^2, xy, 2z^2) \cdot ds = \int_0^{\pi/2} (\sin^2 t, \sin t \cos t, 2t^4) \cdot (\cos t, -\sin t, 2t)\, dt$$

$$= \int_0^{\pi/2} 4t^5\, dt = \tfrac{2}{3} t^6 \Big|_0^{\pi/2} = \tfrac{2}{3} \left(\tfrac{\pi}{2}\right)^6 \approx 10.0145.$$

11. The value of $\mathbf{F}$ along $\mathbf{c}$ is $\mathbf{F}(\mathbf{c}(t)) = \mathbf{F}(4t^3\mathbf{i} + t^2\mathbf{j}) = 8t^5\mathbf{i} + e^{t^2}\mathbf{j}$. Since $\mathbf{c}'(t) = 12t^2\mathbf{i} + 2t\mathbf{j}$, we get

$$\int_{\mathbf{c}} (2xy\mathbf{i} + e^y\mathbf{j}) \cdot ds = \int_0^1 (8t^5\mathbf{i} + e^{t^2}\mathbf{j}) \cdot (12t^2\mathbf{i} + 2t\mathbf{j})\, dt$$

$$= \int_0^1 \left(96t^7 + 2te^{t^2}\right) dt = \left(12t^8 + e^{t^2}\right)\Big|_0^1 = 12 + e - 1 = 11 + e$$

(we used the substitution $u = t^2$, $du = 2t\,dt$, to evaluate $\int 2te^{t^2}\,dt = e^{t^2}$).

13. Recall that

$$\int_{\mathbf{c}} \mathbf{F} \cdot ds = \int_a^b \mathbf{F}(\mathbf{c}(t)) \cdot \mathbf{c}'(t)\, dt = \int_a^b \|\mathbf{F}(\mathbf{c}(t))\|\, \|\mathbf{c}'(t)\| \cos\theta(t)\, dt,$$

where $\theta(t)$ is the angle between $\mathbf{F}(\mathbf{c}(t))$ and $\mathbf{c}'(t)$. By assumption, $\theta(t) = 0$, and thus

$$\int_{\mathbf{c}} \mathbf{F} \cdot ds = k \int_a^b \|\mathbf{c}'(t)\|\, dt = k\ell(\mathbf{c}),$$

where $\ell(\mathbf{c})$ is the length of $\mathbf{c}$.

15. The work is given by $W = \int_{\mathbf{c}} \mathbf{F} \cdot ds$, where $\mathbf{F}(x, y) = x^3\mathbf{i} + (x+y)\mathbf{j}$ and $\mathbf{c}(t) = \sin t\,\mathbf{i} + t^2\mathbf{j}$, $t \in [0, \pi/2]$. Since $\mathbf{F}(\mathbf{c}(t)) = \sin^3 t\,\mathbf{i} + (\sin t + t^2)\mathbf{j}$ and $\mathbf{c}'(t) = \cos t\,\mathbf{i} + 2t\mathbf{j}$, it follows that

$$W = \int_0^{\pi/2} (\sin^3 t\,\mathbf{i} + (\sin t + t^2)\mathbf{j}) \cdot (\cos t\,\mathbf{i} + 2t\mathbf{j})\, dt$$

$$= \int_0^{\pi/2} (\sin^3 t \cos t + 2t \sin t + 2t^3)\, dt.$$

Using the substitution $u = \sin t$, $du = \cos t\,dt$, we get

$$\int_0^{\pi/2} \sin^3 t \cos t\, dt = \int_0^1 u^3\, du = \tfrac{1}{4}.$$

Using integration by parts with $u = t$ and $dv = \sin t\,dt$, we get ($du = dt$, $v = -\cos t$)

$$\int_0^{\pi/2} t \sin t\, dt = -t \cos t \Big|_0^{\pi/2} + \int_0^{\pi/2} \cos t\, dt = \int_0^{\pi/2} \cos t\, dt = \sin t \Big|_0^{\pi/2} = 1.$$

Finally,

$$\int_0^{\pi/2} 2t^3\, dt = \tfrac{1}{2} t^4 \Big|_0^{\pi/2} = \tfrac{1}{32}\pi^4,$$

and so

$$W = \tfrac{1}{4} + 2 + \tfrac{1}{32}\pi^4 = \tfrac{9}{4} + \tfrac{1}{32}\pi^4 \approx 5.2940.$$

17. There is no need to parametrize a circle or to compute a path integral. Just notice that the direction of $\mathbf{F}$ is radial (i.e., along lines through the origin), and therefore orthogonal to any circle centered at the origin. Consequently, the work must be zero.

19. Use W_a, W_b, etc. to denote the work along the curves $\mathbf{c}_a$, $\mathbf{c}_b$, etc. in parts (a), (b), etc.

(a) The work is
$$W_a = \int_{\mathbf{c}_{a1}} \mathbf{F} \cdot d\mathbf{s} + \int_{\mathbf{c}_{a2}} \mathbf{F} \cdot d\mathbf{s},$$
where $\mathbf{c}_{a1}(t) = (t,0)$, $t \in [0,1]$ and $\mathbf{c}_{a2}(t) = (1,t)$, $t \in [0,1]$. Since $\mathbf{c}'_{a1}(t) = (1,0)$ and $\mathbf{c}'_{a2}(t) = (0,1)$, it follows that
$$W_a = \int_{\mathbf{c}_{a1}} \mathbf{F} \cdot d\mathbf{s} + \int_{\mathbf{c}_{a2}} \mathbf{F} \cdot d\mathbf{s} = \int_0^1 (0,0) \cdot (1,0)\, dt + \int_0^1 (t,0) \cdot (0,1)\, dt = 0 + 0 = 0.$$
So the work along $\mathbf{c}_a$ is zero.

(b) Parametrize the given path by $\mathbf{c}_b(t) = (t,t^2)$, $t \in [0,1]$. Then $\mathbf{c}'_b(t) = (1,2t)$, $\mathbf{F}(\mathbf{c}_b(t)) = (t^2,0)$ and
$$W_b = \int_0^1 (t^2,0) \cdot (1,2t)\, dt = \int_0^1 t^2\, dt = \tfrac{1}{3}.$$

(c) Use a parametrization $\mathbf{c}_c(t) = (t,t^4)$, $t \in [0,1]$. Then $\mathbf{c}'_c(t) = (1,4t^3)$, $\mathbf{F}(\mathbf{c}_c(t)) = (t^4,0)$ and
$$W_c = \int_{\mathbf{c}_c} \mathbf{F} \cdot d\mathbf{s} = \int_0^1 (t^4,0) \cdot (1,4t^3)\, dt = \int_0^1 t^4\, dt = \tfrac{1}{5}.$$

(d) Parametrize the given path by $\mathbf{c}_d(t) = (t,t)$, $t \in [0,1]$. Then $\mathbf{c}'_d(t) = (1,1)$, $\mathbf{F}(\mathbf{c}_d(t)) = (t,0)$ and
$$W_d = \int_{\mathbf{c}_d} \mathbf{F} \cdot d\mathbf{s} = \int_0^1 (t,0) \cdot (1,1)\, dt = \int_0^1 t\, dt = \tfrac{1}{2}.$$

(e) Use $\mathbf{c}_e(t) = (t,\sin(\pi t/2))$, $t \in [0,1]$. Then $\mathbf{c}'_e(t) = (1,(\pi/2)\cos(\pi t/2))$, $\mathbf{F}(\mathbf{c}_e(t)) = (\sin(\pi t/2),0)$ and
$$W_e = \int_{\mathbf{c}_e} \mathbf{F} \cdot d\mathbf{s} = \int_0^1 (\sin(\pi t/2),0) \cdot (1,(\pi/2)\cos(\pi t/2))\, dt$$
$$= \int_0^1 \sin(\pi t/2)\, dt = -\tfrac{2}{\pi}\cos(\pi t/2)\Big|_0^1 = \tfrac{2}{\pi} \approx 0.6366.$$

(f) The work is
$$W_f = \int_{\mathbf{c}_{f1}} \mathbf{F} \cdot d\mathbf{s} + \int_{\mathbf{c}_{f2}} \mathbf{F} \cdot d\mathbf{s},$$
where $\mathbf{c}_{f1}(t) = (0,t)$, $t \in [0,1]$ and $\mathbf{c}_{f2}(t) = (t,1)$, $t \in [0,1]$. Then $\mathbf{c}'_{f1}(t) = (0,1)$, $\mathbf{c}'_{f2}(t) = (1,0)$, and
$$W_f = \int_{\mathbf{c}_{f1}} \mathbf{F} \cdot d\mathbf{s} + \int_{\mathbf{c}_{f2}} \mathbf{F} \cdot d\mathbf{s} = \int_0^1 (t,0) \cdot (0,1)\, dt + \int_0^1 (1,0) \cdot (1,0)\, dt = 0 + 1 = 1.$$

See Figure 19; the vector field $\mathbf{F}$ becomes stronger as the y-coordinate increases. The curves $\mathbf{c}_f$, $\mathbf{c}_e$ and $\mathbf{c}_d$ are "aligned better" than $\mathbf{c}_a$, $\mathbf{c}_b$ and $\mathbf{c}_c$ with $\mathbf{F}$ in regions where $\mathbf{F}$ is stronger. That is why in these cases the corresponding work (i.e., W_f, W_e and W_d) is larger. Using similar argument we can compare any pair of paths.

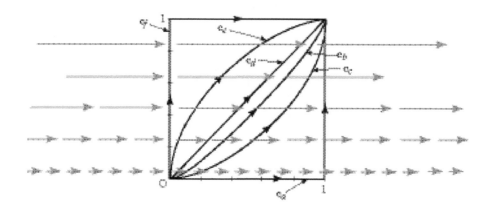

Figure 19

21. Parametrize $\mathbf{c}$ by $\mathbf{c}(t) = (2\cos t, 2\sin t)$, $t \in [0, 2\pi]$. Then $x^2 + y^2 = 4$ and

$$\int_{\mathbf{c}} \frac{ydx + xdy}{x^2 + y^2} = \tfrac{1}{4} \int_0^{2\pi} (2\sin t(-2\sin t) + 2\cos t(2\cos t)) \, dt$$

$$= \int_0^{2\pi} (\cos^2 t - \sin^2 t) \, dt = \int_0^{2\pi} \cos 2t \, dt = \tfrac{1}{2} \sin 2t \Big|_0^{2\pi} = 0.$$

23. (a) Since $x = t$ and $y = t$, it follows that

$$\int_{\mathbf{c}_1} (xy dx + 2y dy) = \int_0^1 \left(xy \frac{dx}{dt} + 2y \frac{dy}{dt} \right) dt$$

$$= \int_0^1 (t^2 + 2t) \, dt = \left(\tfrac{1}{3} t^3 + t^2 \right) \Big|_0^1 = \tfrac{4}{3}.$$

(b) In this case, $x = \sin t$ and $y = \sin t$, and

$$\int_{\mathbf{c}_2} (xy dx + 2y dy) = \int_0^{\pi/2} \left(xy \frac{dx}{dt} + 2y \frac{dy}{dt} \right) dt$$

$$= \int_0^{\pi/2} (\sin^2 t \cos t + 2 \sin t \cos t) \, dt$$

$$= \left(\tfrac{1}{3} \sin^3 t + \sin^2 t \right) \Big|_0^{\pi/2} = \tfrac{1}{3} + 1 = \tfrac{4}{3}.$$

(The substitution $u = \sin t$ is used to solve both integrals.)

(c) From $x = \cos t$ and $y = \cos t$ it follows that

$$\int_{\mathbf{c}_3} (xy dx + 2y dy) = \int_0^{\pi/2} \left(xy \frac{dx}{dt} + 2y \frac{dy}{dt} \right) dt$$

$$= \int_0^{\pi/2} (\cos^2 t(-\sin t) + 2\cos t(-\sin t)) \, dt$$

$$= \left(\tfrac{1}{3}\cos^3 t + \cos^2 t\right)\Big|_0^{\pi/2} = -1 - \tfrac{1}{3} = -\tfrac{4}{3}.$$

(The substitution $u = \cos t$ is used to solve both integrals.)

As a path integral of a vector field, a path integral of a 1-form depends only on the orientation of the parametrization that is used. The paths $\mathbf{c}_1$ and $\mathbf{c}_2$ are oriented from $(0,0)$ to $(1,1)$, whereas $\mathbf{c}_3$ is oriented from $(1,1)$ to $(0,0)$. This is the reason why the results in (a) and (b) are the same, and the result in (c) is the negative of the results in (a) and (b).

25. Let $\mathbf{c}\colon [a,b] \to \mathbb{R}^2$ (or $\mathbb{R}^3$) Assume, for simplicity, that $\mathbf{c}$ consists of two C^1 paths (the proof in the general case of n C^1 paths is analogous); so assume that the restriction $\mathbf{c}_1 = \mathbf{c}|^{[a,d]}$ of $\mathbf{c}$ to the interval $[a,d]$ and the restriction $\mathbf{c}_2 = \mathbf{c}|^{[d,b]}$ of $\mathbf{c}$ to the interval $[d,b]$ are C^1 paths. It follows that

$$\int_{\mathbf{c}} \mathbf{F} \cdot d\mathbf{s} = \int_{\mathbf{c}_1} \mathbf{F} \cdot d\mathbf{s} + \int_{\mathbf{c}_2} \mathbf{F} \cdot d\mathbf{s}.$$

Let $\boldsymbol{\gamma}(t) = \mathbf{c}(\phi(t))$ be a reparametrization of $\mathbf{c}(t)$, and assume that it preserves its orientation. By Theorem 5.3,

$$\int_{\mathbf{c}_1} \mathbf{F} \cdot d\mathbf{s} = \int_{\boldsymbol{\gamma}_1} \mathbf{F} \cdot d\mathbf{s} \quad \text{and} \quad \int_{\mathbf{c}_2} \mathbf{F} \cdot d\mathbf{s} = \int_{\boldsymbol{\gamma}_2} \mathbf{F} \cdot d\mathbf{s},$$

where $\boldsymbol{\gamma}_1 = \mathbf{c}_1(\phi(t))$ and $\boldsymbol{\gamma}_2 = \mathbf{c}_2(\phi(t))$. Therefore,

$$\int_{\mathbf{c}} \mathbf{F} \cdot d\mathbf{s} = \int_{\mathbf{c}_1} \mathbf{F} \cdot d\mathbf{s} + \int_{\mathbf{c}_2} \mathbf{F} \cdot d\mathbf{s} = \int_{\boldsymbol{\gamma}_1} \mathbf{F} \cdot d\mathbf{s} + \int_{\boldsymbol{\gamma}_2} \mathbf{F} \cdot d\mathbf{s} = \int_{\boldsymbol{\gamma}} \mathbf{F} \cdot d\mathbf{s}.$$

The proof in the orientation-reversing case is analogous.

27. Parametrize the circle by $\mathbf{c}(t) = r\cos t\, \mathbf{i} + r\sin t\, \mathbf{j}$, $t \in [0, 2\pi]$. Then $\mathbf{c}'(t) = -r\sin t\, \mathbf{i} + r\cos t\, \mathbf{j}$ and the circulation of $\mathbf{F}$ around $\mathbf{c}$ is

$$\int_{\mathbf{c}} \mathbf{F} \cdot d\mathbf{s} = \int_0^{2\pi} \mathbf{F}(\mathbf{c}(t)) \cdot \mathbf{c}'(t)\, dt$$
$$= \int_0^{2\pi} (r\cos t\, \mathbf{i} + r\sin t\, \mathbf{j}) \cdot (-r\sin t\, \mathbf{i} + r\cos t\, \mathbf{j})\, dt = \int_0^{2\pi} 0\, dt = 0.$$

Next, we need the outward unit normal vector $\mathbf{n}$. The dot product of the vector $\mathbf{N} = \cos t\, \mathbf{i} + \sin t\, \mathbf{j}$ with $\mathbf{c}'(t)$ is zero, hence $\mathbf{N}$ is normal to $\mathbf{c}$. Since $\mathbf{N}$ is a unit vector, it follows that $\mathbf{n} = \pm(\cos t\, \mathbf{i} + \sin t\, \mathbf{j})$. In the first quadrant, $\mathbf{n}$ has positive $\mathbf{i}$ and $\mathbf{j}$ components, and so $\mathbf{n} = \cos t\, \mathbf{i} + \sin t\, \mathbf{j}$ (check that this choice works for all quadrants; for example, in the third quadrant both components of $\mathbf{n}$ are negative — and the outward normal also has negative components). The flux of $\mathbf{F}$ across $\mathbf{c}$ is

$$\int_{\mathbf{c}} \mathbf{F} \cdot \mathbf{n}\, ds = \int_0^{2\pi} (r\cos t\, \mathbf{i} + r\sin t\, \mathbf{j}) \cdot (\cos t\, \mathbf{i} + \sin t\, \mathbf{j})\, dt = \int_0^{2\pi} r\, dt = 2\pi r.$$

29. Parametrize the semi-circular part of $\mathbf{c}$ by $\mathbf{c}_1(t) = r\cos t\, \mathbf{i} + r\sin t\, \mathbf{j}$, $t \in [0, \pi]$, and the straight line part by $\mathbf{c}_2(t) = t\mathbf{i}$, $t \in [-r, r]$. Then $\mathbf{c}_1'(t) = -r\sin t\, \mathbf{i} + r\cos t\, \mathbf{j}$ and $\mathbf{c}_2'(t) = \mathbf{i}$; so

$$\int_{\mathbf{c}_1} \mathbf{F} \cdot d\mathbf{s} = \int_{-r}^{r} (r\cos t\, \mathbf{i} + r\sin t\, \mathbf{j}) \cdot (-r\sin t\, \mathbf{i} + r\cos t\, \mathbf{j})\, dt = \int_{-r}^{r} 0\, dt = 0$$

and

$$\int_{\mathbf{c}_2} \mathbf{F} \cdot d\mathbf{s} = \int_{-r}^{r} t\mathbf{i} \cdot \mathbf{i} \, dt = \int_{-r}^{r} t \, dt = 0.$$

It follows that the circulation of $\mathbf{F}$ along $\mathbf{c}$ is

$$\int_{\mathbf{c}} \mathbf{F} \cdot d\mathbf{s} = \int_{\mathbf{c}_1} \mathbf{F} \cdot d\mathbf{s} + \int_{\mathbf{c}_2} \mathbf{F} \cdot d\mathbf{s} = 0.$$

The outward unit normal to $\mathbf{c}_1$ is $\mathbf{n}_1 = \cos t\, \mathbf{i} + \sin t\, \mathbf{j}$ (see the solution of Exercise 27). Therefore,

$$\int_{\mathbf{c}_1} \mathbf{F} \cdot \mathbf{n}_1 \, ds = \int_0^{\pi} (r\cos t\, \mathbf{i} + r\sin t\, \mathbf{j}) \cdot (\cos t\, \mathbf{i} + \sin t\, \mathbf{j}) \, dt = \int_0^{\pi} r \, dt = \pi r.$$

The outward unit normal to $\mathbf{c}_2$ is $\mathbf{n}_2 = -\mathbf{j}$. It follows that

$$\int_{\mathbf{c}_2} \mathbf{F} \cdot \mathbf{n}_2 \, ds = \int_0^{\pi} t\mathbf{i} \cdot (-\mathbf{j}) \, dt = \int_0^{\pi} 0 \, dt = 0.$$

Consequently, the flux of $\mathbf{F}$ across $\mathbf{c}$ is

$$\int_{\mathbf{c}} \mathbf{F} \cdot \mathbf{n} \, ds = \int_{\mathbf{c}_1} \mathbf{F} \cdot \mathbf{n}_1 \, ds + \int_{\mathbf{c}_2} \mathbf{F} \cdot \mathbf{n}_2 \, ds = \pi r.$$

31. Consider the flow given by $\mathbf{F}(x,y,z) = (F_1(x,y,z), F_2(x,y,z), F_3(x,y,z))$ in $\mathbb{R}^3$. We are interested in the behavior of the flow at some point P, so, for simplicity, we choose a coordinate system in which P is located at the origin. Take a small number $h > 0$. The circulation of $\mathbf{F}$ around $\mathbf{c}(t) = (h\cos t, h\sin t, 0)$, $t \in [0, 2\pi]$, has been discussed in the text (in the case of a two-dimensional vector field, and the curve $\mathbf{c}(t) = (h\cos t, h\sin t)$). Here, we consider $\mathbf{c}(t) = (h\cos t, 0, h\sin t)$, $t \in [0, 2\pi]$.

In order to find $\int_{\mathbf{c}} \mathbf{F} \cdot d\mathbf{s}$, we use the linear approximation for the components of $\mathbf{F}$,

$$\begin{aligned}
\mathbf{F}(x,y,z) &= (F_1(x,y,z), F_2(x,y,z), F_3(x,y,z)) \\
&= \Big(F_1(0,0,0) + \frac{\partial F_1}{\partial x}(0,0,0)\, x + \frac{\partial F_1}{\partial y}(0,0,0)\, y + \frac{\partial F_1}{\partial z}(0,0,0)\, z, \\
&\qquad F_2(0,0,0) + \frac{\partial F_2}{\partial x}(0,0,0)\, x + \frac{\partial F_2}{\partial y}(0,0,0)\, y + \frac{\partial F_2}{\partial z}(0,0,0)\, z, \\
&\qquad F_3(0,0,0) + \frac{\partial F_3}{\partial x}(0,0,0)\, x + \frac{\partial F_3}{\partial y}(0,0,0)\, y + \frac{\partial F_3}{\partial z}(0,0,0)\, z \Big).
\end{aligned}$$

Thus (keeping in mind that F_1, F_2 and F_3 and their partial derivatives are computed at $(0,0,0)$)

$$\begin{aligned}
\mathbf{F}(\mathbf{c}(t)) &= \Big(F_1 + \frac{\partial F_1}{\partial x}\, h\cos t + \frac{\partial F_1}{\partial z}\, h\sin t, \\
&\qquad F_2 + \frac{\partial F_2}{\partial x}\, h\cos t + \frac{\partial F_2}{\partial z}\, h\sin t, \; F_3 + \frac{\partial F_3}{\partial x}\, h\cos t + \frac{\partial F_3}{\partial z}\, h\sin t \Big),
\end{aligned}$$

Since $\mathbf{c}'(t) = (-h\sin t, 0, h\cos t)$, we get

$$\mathbf{F}(\mathbf{c}(t)) \cdot \mathbf{c}'(t) = -F_1 h\sin t - \frac{\partial F_1}{\partial x} h^2 \sin t \cos t - \frac{\partial F_1}{\partial z} h^2 \sin^2 t + F_3 h\cos t$$

$$+ \frac{\partial F_3}{\partial x} h^2 \cos^2 t + \frac{\partial F_3}{\partial z} h^2 \sin t \cos t$$

and

$$\int_{\mathbf{c}} \mathbf{F} \cdot d\mathbf{s} = \int_0^{2\pi} \mathbf{F}(\mathbf{c}(t)) \cdot \mathbf{c}'(t) \, dt = \left(\frac{\partial F_3}{\partial x} - \frac{\partial F_1}{\partial z} \right) h^2 \pi,$$

since $\int_0^{2\pi} \sin t \, dt = \int_0^{2\pi} \cos t \, dt = \int_0^{2\pi} \sin t \cos t \, dt = 0$ and $\int_0^{2\pi} \sin^2 t \, dt = \int_0^{2\pi} \cos^2 t \, dt = \pi$.

Thus, if $\mathbf{c}(t) = (h \cos t, 0, h \sin t)$, $t \in [0, 2\pi]$, then $\int_{\mathbf{c}} \mathbf{F} \cdot d\mathbf{s} = -h^2 \pi \cdot (\mathbf{j}$ component of *curl* $\mathbf{F}$). In the text it has been shown that if $\mathbf{c}(t) = (h \cos t, h \sin t, 0)$, $t \in [0, 2\pi]$, then $\int_{\mathbf{c}} \mathbf{F} \cdot d\mathbf{s} = h^2 \pi \cdot (\mathbf{k}$ component of *curl* $\mathbf{F}$). Analogously, we prove that if $\mathbf{c}(t) = (0, h \cos t, h \sin t)$, $t \in [0, 2\pi]$, then $\int_{\mathbf{c}} \mathbf{F} \cdot d\mathbf{s} = h^2 \pi \cdot (\mathbf{i}$ component of *curl* $\mathbf{F}$).

5.4. Path Integrals Independent of Path

1. Let $U = \mathbb{R}^2 - \{(x, y) \mid x^2 + y^2 = 1\}$. U is not connected, since any continuous curve joining a point inside to a point outside the circle $x^2 + y^2 = 1$ would have to cross the circle, and so would not lie entirely in U. Since U is not connected, it cannot be simply-connected. It is not star-shaped either; if it were star-shaped, it would have been connected.

3. Let U denote the given set and let $\mathbf{c}$ be the circle $x^2 + y^2 = 1$ in the plane $z = 1$; then $U = \mathbb{R}^3 - \mathbf{c}$. The set U is connected: any two points in U can be joined by a continuous curve that is completely contained in U (compare with Exercise 1: this time, we have a lot of "space" to avoid the points on the circle $\mathbf{c}$). U is not simply-connected; see Figure 3. The curve $\mathbf{c}_1$ that goes around $\mathbf{c}$ cannot be deformed (without breaking) to a point. U is not star-shaped either: let A be any point and let ℓ be the line through A and any point B on $\mathbf{c}$; see Figure 3. The line segment $\overline{AP}$, where P is a point on the half-line of ℓ starting at B that does not contain A is not completely contained in U, since B does not belong to U.

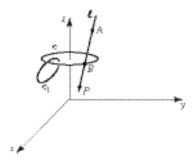

Figure 3

5. Let U be the given set. U is not connected: any continuous curve connecting a point inside the sphere to a point outside of it would have to cross the sphere, and would not be completely contained in U. Since U is not connected, it cannot be simply-connected or star-shaped either.

Notice that this exercise is a higher-dimensional analogue of Exercise 1.

7. Let $U = \mathbb{R}^3 - \mathbf{c}$. The set U is connected: there is a continuous curve joining any two points in U. U is also simply-connected: any simple closed curve can be continuously deformed to a point; see Figure 7. U is not star-shaped: take any point A and let ℓ be the line through A and through any point B on $\mathbf{c}$. A point P on the half-line of ℓ starting at B that does not contain A cannot be joined to A by a straight line that is entirely in U; see Figure 7.

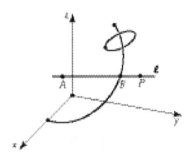

Figure 7

9. Let U be the given region, and let $A = (1, 4)$. Any point P that belongs to U can be joined with A by a straight line segment that is in U. So U is star-shaped; see Figure 9.

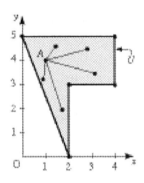

Figure 9

Since U is star-shaped, it is also connected. Since any simple closed curve can be deformed to a point (there are no "holes" in U), it follows that U is simply-connected.

11. Every simple closed curve on a sphere can be shrunk (without breaking) to a point, by moving along the surface of the sphere only. This is also true if a curve encloses a hole. If it does, all we have to do is to push it around the other side.

Suggestion: draw a sphere and a simple closed curve on it, and visualize how the deformation goes. Do the same for a sphere with a hole.

13. The domain of $\mathbf{F}$ is $\mathbb{R}^2$, which is an open and simply-connected set. Since

$$curl\,\mathbf{F} = \begin{vmatrix} \mathbf{i} & \mathbf{j} & \mathbf{k} \\ \partial/\partial x & \partial/\partial y & \partial/\partial z \\ y^2\cos x & 2y\sin x & 0 \end{vmatrix} = (2y\cos x - 2y\cos x)\mathbf{k} = \mathbf{0},$$

it follows from Theorem 5.4 that $\mathbf{F}$ is a gradient vector field. Consequently, any path starting at $(1, 1)$ and ending at $(1, 3)$ can be used in the evaluation of the path integral of $\mathbf{F}$. Let us take the straight line segment and parametrize it as $\mathbf{c}(t) = \mathbf{i} + t\mathbf{j}$, $t \in [1, 3]$. Then $\mathbf{c}'(t) = \mathbf{j}$ and

$$\int_{\mathbf{c}} \mathbf{F} \cdot d\mathbf{s} = \int_1^3 (t^2 \cos 1\, \mathbf{i} + 2t \sin 1\, \mathbf{j}) \cdot \mathbf{j}\, dt = 2 \sin 1 \int_1^3 t\, dt = 8 \sin 1 \approx 6.7318.$$

Here is an alternative: first, we find a function f such that $\nabla f = \mathbf{F}$. From $y^2 \cos x\, \mathbf{i} + 2y \sin x\, \mathbf{j} = f_x \mathbf{i} + f_y \mathbf{j}$ it follows that $f_x = y^2 \cos x$ and $f_y = 2y \sin x$. Integrate the first equation with respect to x to obtain $f = y^2 \sin x + C(y)$. Now substitute this into the expression for f_y, thus getting $2y \sin x + C'(y) = 2y \sin x$; i.e., $C'(y) = 0$ and $C(y) = C$ (where C is a real number). Hence $f = y^2 \sin x + C$. Now use the Fundamental Theorem 5.5:

$$\int_{\mathbf{c}} \mathbf{F} \cdot d\mathbf{s} = \int_{\mathbf{c}} \nabla f \cdot d\mathbf{s} = f(1, 3) - f(1, 1) = 9 \sin 1 - \sin 1 = 8 \sin 1.$$

15. By definition,

$$curl\, \mathbf{F} = \begin{vmatrix} \mathbf{i} & \mathbf{j} & \mathbf{k} \\ \partial/\partial x & \partial/\partial y & \partial/\partial z \\ -y/(x^2 + y^2) & x/(x^2 + y^2) & 1 \end{vmatrix} = \left(\frac{x^2 + y^2 - 2x^2}{(x^2 + y^2)^2} + \frac{x^2 + y^2 - 2y^2}{(x^2 + y^2)^2} \right) \mathbf{k} = \mathbf{0}.$$

From $\mathbf{c}_1(t) = (\cos t, \sin t, 0)$ it follows that $\mathbf{c}_1'(t) = (-\sin t, \cos t, 0)$ and

$$\int_{\mathbf{c}_1} \mathbf{F} \cdot d\mathbf{s} = \int_0^\pi (-\sin t, \cos t, 1) \cdot (-\sin t, \cos t, 0)\, dt = \int_0^\pi 1\, dt = \pi.$$

From $\mathbf{c}_2(t) = (\cos t, -\sin t, 0)$, we get $\mathbf{c}_2'(t) = (-\sin t, -\cos t, 0)$ and

$$\int_{\mathbf{c}_2} \mathbf{F} \cdot d\mathbf{s} = \int_0^\pi (\sin t, \cos t, 0) \cdot (-\sin t, -\cos t, 0)\, dt = \int_0^\pi (-1)\, dt = -\pi.$$

Obviously, the path integral of $\mathbf{F}$ depends on the path, so $\mathbf{F}$ cannot be conservative. The domain of $\mathbf{F}$ is the set $\mathbb{R}^3 - \{(0, 0, 0)\}$, which is not simply-connected; consequently, Theorem 5.8 does not apply and there is no contradiction.

17. Since

$$curl\, \mathbf{F} = \begin{vmatrix} \mathbf{i} & \mathbf{j} & \mathbf{k} \\ \partial/\partial x & \partial/\partial y & \partial/\partial z \\ 4x^2 - 4y^2 + x & 7xy + \ln y & 0 \end{vmatrix} = (7y + 8y)\mathbf{k} \neq \mathbf{0},$$

it follows that $\mathbf{F}$ is not a gradient vector field.

19. The *curl* of $\mathbf{F}$ is computed to be

$$curl\, \mathbf{F} = \begin{vmatrix} \mathbf{i} & \mathbf{j} & \mathbf{k} \\ \partial/\partial x & \partial/\partial y & \partial/\partial z \\ 2x \ln y & 2y + x^2/y & 0 \end{vmatrix} = \left(\frac{2x}{y} - \frac{2x}{y} \right) \mathbf{k} = \mathbf{0}.$$

The domain of $\mathbf{F}$ is the set $\{(x, y) \mid y > 0\}$; i.e., it is the upper half-plane (without the x-axis). It is simply-connected, and consequently, $\mathbf{F}$ is a gradient vector field.

From $\mathbf{F} = \nabla f$ it follows that

$$2x \ln y\, \mathbf{i} + 2y + \frac{x^2}{y}\mathbf{j} = f_x \mathbf{i} + f_y \mathbf{j};$$

i.e., $f_x = 2x \ln y$ and $f_y = 2y + x^2/y$. Integrating the equation for f_x with respect to x we get $f = x^2 \ln y + C(y)$; the equation for f_y then implies that $x^2/y + C'(y) = 2y + x^2/y$, so that $C'(y) = 2y$ and $C(y) = y^2 + C$. So all functions of the form

$$f(x, y) = x^2 \ln y + y^2 + C,$$

where C is a constant, satisfy $\mathbf{F} = \nabla f$.

21. The *curl* of $\mathbf{F}$ is computed to be

$$curl\, \mathbf{F} = \begin{vmatrix} \mathbf{i} & \mathbf{j} & \mathbf{k} \\ \partial/\partial x & \partial/\partial y & \partial/\partial z \\ y\cos(xy) & x\cos(xy) - z\sin y & \cos y \end{vmatrix}$$

$$= (-\sin y + \sin y)\mathbf{i} + (\cos(xy) - xy\sin(xy) - \cos(xy) + xy\sin(xy))\mathbf{k} = \mathbf{0}$$

The domain of $\mathbf{F}$ is $\mathbb{R}^3$, which is a simply-connected set. Hence, $\mathbf{F}$ is a gradient vector field.

From $\mathbf{F} = \nabla f$ it follows that

$$y\cos(xy)\mathbf{i} + (x\cos(xy) - z\sin y)\mathbf{j} + \cos y\,\mathbf{k} = f_x \mathbf{i} + f_y \mathbf{j} + f_z \mathbf{k};$$

i.e., $f_x = y\cos(xy)$, $f_y = x\cos(xy) - z\sin y$ and $f_z = \cos y$. Integrating the equation for f_x with respect to x we get $f = \sin(xy) + C(y, z)$; the equation for f_y implies that

$$x\cos(xy) + \frac{\partial C(y, z)}{\partial y} = x\cos(xy) - z\sin y,$$

so that $\partial C(y, z)/\partial y = -z\sin y$; it follows that $C(y, z) = z\cos y + C(z)$. Hence

$$f = \sin(xy) + z\cos y + C(z).$$

Finally, substituting this formula for f into the expression for f_z, we get $\cos y + C'(z) = \cos y$; i.e., $C'(z) = 0$ and $C(z) = C$. So all functions of the form

$$f(x, y, z) = \sin(xy) + z\cos y + C,$$

where C is a constant, satisfy $\mathbf{F} = \nabla f$.

23. Given integral can be interpreted as $\int_c \mathbf{F} \cdot d\mathbf{s}$, where $\mathbf{F} = (\cos x \tan z, 1, \sin x \sec^2 z)$, and $\mathbf{c}$ is a curve joining $(0, 0, 0)$ and $(\pi, \pi/2, \pi/3)$. The curl of $\mathbf{F}$ is computed to be

$$curl\, \mathbf{F} = \begin{vmatrix} \mathbf{i} & \mathbf{j} & \mathbf{k} \\ \partial/\partial x & \partial/\partial y & \partial/\partial z \\ \cos x \tan z & 1 & \sin x \sec^2 z \end{vmatrix} = -(\cos x \sec^2 z - \cos x \sec^2 z)\mathbf{j} = \mathbf{0}.$$

Take the domain of $\mathbf{F}$ to be the set $U = \{(x, y, z) \mid -\pi/2 < z < \pi/2\}$, since both endpoints $(0, 0, 0)$ and $(\pi, \pi/2, \pi/3)$ belong to it. U is a simply-connected set and so $\mathbf{F}$ is a gradient vector field on U.

From $\mathbf{F} = \nabla f$ it follows that

$$(\cos x \tan z, 1, \sin x \sec^2 z) = (f_x, f_y, f_z);$$

i.e., $f_x = \cos x \tan z$, $f_y = 1$ and $f_z = \sin x \sec^2 z$. Integrating the equation for f_x with respect to x we get $f = \sin x \tan z + C(y, z)$; substituting f into the equation for f_y we get $\partial C(y, z)/\partial y = 1$, so that $C(y, z) = y + C(z)$. Hence

$$f = \sin x \tan z + y + C(z).$$

Finally, substituting this formula for f into the expression for f_z, we get $\sin x \sec^2 z + C'(z) = \sin x \sec^2 z$; i.e., $C'(z) = 0$ and $C(z) = C$. So $\mathbf{F} = \nabla f$, where $f(x, y, z) = \sin x \tan z + y + C$, and C is a constant. By the Fundamental Theorem 5.5,

$$\int_{\mathbf{c}} \mathbf{F} \cdot d\mathbf{s} = \int_{\mathbf{c}} \nabla f \cdot d\mathbf{s} = f(\pi, \pi/2, \pi/3) - f(0, 0, 0) = \sin \pi \tan(\pi/3) + \frac{\pi}{2} - 0 = \frac{\pi}{2}.$$

25. Let $\overline{\mathbf{c}}_1(t) = (0, t)$, $t \in [0, y]$; i.e., $\overline{\mathbf{c}}_1$ parametrizes the part of $\overline{\mathbf{c}}$ that goes along the y-axis. Then

$$\int_{\overline{\mathbf{c}}_1} \mathbf{F} \cdot d\mathbf{s} = \int_0^y \mathbf{F}(\overline{\mathbf{c}}_1(t)) \cdot \overline{\mathbf{c}}_1'(t) \, dt$$

$$= \int_0^y \left(F_1(0, t), F_2(0, t) \right) \cdot (0, 1) \, dt = \int_0^y F_2(0, t) \, dt.$$

Parametrize the horizontal part of $\overline{\mathbf{c}}$ by $\overline{\mathbf{c}}_2(t) = (t, y)$, $t \in [0, x]$. Then $\overline{\mathbf{c}}_2'(t) = (1, 0)$ and

$$\int_{\overline{\mathbf{c}}_2} \mathbf{F} \cdot d\mathbf{s} = \int_0^x \mathbf{F}(\overline{\mathbf{c}}_2(t)) \cdot \overline{\mathbf{c}}_2'(t) \, dt$$

$$= \int_0^x \left(F_1(t, y), F_2(t, y) \right) \cdot (1, 0) \, dt = \int_0^x F_1(t, y) \, dt.$$

Hence

$$f(x, y) = \int_0^y F_2(0, t) \, dt + \int_0^x F_1(t, y) \, dt.$$

It follows that

$$\frac{\partial f}{\partial x}(x, y) = 0 + \frac{\partial}{\partial x} \left(\int_0^x F_1(t, y) \, dt \right) = F_1(x, y),$$

where the right side was computed using the Fundamental Theorem of Calculus.

27. (a) The *curl* of $\mathbf{F}$ is

$$curl\, \mathbf{F} = \begin{vmatrix} \mathbf{i} & \mathbf{j} & \mathbf{k} \\ \partial/\partial x & \partial/\partial y & \partial/\partial z \\ -(1 + x)ye^x & -xe^x & 0 \end{vmatrix} = (-e^x - xe^x + (1 + x)e^x)\mathbf{k} = \mathbf{0}.$$

Since the domain of $\mathbf{F}$ is $\mathbb{R}^3$ (which is simply-connected), it follows that $\mathbf{F}$ is conservative.

(b) By the Fundamental Theorem (see Theorem 5.5),

$$\int_{\mathbf{c}} \mathbf{F} \cdot d\mathbf{s} = -\int_{\mathbf{c}} \nabla V \cdot d\mathbf{s} = -V(x, y) \Big|_{(0,0)}^{(x,0)} = -V(x, 0) + V(0, 0) = -V(x, 0),$$

since $V(0,0) = 0$ by assumption. Now let $\mathbf{c}$ be the straight line segment from $(0,0)$ to $(x,0)$; parametrize it by $\mathbf{c}(t) = t\mathbf{i}$, $t \in [0, x]$. Then $\mathbf{F}(\mathbf{c}(t)) = -te^t\mathbf{j}$, $\mathbf{c}'(t) = \mathbf{i}$ and hence

$$\int_{\mathbf{c}} \mathbf{F} \cdot d\mathbf{s} = \int_0^x \mathbf{F}(\mathbf{c}(t)) \cdot \mathbf{c}'(t) \, dt = \int_0^x (-te^t\mathbf{j}) \cdot \mathbf{i} \, dt = \int_0^x 0 \, dt = 0.$$

Comparing the two expressions for $\int_{\mathbf{c}} \mathbf{F} \cdot d\mathbf{s}$, we get that $V(x,0) = 0$ for every x.

(c) The computation is analogous to (b). By the Fundamental Theorem 5.5,

$$\int_{\mathbf{c}} \mathbf{F} \cdot d\mathbf{s} = -\int_{\mathbf{c}} \nabla V \cdot d\mathbf{s} = -V(x,y)\Big|_{(0,0)}^{(0,y)} = -V(0,y) + V(0,0) = -V(0,y),$$

since $V(0,0) = 0$. Take $\mathbf{c}$ to be the straight line segment from $(0,0)$ to $(0,y)$; parametrize it by $\mathbf{c}(t) = t\mathbf{j}$, $t \in [0, y]$. Then $\mathbf{F}(\mathbf{c}(t)) = -t\mathbf{i}$, $\mathbf{c}'(t) = \mathbf{j}$ and

$$\int_{\mathbf{c}} \mathbf{F} \cdot d\mathbf{s} = \int_0^x \mathbf{F}(\mathbf{c}(t)) \cdot \mathbf{c}'(t) \, dt = \int_0^y (-t\mathbf{i}) \cdot \mathbf{j} \, dt = \int_0^y 0 \, dt = 0.$$

Comparing the two expressions for $\int_{\mathbf{c}} \mathbf{F} \cdot d\mathbf{s}$, we get that $V(0,y) = 0$ for every y.

The reader is invited to check that $V(x,y) = xye^x$.

29. Parametrize $\mathbf{c}$ by $\mathbf{c}(t) = 3t\mathbf{i} - 2t\mathbf{j}$, $t \in [0, 1]$. Then

$$\mathbf{F}(\mathbf{c}(t)) = \mathbf{F}(3t\mathbf{i} - 2t\mathbf{j}) = -12t^2e^{-2t}\mathbf{i} + (9t^2 - 18t^3)e^{-2t}\mathbf{j},$$

$\mathbf{c}'(t) = 3\mathbf{i} - 2\mathbf{j}$ and

$$\int_{\mathbf{c}} \mathbf{F} \cdot d\mathbf{s} = \int_0^1 \mathbf{F}(\mathbf{c}(t)) \cdot \mathbf{c}'(t) \, dt = \int_0^1 (-12t^2e^{-2t}\mathbf{i} + (9t^2 - 18t^3)e^{-2t}\mathbf{j}) \cdot (3\mathbf{i} - 2\mathbf{j}) \, dt$$

$$= \int_0^1 (-54t^2e^{-2t} + 36t^3e^{-2t}) \, dt.$$

From integration tables (or, using integration by parts) we get

$$\int_0^1 t^2e^{-2t} \, dt = -\tfrac{1}{4}e^{-2t}(2t^2 + 2t + 1)\Big|_0^1 = \tfrac{1}{4}(1 - 5e^{-2}).$$

Similarly,

$$\int_0^1 t^3e^{-2t} \, dt = -\tfrac{1}{8}e^{-2t}(4t^3 + 6t^2 + 6t + 3)\Big|_0^1 = \tfrac{1}{8}(3 - 19e^{-2}).$$

Finally,

$$\int_{\mathbf{c}} \mathbf{F} \cdot d\mathbf{s} = -54\left(\tfrac{1}{4}(1 - 5e^{-2})\right) + 36\left(\tfrac{1}{8}(3 - 19e^{-2})\right) = -18e^{-2} \approx -2.4360.$$

(b) Since $curl\,\mathbf{F} = \mathbf{0}$ and the domain of $\mathbf{F}$ is $\mathbb{R}^2$, which is a simply-connected set, it follows that $\mathbf{F}$ is a gradient vector field. From $\mathbf{F} = \nabla f$ it follows that

$$2xye^y\mathbf{i} + x^2e^y(1 + y)\mathbf{j} = f_x\mathbf{i} + f_y\mathbf{j};$$

i.e., $f_x = 2xye^y$ and $f_y = x^2e^y(1 + y)$. Integrating the equation for f_x with respect to x we get $f = x^2ye^y + C(y)$; the equation for f_y implies that $x^2(e^y + ye^y) + C'(y) = x^2e^y(1 + y)$, so that $C'(y) = 0$ and $C(y) = C$. So all functions of the form $f(x, y) = x^2ye^y + C$, where C is a constant, satisfy $\mathbf{F} = \nabla f$. Hence, by the Fundamental Theorem,

$$\int_c \mathbf{F} \cdot d\mathbf{s} = \int_c \nabla f \cdot d\mathbf{s} = f(3, -2) - f(0, 0) = -18e^{-2}.$$

31. By a product rule for *curl*,

$$curl\, \mathbf{F}(\mathbf{r}) = curl\,(\|\mathbf{r}\|^2\mathbf{r}) = \|\mathbf{r}\|^2 curl\, \mathbf{r} + grad\,\|\mathbf{r}\|^2 \times \mathbf{r}.$$

Now

$$curl\, \mathbf{r} = \begin{vmatrix} \mathbf{i} & \mathbf{j} & \mathbf{k} \\ \partial/\partial x & \partial/\partial y & \partial/\partial z \\ x & y & z \end{vmatrix} = \mathbf{0}$$

and

$$\nabla\|\mathbf{r}\|^2 = \nabla(x^2 + y^2 + z^2) = 2x\mathbf{i} + 2y\mathbf{j} + 2z\mathbf{k} = 2\mathbf{r},$$

and therefore

$$curl\, \mathbf{F}(\mathbf{r}) = \|\mathbf{r}\|^2 \mathbf{0} + 2\mathbf{r} \times \mathbf{r} = \mathbf{0}.$$

From $\mathbf{F} = \nabla f$ it follows that

$$((x^2 + y^2 + z^2)x, (x^2 + y^2 + z^2)y, (x^2 + y^2 + z^2)z) = (f_x, f_y, f_z);$$

i.e., $f_x = (x^2 + y^2 + z^2)x$, $f_y = (x^2 + y^2 + z^2)y$ and $f_z = (x^2 + y^2 + z^2)z$. Integrating the equation for f_x with respect to x we get

$$f = \tfrac{1}{4}x^4 + \tfrac{1}{2}x^2(y^2 + z^2) + C(y, z);$$

substituting f into the equation for f_y we get $yx^2 + \partial C(y, z)/\partial y = yx^2 + y^3 + z^2y$, so that $\partial C(y, z)/\partial y = y^3 + z^2y$, and $C(y, z) = y^4/4 + y^2z^2/2 + C(z)$. Hence

$$f = \tfrac{1}{4}x^4 + \tfrac{1}{4}y^4 + \tfrac{1}{2}x^2y^2 + \tfrac{1}{2}x^2z^2 + \tfrac{1}{2}y^2z^2 + C(z).$$

Finally, substituting this formula for f into the expression for f_z, we get $C'(z) = z^3$; i.e., $C(z) = z^4/4 + C$. So $\mathbf{F} = \nabla f$, where

$$f(x, y, z) = \tfrac{1}{4}(x^4 + y^4 + z^4) + \tfrac{1}{2}(x^2y^2 + x^2z^2 + y^2z^2) + C$$
$$= \tfrac{1}{4}(x^2 + y^2 + z^2)^2 + C = \tfrac{1}{4}\|\mathbf{r}\|^4 + C.$$

33. The *curl* of $\mathbf{F}$ is

$$curl\, \mathbf{F} = \begin{vmatrix} \mathbf{i} & \mathbf{j} & \mathbf{k} \\ \partial/\partial x & \partial/\partial y & \partial/\partial z \\ \ln(x + y^2) + x/(x + y^2) & 2xy/(x + y^2) & 0 \end{vmatrix}$$

$$= \left(\frac{2y(x + y^2) - 2xy}{(x + y^2)^2} - \frac{2y}{(x + y^2)} + \frac{2xy}{(x + y^2)^2}\right)\mathbf{k} = \mathbf{0}.$$

The domain of $\mathbf{F}$ is the set $U = \{(x,y) \mid x + y^2 > 0\}$. We do not have to identify it precisely, but only note that it contains the first quadrant. That part is simply-connected, and contains the curve; it follows that $\mathbf{F}$ is a gradient vector field on that set.

From $\mathbf{F} = \nabla f$ it follows that

$$(\ln(x + y^2) + x/(x + y^2))\mathbf{i} + (2xy/(x + y^2))\mathbf{j} = f_x\mathbf{i} + f_y\mathbf{j};$$

i.e., $f_x = \ln(x + y^2) + x/(x + y^2)$ and $f_y = 2xy/(x + y^2)$. Integrating the equation for f_y with respect to y we get

$$f = x\ln(x + y^2) + C(x);$$

the equation for f_x implies that

$$\ln(x + y^2) + \frac{x}{x + y^2} + C'(x) = \ln(x + y^2) + \frac{x}{x + y^2},$$

so that $C'(x) = 0$ and $C(x) = C$. Hence $f(x,y) = x\ln(x + y^2) + C$, and, by the Fundamental Theorem,

$$\int_c \mathbf{F} \cdot d\mathbf{s} = \int_c \nabla f \cdot d\mathbf{s} = x\ln(x + y^2)\Big|_{(1,1)}^{(2,8)} = 2\ln 66 - \ln 2 = \ln 2178 \approx 7.6862.$$

Chapter Review

True/false Quiz

1. True. The assumption implies that $\int_c \nabla f \cdot d\mathbf{s} = 0$, where $\mathbf{c}$ joins P and Q. By the Fundamental Theorem of Calculus (see Theorem 5.5), it follows that

$$\int_c \nabla f \cdot d\mathbf{s} = f(\mathbf{c}(b)) - f(\mathbf{c}(a)) = f(Q) - f(P) = 0,$$

and thus $f(P) = f(Q)$.

3. False. Take, for instance, $\mathbf{F}(x,y,z) = (\sin x, \cos x, 1)$. Clearly, $\|\mathbf{F}\| = \sqrt{2}$. However, $\mathbf{F}$ is not path-independent (by Theorem 5.8) since $curl\, \mathbf{F} = -\sin x\, \mathbf{k} \neq \mathbf{0}$

5. True. Follows from Theorem 5.9.

7. True. Follows from the fact that $\mathbf{F}$ is perpendicular to any vertical path at all its points.

9. False. Take the circle $\mathbf{c}(t) = (\cos t, \sin t)$, $t \in [0, 2\pi]$. Let $\mathbf{F}(x,y) = (1,0)$ and $\mathbf{G}(x,y) = (0,1)$. Then

$$\int_c \mathbf{F} \cdot d\mathbf{s} = \int_0^{2\pi} (1,0) \cdot (-\sin t, \cos t)\, dt = -\int_0^{2\pi} \sin t\, dt = 0$$

and

$$\int_c \mathbf{G} \cdot d\mathbf{s} = \int_0^{2\pi} (0,1) \cdot (-\sin t, \cos t)\, dt = \int_0^{2\pi} \cos t\, dt = 0.$$

So, $\mathbf{F}$ and $\mathbf{G}$ have the same circulation along $\mathbf{c}$, but $\mathbf{F} \neq \mathbf{G}$.

11. True. Theorem 5.8 applies, since $\mathbb{R}^2$ is open and connected.

Review Exercises and Problems

1. Define $\boldsymbol{\gamma}(t) = \mathbf{c}(\phi(t)) = (\phi(t)^{3/2}, \phi(t))$. Drop t for simplicity of notation. Then $\boldsymbol{\gamma}' = (3\phi^{1/2}\phi'/2, \phi')$, and from $\|\boldsymbol{\gamma}'\| = C$ ($C > 0$ is a constant) it follows that

$$\sqrt{\left(\tfrac{3}{2}\phi^{1/2}\phi'\right)^2 + (\phi')^2} = C;$$

i.e.,

$$|\phi'|\sqrt{\tfrac{9}{4}\phi + 1} = C.$$

Since we are looking for an orientation-preserving parametrization, $\phi' > 0$ and hence $|\phi'| = \phi'$; so we have to solve the differential equation

$$\phi'\sqrt{\tfrac{9}{4}\phi + 1} = C,$$

where $C > 0$. Integrating both sides, we get

$$\tfrac{2}{3}\left(\tfrac{9}{4}\phi + 1\right)^{3/2}\tfrac{4}{9} = Ct + D$$

and

$$\left(\tfrac{9}{4}\phi + 1\right)^{3/2} = \tfrac{27}{8}(Ct + D).$$

Solving for ϕ, we get $\phi(t) = (Ct + D)^{2/3} - 4/9$.

3. Parametrize $\mathbf{c}$ by $\mathbf{c}(t)$, $t \in [a, b]$. Since $\mathbf{T}(t) = \mathbf{c}'(t)/\|\mathbf{c}'(t)\|$, it follows that

$$\int_{\mathbf{c}} \mathbf{T} \cdot d\mathbf{s} = \int_a^b \frac{\mathbf{c}'(t)}{\|\mathbf{c}'(t)\|} \cdot \mathbf{c}'(t)\, dt = \int_a^b \|\mathbf{c}'(t)\|\, dt = \ell(\mathbf{c}),$$

where $\ell(\mathbf{c})$ is the length of $\mathbf{c}$.

If $\mathbf{c}$ is a contour curve of $\mathbf{F}$, then ∇f is perpendicular to $\mathbf{c}$ and thus $\int_{\mathbf{c}} \nabla f \cdot d\mathbf{s} = 0$.

5. The domain of $\mathbf{F}$ is $\mathbb{R}^2$ (which is simply-connected), so it suffices to show that the scalar curl of $\mathbf{F}$ is zero. This is indeed true, since $\partial F_2/\partial x = 8x$ and $\partial F_1/\partial y = 8x$.

From $\nabla f = \mathbf{F}$ we conclude that $\partial f/\partial x = 3x^2 + 8xy$ and $\partial f/\partial y = 4x^2$. Integrating the second equation with respect to y, we get $f = 4x^2y + C(x)$. Substituting this expression for f into $\partial f/\partial x = 3x^2 + 8xy$ yields $8xy + C'(x) = 3x^2 + 8xy$. So, $C'(x) = 3x^2$ and $C(x) = x^3 + C$ (where $C \in \mathbb{R}$). Thus, $f(x, y) = 4x^2y + x^3 + C$.

7. Take a path $\mathbf{c}(t)$, $t \in [a, b]$, from $(1, 2)$ to $(-2, 1)$; in that case, $\mathbf{c}(a) = (1, 2)$ and $\mathbf{c}(b) = (-2, 1)$. The required change in f can be written as $f(-2, 1) - f(1, 2) = f(\mathbf{c}(b)) - f(\mathbf{c}(a))$. Using the Fundamental Theorem of Calculus (see Theorem 5.5)

$$\int_{\mathbf{c}} \nabla f \cdot d\mathbf{s} = f(\mathbf{c}(b)) - f(\mathbf{c}(a)),$$

we see that the change in f is given by $\int_{\mathbf{c}} \nabla f \cdot d\mathbf{s}$.

According to Theorem 5.8, we can take any path from $(1, 2)$ to $(-2, 1)$ to calculate the above path integral. So, let $\mathbf{c}$ be the straight line segment, parametrized by

$$\mathbf{c}(t) = (1, 2) + t(-3, -1) = (1 - 3t, 2 - t), \qquad t \in [0, 1].$$

Thus

$$\int_{\mathbf{c}} \nabla f \cdot d\mathbf{s} = \int_0^1 ((1 - 3t)^2, -(2 - t)^3) \cdot (-3, -1) \, dt$$

$$= \int_0^1 (-t^3 - 21t^2 + 6t + 5) \, dt = \frac{3}{4}.$$

9. We will compute the work of $\mathbf{F}$ in two different ways. Since $\mathbf{F}$ is conservative, we can choose any path joining $(0, 0, 0)$ and (x, y, z). Take the line $\mathbf{c}(t) = (tx, ty, tz)$, $t \in [0, 1]$. Then

$$\mathbf{F}(\mathbf{c}(t)) = a\|\mathbf{c}(t)\|\mathbf{c}(t) = a\sqrt{t^2(x^2 + y^2 + z^2)} \ (tx, ty, tz) = at^2\sqrt{x^2 + y^2 + z^2} \ (x, y, z)$$

and $\mathbf{c}'(t) = (x, y, z)$, and therefore

$$W = \int_{\mathbf{c}} \mathbf{F} \cdot d\mathbf{s} = \int_0^1 at^2\sqrt{x^2 + y^2 + z^2} \ (x, y, z) \cdot (x, y, z) \, dt$$

$$= a(x^2 + y^2 + z^2)^{3/2} \ \tfrac{1}{3}t^3 \Big|_0^1 = \tfrac{1}{3}a(x^2 + y^2 + z^2)^{3/2}.$$

Since $\mathbf{F}$ is conservative, there is a potential function V such that $\mathbf{F} = -\nabla V$. Hence

$$W = \int_{\mathbf{c}} \mathbf{F} \cdot d\mathbf{s} = -\int_{\mathbf{c}} \nabla V \cdot d\mathbf{s} = -V \Big|_{(0,0,0)}^{(x,y,z)} = -V(x, y, z) + V(0, 0, 0) = -V(x, y, z),$$

since, by assumption, $V(0, 0, 0) = 0$. Comparing the two expressions for the work, we get

$$V(x, y, z) = -\tfrac{1}{3}a(x^2 + y^2 + z^2)^{3/2};$$

i.e., $V(\mathbf{r}) = -a\|\mathbf{r}\|^3/3$.

11. Let $E(t) = f(\mathbf{c}(t)) + m\|\mathbf{c}'(t)\|^2/2$. We will show that $E'(t) = 0$ (and consequently, that $E(t)$ is a constant). By the chain rule,

$$\frac{d}{dt}(f(\mathbf{c}(t))) = Df(\mathbf{c}(t)) \cdot D\mathbf{c}(t) = \nabla f(\mathbf{c}(t)) \cdot \mathbf{c}'(t).$$

Since $\|\mathbf{c}'(t)\|^2 = \mathbf{c}'(t) \cdot \mathbf{c}'(t)$, it follows that $(d/dt)\|\mathbf{c}'(t)\|^2 = 2\mathbf{c}'(t) \cdot \mathbf{c}''(t)$. Therefore,

$$\frac{d}{dt}E(t) = \nabla f(\mathbf{c}(t)) \cdot \mathbf{c}'(t) + m\mathbf{c}'(t) \cdot \mathbf{c}''(t).$$

Since $\mathbf{F} = -\nabla f$ and $\mathbf{F}(t) = m\mathbf{c}''(t)$ by Newton's Second Law, it follows that

$$\frac{d}{dt}E(t) = -\mathbf{F}(\mathbf{c}(t)) \cdot \mathbf{c}'(t) + \mathbf{c}'(t) \cdot \mathbf{F}(\mathbf{c}(t)) = 0.$$

$E(t)$ is the sum of the kinetic energy and the potential energy. The fact that $E'(t) = 0$ is the Law of Conservation of Energy (i.e., total energy remains constant throughout the motion).

13. Let $\mathbf{c}(t) = (x(t), y(t))$, $t \in [a, b]$, be the trajectory of the satellite as it is moved from its location at height h_1 to a new location at height h_2. Then

$$W = \int_{\mathbf{c}} \mathbf{F} \cdot d\mathbf{s} = \int_a^b (0, -mg) \cdot (x'(t), y'(t)) \, dt$$

$$= -mg \int_a^b y'(t) \, dt = -mg(y(b) - y(a)) = -mg(h_2 - h_1).$$

6. DOUBLE AND TRIPLE INTEGRALS

6.1. Double Integrals: Definition and Properties

1. Dividing $[1, 5]$ into subintervals $[1, 2], [2, 3], [3, 4]$ and $[4, 5]$, and interval $[0, 2]$ into subintervals $[0, 0.5], [0.5, 1], [1, 1.5]$ and $[1.5, 2]$, we have subdivided $R = [1, 5] \times [0, 2]$ into 16 rectangles R_{ij}, of area $\Delta A_{ij} = 0.5$, for $i, j = 1, 2, 3, 4$. We form $\mathcal{R}_4 = \sum_{i=1}^{4} \sum_{j=1}^{4} f(x_i^*, y_j^*) \Delta A_{ij}$, where we read the values $f(x_i^*, y_j^*)$ from the contour diagram. It follows that (we start at the upper left rectangle, and go down row by row, taking (x_i^*, y_j^*) to be (approximately) the center)

$$\mathcal{R}_4 = 0.5(11 + 10 + 7 + 6 + 10 + 8 + 6 + 5 + 9 + 7 + 5.5 + 4.5 + 9 + 7 + 5 + 4) = 56.5.$$

3. Form the partition of R into 9 rectangles $R_{11}, R_{12}, \ldots, R_{33}$ using vertical lines $x = 0$, $x = 1$, $x = 2$ and $x = 3$ and horizontal lines $y = 0$, $y = 2/3$, $y = 4/3$ and $y = 2$. Taking (x_i^*, y_j^*) to be the upper-left corner of a subrectangle R_{ij}, $i, j = 1, 2, 3$, we get

$$\mathcal{R}_3 = f(0, 2/3) \Delta A_{11} + f(0, 4/3) \Delta A_{12} + f(0, 2) \Delta A_{13} + f(1, 2/3) \Delta A_{21} + f(1, 4/3) \Delta A_{22}$$
$$+ f(1, 2) \Delta A_{23} + f(2, 2/3) \Delta A_{31} + f(2, 4/3) \Delta A_{32} + f(2, 2) \Delta A_{33},$$

where $\Delta A_{ij} = 2/3$ (for all i and j) is the area of R_{ij}. It follows that (factor out $\Delta A_{ij} = 2/3$)

$$\mathcal{R}_3 = \tfrac{2}{3} \left(\tfrac{7}{9} + \tfrac{55}{9} + 15 + \tfrac{16}{9} + \tfrac{64}{9} + 16 + \tfrac{25}{9} + \tfrac{73}{9} + 17 \right)$$
$$= \tfrac{2}{3} \left(\tfrac{240}{9} + 48 \right) = \tfrac{448}{9} \approx 49.7778.$$

5. Form the partition of R into 4 rectangles $R_{11}, R_{12}, R_{21}, R_{22}$, using vertical lines $x = 0$, $x = 5/2$ and $x = 5$ and horizontal lines $y = 0$, $y = 3/2$ and $y = 3$. Taking (x_i^*, y_j^*) to be the upper left corner of each subrectangle R_{ij}, $i, j = 1, 2$, we get

$$\mathcal{R}_2 = f(0, 3/2) \Delta A_{11} + f(0, 3) \Delta A_{12} + f(5/2, 3/2) \Delta A_{21} + f(5/2, 3) \Delta A_{22},$$

where $\Delta A_{ij} = (3/2)(5/2) = 15/4$ (for all i, j) is the area of R_{ij}. It follows that (factor out $\Delta A_{ij} = 15/4$)

$$\mathcal{R}_2 = \tfrac{15}{4} \left(-\tfrac{9}{4} - 9 + 4 - \tfrac{11}{4} \right) = \tfrac{15}{4} (-10) = -75/2.$$

7. Form the partition of R into 4 rectangles $R_{11}, R_{12}, R_{21}, R_{22}$, using vertical lines $x = -1$, $x = 0$ and $x = 1$ and horizontal lines $y = 0$, $y = 1/2$ and $y = 1$. Taking (x_i^*, y_j^*) to be the lower right corner of each subrectangle R_{ij}, $i, j = 1, 2$, we get

$$\mathcal{R}_2 = f(0, 0) \Delta A_{11} + f(0, 1/2) \Delta A_{12} + f(1, 0) \Delta A_{21} + f(1, 1/2) \Delta A_{22},$$

where $\Delta A_{ij} = 1/2$ (for all i, j) is the area of R_{ij}. It follows that

$$\mathcal{R}_2 = \tfrac{1}{2} \left(0 - \tfrac{1}{2} + 1 + \tfrac{1}{2} e^{1/2} \right) = \tfrac{1}{4} (1 + e^{1/2}) \approx 0.6622.$$

9. Using (c) in Theorem 6.2, we write

$$\iint_{[-2,1] \times [-1,1]} xy^2 \, dA = \iint_{[-2,-1] \times [-1,1]} xy^2 \, dA + \iint_{[-1,1] \times [-1,1]} xy^2 \, dA.$$

Since $xy^2 < 0$ on $[-2, -1] \times [-1, 1]$, it follows that $\iint_{[-2,-1]\times[-1,1]} xy^2 dA < 0$. The integral $\iint_{[-1,1]\times[-1,1]} xy^2 dA$ is equal to zero, since the contribution to a Riemann sum coming from a rectangle R_{ij} cancels with the contribution coming from the rectangle symmetric to R_{ij} with respect to the y-axis.

11. Since $z = 3 - x \geq 0$ on $R = [0, 1] \times [0, 2]$, it follows that the double integral $\iint_R (3 - x)\, dA$ is the volume of the solid V under $z = 3 - x$ and over $[0, 1] \times [0, 2]$; see Figure 11. The solid V consists of the rectangular box over $[0, 1] \times [0, 2]$ of height 2 with a triangular prism on top (the triangle is the right triangle with sides 1) of length 2. Therefore, the volume of V is $4 + 1 = 5$, and so $\iint_R (3 - x)\, dA = 5$.

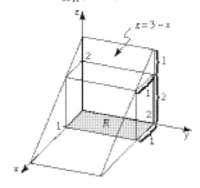

Figure 11

13. Since $-1 \leq x, y \leq 1$, it follows that $0 \leq x^2 + y^2 \leq 2$, and so $1 \leq e^{x^2+y^2} \leq e^2$. Consequently,

$$\iint_R 1\, dA \leq \iint_R e^{x^2+y^2}\, dA \leq \iint_R e^2\, dA.$$

Since $\iint_R 1\, dA = \text{area}(R) = 4$ and $\iint_R e^2\, dA = e^2 \iint_R 1\, dA = 4e^2$, we get the desired inequality.

15. Since $\sin a \leq 1$ and $\cos a \leq 1$ for all $a \in R$, it follows that $\sin(2x)\cos(5y) \leq 1$ and therefore

$$\iint_R \sin(2x)\cos(5y)\, dA \leq \iint_R 1\, dA = \text{area}(R) = 2\pi^2,$$

since the dimensions of R are 2π and π.

17. The function f is continuous at all points in $R = [-1, 1] \times [0, 2]$ (notice that $\lim_{x \to 0} f(x, y) = 3$ for all y), and is hence integrable on R.

Since $f(x, y) > 0$ on R, we can interpret $\iint_R f\, dA$ as the volume of the solid V below $f(x, y)$ and above R. The part of V over $[0, 1] \times [0, 2]$ is a rectangular box of height 3; its volume is 6. The part of V over $[-1, 0] \times [0, 2]$ consists of a rectangular box of height 2 with a triangular prism on top (the triangle is the right triangle with sides 1), of height (or length) 2. The volume of the box and the prism is $4 + 1 = 5$. It follows that $\iint_R f\, dA = 11$.

19. The inequality $x^2 + y^2 \leq x^3 + y^3$ does not hold on $[0,1] \times [0,1]$; take, for example, $x = y = 1/2$: then $x^2 + y^2 = 1/2$, but $x^3 + y^3 = 1/4$. As a matter of fact, if $0 \leq x, y \leq 1$, then $x^2 + y^2 \geq x^3 + y^3$, and $\iint_R (x^2 + y^2)\, dA > \iint_R (x^3 + y^3)\, dA$, if $R = [0,1] \times [0,1]$.

However, if $x, y \geq 1$, then $x^2 \leq x^3$ and $y^2 \leq y^3$, and so $x^2 + y^2 \leq x^3 + y^3$, and the inequality $\iint_R (x^2 + y^2)\, dA \leq \iint_R (x^3 + y^3)\, dA$ holds for $R = [1,2] \times [2,3]$.

21. The function $f(x,y) = e^{1/(x^2+y^2)}$ is not bounded on $[-1,1] \times [-1,1]$; its value approaches ∞ as $(x,y) \to (0,0)$. Consequently, f is not integrable on $[-1,1] \times [-1,1]$.

6.2. Double Integrals over General Regions

1. D is of type 1. It is bounded by $\phi(x) = -\sqrt{1-x^2}$ and $\psi(x) = \sqrt{1-x^2}$, defined on the interval $[a = -1, b = 1]$. It is also of type 2: take $\phi(y) = -\sqrt{1-y^2}$ and $\psi(y) = \sqrt{1-y^2}$, defined on the interval $[c = -1, d = 1]$. Thus, D is of type 3.

3. Equations of lines that form the boundary of D are $y = x$, $y = 2$ and $y = 2x$. D is of type 1: it is bounded by $\phi(x) = x$ and by

$$\psi(x) = \begin{cases} 2x & \text{if } 0 \leq x \leq 1 \\ 2 & \text{if } 1 \leq x \leq 2 \end{cases}$$

on the interval $[a = 0, b = 2]$. D is also of type 2: take $\phi(y) = y/2$ and $\psi(y) = y$ defined on the interval $c = 0$ and $d = 2$. Thus, D is of type 3.

5. Let $g(x,y) = -f(x,y)$. Then $g(x,y) \geq 0$ and the solid region under g and above D is the mirror image (i.e., symmetric with respect to the xy-plane) of the solid region below D and above $f(x,y)$. Since

$$\iint_D g\, dA = -\iint_D f\, dA,$$

it follows that $\iint_D f\, dA$ is the negative of the volume of the solid region below D and above f.

If $f(x,y) \geq g(x,y)$ and $f(x,y) \geq 0$, $g(x,y) \geq 0$, then

$$\iint_D (f - g)\, dA = \iint_D f\, dA - \iint_D g\, dA$$

is the difference of the volume of the solid under f and above D and the volume of the solid under g and above D. In other words, $\iint_D (f - g)\, dA$ is the volume of the solid region between f and g and inside the vertical "cylinder" whose cross-sections all look like $D \subseteq \mathbb{R}^2$; see Figure 5.

Notice that this interpretation holds no matter whether f and g are positive or not (but as long as $f(x,y) \geq g(x,y)$). Here is a reason: by definition, $(R_{ij}, i,j = 1, \ldots, n$, is a rectangle forming a subdivision and ΔA_{ij} is its area)

$$\iint_D (f - g)\, dA = \lim_{n \to \infty} \sum_{i=1}^{n} \sum_{j=1}^{n} (f(x_i^*, y_j^*) - g(x_i^*, y_j^*)) \Delta A_{ij}.$$

Since $f(x,y) \geq g(x,y)$ (no additional assumptions!), $f(x_i^*, y_j^*) - g(x_i^*, y_j^*) \geq 0$, and

$$(f(x_i^*, y_j^*) - g(x_i^*, y_j^*))\Delta A_{ij}$$

represents the volume of the rectangular box that approximates the volume of the solid between f and g and above R_{ij}. Adding up all such rectanglular boxes and taking the limit as R_{ij}'s shrink to a point, we get the volume of the solid between f and g defined over $D \subseteq \mathbb{R}^2$.

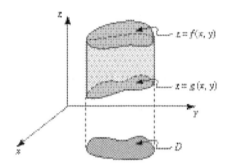

Figure 5

7. Place the cone so that its base lies in the xy-plane and its axis of rotation coincides with the z-axis. At a height z, $0 \leq z \leq h$, the cross-section is a circle of radius $\rho = r(h-z)/h$; see Figure 7. The radius was determined from the similar triangles ratio $h/r = (h-z)/\rho$. It follows that the cross-sectional area at height z is $\pi\rho^2 = \pi r^2(h-z)^2/h^2$ and the volume of the cone is equal to the "sum" of all such cross-sections from $z = 0$ (where the base of the cone is) to $z = h$ (where its vertex is); i.e.,

$$v(V) = \int_0^h \pi \frac{r^2(h-z)^2}{h^2}\,dz = \frac{\pi r^2}{h^2}\int_0^h (h-z)^2\,dz$$

$$= \frac{\pi r^2}{h^2}\left(-\tfrac{1}{3}\right)(h-z)^3\Big|_0^h = \frac{\pi r^2}{h^2}\frac{h^3}{3} = \tfrac{1}{3}\pi r^2 h.$$

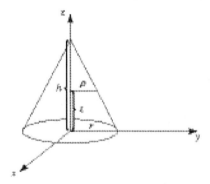

Figure 7

9. Consider cross-sections parallel to the yz-plane. At x, $0 \leq x \leq 2$, the cross-section is a rectangle of length 1 and height $e^{-x+2} - 1$. The volume of V is the "sum" of all such rectangles as x takes on values between 0 and 2; so

$$v(V) = \int_0^2 \left(e^{-x+2} - 1\right) dx = \left(-e^{-x+2} - x\right)\Big|_0^2 = e^2 - 3.$$

11. The region D can be broken into two regions, D_1 and D_2: D_1 is bounded by $x = 0$, $x = 4$, $y = -\sqrt{x}$ and $y = \sqrt{x}$; D_2 is bounded by $x = 4$, $x = 9$, $y = -x + 6$ and $y = \sqrt{x}$. Thus,

$$\iint_D 2y\,dA = \iint_{D_1} 2y\,dA + \iint_{D_2} 2y\,dA = \int_0^4 \left(\int_{-\sqrt{x}}^{\sqrt{x}} 2y\,dy\right) dx + \int_4^9 \left(\int_{x-6}^{\sqrt{x}} 2y\,dy\right) dx$$

$$= \int_0^4 \left(y^2\Big|_{-\sqrt{x}}^{\sqrt{x}}\right) dx + \int_4^9 \left(y^2\Big|_{x-6}^{\sqrt{x}} dx\right)$$

$$= 0 + \int_4^9 \left(x - (x-6)^2\right) dx = \frac{x^2}{2} - \frac{(x-6)^3}{3}\Big|_4^9 = \frac{125}{6}.$$

13. Integrating with respect to y first, we get

$$\int_0^2 \left(\int_{2-x}^{x+1} (xe^y - 2y - 1)dy\right) dx = \int_0^2 \left(xe^y - y^2 - y\right)\Big|_{2-x}^{x+1} dx$$

$$= \int_0^2 \left(xe^{x+1} - (x+1)^2 - (x+1)\right) - \left(xe^{2-x} - (2-x)^2 - (2-x)\right) dx$$

$$= \int_0^2 (xe^{x+1} - xe^{2-x} - 8x + 4)\,dx.$$

Using integration by parts with $u = x$ and $dv = e^x dx$, we get

$$\int_0^2 xe^{x+1}\,dx = e\int_0^2 xe^x\,dx = e(x-1)e^x\Big|_0^2 = e^3 + e.$$

Similarly (by integration by parts with $u = x$ and $dv = e^{-x}dx$),

$$\int_0^2 xe^{2-x}\,dx = e^2\int_0^2 xe^{-x}\,dx = e^2(-x-1)e^{-x}\Big|_0^2 = e^2 - 3.$$

Since

$$\int_0^2 (-8x+4)\,dx = (-4x^2 + 4x)\Big|_0^2 = -8,$$

it follows that the value of the integral is $e^3 + e - (e^2 - 3) - 8 = e^3 - e^2 + e - 5 \approx 10.4148$.

15. Integrating with respect to ρ first, we get

$$\int_0^\pi \left(\int_0^{\cos\theta} \rho^2 \sin\theta d\rho\right) d\theta = \int_0^\pi \left(\tfrac{1}{3}\rho^3 \sin\theta\right)\Big|_0^{\cos\theta} d\theta$$

$$= \tfrac{1}{3}\int_0^\pi \cos^3\theta \sin\theta\,d\theta$$

$$= \tfrac{1}{3}\left(-\tfrac{1}{4}\cos^4\theta\right)\Big|_0^\pi = -\tfrac{1}{12}(1 - 1) = 0.$$

(To solve $\int \cos^3\theta \sin\theta d\theta$, use the substitution $u = \cos\theta$, $du = -\sin\theta d\theta$.)

17. Integrating with respect to x first, we get

$$\int_0^{\pi/2} \left(\int_0^{\sin y} x \cos y \, dx \right) dy = \int_0^{\pi/2} \left(\tfrac{1}{2} x^2 \cos y \right) \Big|_0^{\sin y} dy$$

$$= \tfrac{1}{2} \int_0^{\pi/2} \sin^2 y \cos y \, dy = \tfrac{1}{6} \sin^3 y \Big|_0^{\pi/2} = \tfrac{1}{6}.$$

(To solve $\int \sin^2 y \cos y \, dy$, use the substitution $u = \sin y$, $du = \cos y \, dy$.)

19. The given integral can be expressed as the iterated integral

$$\iint_D f \, dA = \int_3^4 \left(\int_0^2 (xy^{-1} - x^2 y^2) dx \right) dy$$

$$= \int_3^4 \left(\tfrac{1}{2} x^2 y^{-1} - \tfrac{1}{3} x^3 y^2 \right) \Big|_0^2 dy = \int_3^4 \left(2y^{-1} - \tfrac{8}{3} y^2 \right) dy$$

$$= \left(2 \ln y - \tfrac{8}{9} y^3 \right) \Big|_3^4 = \left(2 \ln 4 - \tfrac{512}{9} \right) - \left(2 \ln 3 - 24 \right)$$

$$= 2 \ln(4/3) - \tfrac{296}{9} \approx -32.3135.$$

21. The given integral can be evalauted as the iterated integral

$$\iint_D f \, dA = \int_0^1 \left(\int_{-y}^{1+y} (2xy - y) dx \right) dy$$

$$= \int_0^1 \left(x^2 y - xy \right) \Big|_{-y}^{1+y} dy$$

$$= \int_0^1 \left[\left((1+y)^2 y - (1+y)y \right) - \left(y^3 - (-y)y \right) \right] dy$$

$$= \int_0^1 0 \, dy = 0.$$

23. Combining $y = x^2$ and $y = 4 - x^2$ we get $x^2 = 4 - x^2$ or $x^2 = 2$ and $x = \pm\sqrt{2}$. Therefore, the given integral can be evalauted as the iterated integral

$$\iint_D f \, dA = \int_{-\sqrt{2}}^{\sqrt{2}} \left(\int_{x^2}^{4-x^2} x^{-2/3} dy \right) dx$$

$$= \int_{-\sqrt{2}}^{\sqrt{2}} x^{-2/3} y \Big|_{x^2}^{4-x^2} dx = \int_{-\sqrt{2}}^{\sqrt{2}} \left(4x^{-2/3} - 2x^{4/3} \right) dx$$

$$= \left(12x^{1/3} - \tfrac{6}{7} x^{7/3} \right) \Big|_{-\sqrt{2}}^{\sqrt{2}} = 24 \, (2)^{1/6} - \tfrac{12}{7} \, (2)^{7/6} \approx 23.0906.$$

25. Let $f(x, y) = e^{-x-y}$. Since $-1 \leq x \leq 1$ and $0 \leq y \leq 2$, it follows that $-1 \leq -x \leq 1$ and $-2 \leq -y \leq 0$ and so $-3 \leq -x - y \leq 1$. The fact that e^a is an increasing function of a implies that $e^{-3} \leq e^{-x-y} \leq e^1$. Consequently,

$$\iint_D e^{-3} \, dA \leq \iint_D e^{-x-y} \, dA \leq \iint_D e \, dA.$$

The area of D is 4, and so $\iint_D e^{-3} \, dA = 4e^{-3}$ and $\iint_D e \, dA = 4e$ and

$$4e^{-3} \le \iint_D e^{-x-y} \, dA \le 4e.$$

27. The plane can be viewed as the graph of the function $z = 6 - x - y/2$. For points (x, y) in the rectangle $R = [-1, 1] \times [0, 2]$, z is positive, and so the volume is equal to the double integral $\iint_R (6 - x - y/2) \, dA$. Using Fubini's Theorem,

$$\iint_R \left(6 - x - \tfrac{1}{2}y\right) dA = \int_0^2 \left(\int_{-1}^1 \left(6 - x - \tfrac{1}{2}y\right) dx\right) dy$$

$$= \int_0^2 \left(6x - \tfrac{1}{2}x^2 - \tfrac{1}{2}xy\right)\Big|_{-1}^1 \, dy = \int_0^2 (12 - y) \, dy$$

$$= \left(12y - \tfrac{1}{2}y^2\right)\Big|_0^2 = 24 - 2 = 22.$$

29. Write the equations of the given planes as $z = 1 - x - y$ and $z = (1 - x - y)/2$. The two planes intersect when $1 - x - y = (1 - x - y)/2$; i.e., when $x + y = 1$. It follows that the solid in question is bounded from above by $z = 1 - x - y$, from below by $z = (1 - x - y)/2$, and is located over the triangular region D in the xy-plane bounded by the coordinate axes and the line $x + y = 1$. Its volume is

$$v = \iint_D \left(1 - x - y - \tfrac{1}{2}(1 - x - y)\right) dA = \tfrac{1}{2} \int_0^1 \left(\int_0^{1-y} (1 - x - y) dx\right) dy$$

$$= \tfrac{1}{2} \int_0^1 \left(x - \tfrac{1}{2}x^2 - xy\right)\Big|_0^{1-y} \, dy = \tfrac{1}{2} \int_0^1 \left(\tfrac{1}{2} - y + \tfrac{1}{2}y^2\right) dy$$

$$= \tfrac{1}{2} \left(\tfrac{1}{2}y - \tfrac{1}{2}y^2 + \tfrac{1}{6}y^3\right)\Big|_0^1 = \tfrac{1}{12}.$$

31. Let $h = f - g$; the assumption can be written as $\iint_D h \, dA = 0$, and so the average value of h over D is

$$\bar{h} = \frac{1}{\text{area}(D)} \iint_D h \, dA = 0.$$

By Theorem 7 and the remark following (6.2), it follows that there must be a point (x_0, y_0) in D where $h(x_0, y_0) = 0$. Since $h = f - g$, it follows that $f(x_0, y_0) = g(x_0, y_0)$.

33. The equation of the ellipse with semi-axes a and b is $x^2/a^2 + y^2/b^2 = 1$, or $y^2 = b^2(1 - x^2/a^2)$. We will compute the area in the first quadrant and then multiply by 4. Let D be the region in the first quadrant bounded by $y = b\sqrt{1 - x^2/a^2}$. Its area $A(D)$ can be computed as the double integral

$$A(D) = \iint_D 1 \, dA = \int_0^1 \left(\int_0^{b\sqrt{1-x^2/a^2}} 1 \, dy\right) dx$$

$$= b \int_0^a \sqrt{1 - x^2/a^2} \, dx = \frac{b}{a} \int_0^a \sqrt{a^2 - x^2} \, dx.$$

We proceed using the formula for $\int \sqrt{a^2 - x^2}dx$ from integration tables (one could use the trigonometric substitution $x = a \sin t$ to compute this integral):

$$A(D) = \frac{b}{a} \left(\frac{x}{2}\sqrt{a^2 - x^2} + \frac{a^2}{2} \arcsin(x/a) \right) \bigg|_0^a$$

$$= \frac{b}{a} \frac{a^2}{2} (\arcsin 1 - \arcsin 0) = \frac{ab}{2} \frac{\pi}{2} = \frac{ab\pi}{4}.$$

It follows that the area of the given ellipse in $ab\pi$.

35. Combining $y = x^2$ and $y = 4 - x^2$ we get $x^2 = 2$ and $x = \pm\sqrt{2}$. So D is the region in the xy-plane bounded by the y-axis on the left, by $y = 4 - x^2$ from above and by $y = x^2$ from below. Its area is

$$A(D) = \iint_D 1 \, dA = \int_0^{\sqrt{2}} \left(\int_{x^2}^{4-x^2} 1 dy \right) dx$$

$$= \int_0^{\sqrt{2}} (4 - 2x^2) \, dx$$

$$= \left(4x - \tfrac{2}{3}x^3 \right) \bigg|_0^{\sqrt{2}} = 4\sqrt{2} - \tfrac{2}{3}(\sqrt{2})^3 = \tfrac{8}{3}\sqrt{2}.$$

37. The average value of f on R is

$$\overline{f} = \frac{1}{A(R)} \iint_R k(x^2 + y^2) \, dA$$

$$= \tfrac{1}{2}k \int_0^2 \left(\int_0^1 (x^2 + y^2) dx \right) dy = \tfrac{1}{2}k \int_0^2 \left(\tfrac{1}{3}x^3 + xy^2 \right) \bigg|_0^1 dy$$

$$= \tfrac{1}{2}k \int_0^2 \left(\tfrac{1}{3} + y^2 \right) dy = \tfrac{1}{2}k \left(\tfrac{1}{3}y + \tfrac{1}{3}y^3 \right) \bigg|_0^2 = \tfrac{1}{2}k \tfrac{10}{3} = \tfrac{5}{3}k.$$

All points where $f(x, y) = 5k/3$ must satisfy $k(x^2 + y^2) = 5k/3$, or $x^2 + y^2 = 5/3$. Therefore, all points in R that also lie on the circle centered at the origin of radius $\sqrt{5/3}$ satisfy the conclusion of the Mean Value Theorem.

39. (a) By the Extreme Value Theorem, a continuous function defined on a closed interval has a maximum and a minimum. Thus, there exist x_m and x_M in $[a, b]$ such that $f(x_m) \leq f(x)$ and $f(x) \leq f(x_M)$ for all $x \in [a, b]$. Leeting $m = f(x_m)$ and $M = f(x_M)$, we get that $m \leq f(x) \leq M$ for all x in $[a, b]$.

(b) Integrate the inequality in (a) from a to b, to get

$$\int_a^b m \, dx \leq \int_a^b f(x) \, dx \leq \int_a^b M \, dx,$$

i.e., since m and M are constants,

$$m(b - a) \leq \int_a^b f(x) \, dx \leq M(b - a).$$

Dividing all sides by $b - a$ we obtain the desired inequality.

(c) Pick any value N such that $m \le N \le M$. The Intermediate Value Theorem (f is continuous on $[a,b]$) implies that there exists $x_0 \in [a,b]$ such that $f(x_0) = N$. In particular, take $N = \frac{1}{b-a}\int_a^b f(x)\,dx$. Thus, there is an x_0 in $[a,b]$ where $f(x_0) = \frac{1}{b-a}\int_a^b f(x)\,dx$.

(d) There is a point in the interval $[a,b]$ where f attains its average value.

6.3. Examples and Techniques of Evaluation of Double Integrals

1. The given integral can be evaluated as the iterated integral

$$\iint_D f\,dA = \int_0^1 \left(\int_0^{y/2} ye^x\,dx\right)dy$$

$$= \int_0^1 ye^x\Big|_0^{y/2}\,dy = \int_0^1 \left(ye^{y/2} - y\right)dy$$

$$= \left((2y-4)e^{y/2} - \tfrac{1}{2}y^2\right)\Big|_0^1 = \left(-2e^{1/2} - \tfrac{1}{2}\right) - (-4) = \tfrac{7}{2} - 2e^{1/2}.$$

We used the formula $\int ye^{y/2}dy = (2y-4)e^{y/2}$, that can be found in tables of integrals; alternatively, this integral can be computed using integration by parts with $u = y$ and $dv = e^{y/2}dy$.

3. Since $y = 2x$ and $y = 5x$ intersect at the origin, D is the region in the first quadrant: it is bounded from above by the line $y = 5x$, from below by the line $y = 2x$, and is located over the interval $[0,2]$. Hence

$$\iint_D f\,dA = \int_0^2 \left(\int_{2x}^{5x} y^2\,dy\right)dx$$

$$= \int_0^1 \tfrac{1}{3}y^3\Big|_{2x}^{5x}\,dx = \tfrac{1}{3}\int_0^2 117x^3\,dx = \tfrac{39}{4}x^4\Big|_0^2 = 156.$$

5. The solid in question is the solid below $z = \sqrt{4-y^2}$ and above the triangle D in the xy-plane that is bounded by the lines $x = 0$, $y = 0$ and $x + y = 2$. Its volume is

$$v = \iint_D \sqrt{4-y^2}\,dA = \int_0^2 \left(\int_0^{2-y} \sqrt{4-y^2}dx\right)dy$$

$$= \int_0^2 x\sqrt{4-y^2}\Big|_0^{2-y}\,dy = \int_0^2 (2-y)\sqrt{4-y^2}\,dy$$

$$= \int_0^2 2\sqrt{4-y^2}\,dy - \int_0^2 y\sqrt{4-y^2}\,dy$$

From a table of integrals (or using the trigonometric substitution $y = 2\sin\theta$), we get

$$\int_0^2 \sqrt{4-y^2}\,dy = \left(\tfrac{y}{2}\sqrt{4-y^2} + 2\arcsin(y/2)\right)\Big|_0^2 = 2\arcsin 1 = \pi.$$

Using the substitution $u = 4 - y^2$, we get

$$\int_0^2 y\sqrt{4-y^2}\,dy = -\tfrac{1}{3}(4-y^2)^{3/2}\Big|_0^2 = \tfrac{1}{3}4^{3/2} = \tfrac{8}{3}.$$

Therefore, the volume of the solid is $2\pi - 8/3 \approx 3.6165$.

7. The plane $x + y + z = 4$ intersects the xy-plane (i.e., $z = 0$) along the line $x + y = 4$. Combining $x + y = 4$ and $y = 3x$ we get $x = 1$ and $y = 3$. It follows that the solid in question (call it W) is below the plane $z = 4 - x - y$, and above the triangular region D in the xy-plane defined by $x = 0$, $y = 3x$ and $x + y = 4$; see Figure 7. The volume of W is

$$v(W) = \iint_D (4 - x - y)\, dA = \int_0^1 \left(\int_{3x}^{4-x} (4 - x - y)dy \right) dx$$

$$= \int_0^1 \left(4y - xy - \tfrac{1}{2}y^2 \right) \Big|_{3x}^{4-x} dx = \int_0^1 (8x^2 - 16x + 8)\, dx = 8 \int_0^1 (x - 1)^2 \, dx$$

$$= \tfrac{8}{3}(x - 1)^3 \Big|_0^1 = \tfrac{8}{3}.$$

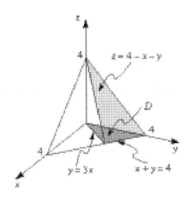

Figure 7

9. The given triangle (call it D) is bounded by $y = 1$, $y = x + 1$ and $x = 1$. The volume of the solid is

$$v = \iint_D xy\, dA = \int_0^1 \left(\int_1^{x+1} xy\, dy \right) dx$$

$$= \tfrac{1}{2} \int_0^1 xy^2 \Big|_1^{x+1} dx = \tfrac{1}{2} \int_0^1 (x^3 + 2x^2)\, dx$$

$$= \tfrac{1}{2} \left(\tfrac{1}{4}x^4 + \tfrac{2}{3}x^3 \right) \Big|_0^1 = \tfrac{11}{24}.$$

11. By separation of variables,

$$\int_0^1 \left(\int_0^2 (1 - x - y + xy)dx \right) dy = \int_0^1 \left(\int_0^2 (1 - x)(1 - y)dx \right) dy$$

$$= \left(\int_0^1 (1 - y)\, dy \right) \left(\int_0^2 (1 - x)\, dx \right) = \left(\left(y - \tfrac{1}{2}y^2 \right) \Big|_0^1 \right) \left(\left(x - \tfrac{1}{2}x^2 \right) \Big|_0^2 \right) = \tfrac{1}{2}(0) = 0.$$

Direct computation gives

$$\int_0^1 \left(\int_0^2 (1 - x - y + xy)dx \right) dy = \int_0^1 \left(x - \tfrac{1}{2}x^2 - xy + \tfrac{1}{2}x^2 y \right) \Big|_0^2 dy$$

$$= \int_0^1 (-2y + 2y)\, dy = 0.$$

13. The given integral is equal to $\iint_D (y^2 - x)\, dA$, where the region of integration D is given by $0 \le x \le 1$ and $\arctan x \le y \le \pi/4$; see Figure 13 (recall that $\arctan(1) = \pi/4$). Reversing the order of integration, we get (from $y = \arctan x$ it follows that $x = \tan y$)

$$\iint_D (y^2 - x)\, dA = \int_0^{\pi/4} \left(\int_0^{\tan y} (y^2 - x) dx \right) dy.$$

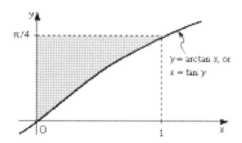

Figure 13

15. The given integral is equal to $\iint_D (\ln x / x)\, dA$, where the region of integration D is given by $1 \le y \le 2$ and $1 \le x \le 2y$; see Figure 15. The integral in the reversed order of integration has to be broken up into two integrals:

$$\iint_D \frac{\ln x}{x}\, dA = \int_1^2 \left(\int_1^2 \frac{\ln x}{x} dy \right) dx + \int_2^4 \left(\int_{x/2}^2 \frac{\ln x}{x} dy \right) dx.$$

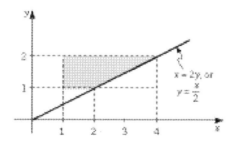

Figure 15

17. The integral in question is equal to the double integral $\iint_D x\, dA$, where the D is the region in the xy-plane defined by $0 \le y \le 1$ and $0 \le x \le \arccos y$; in other words (recall that $x = \arccos y$ is the same curve as $y = \cos x$), D is the region under $y = \cos x$ on $[0, \pi/2]$. Reversing the order of integration, we get

$$\iint_D x\, dA = \int_0^{\pi/2} \left(\int_0^{\cos x} x\, dy \right) dx = \int_0^{\pi/2} \left(xy \Big|_0^{\cos x} \right) dx$$

$$= \int_0^{\pi/2} x \cos x\, dx = (x \sin x + \cos x) \Big|_0^{\pi/2} = \frac{\pi}{2} - 1.$$

The integral $\int x \cos x\, dx$ is solved using integration by parts, with $u = x$ and $dv = \cos x\, dx$.

19. The integral in question is equal to the double integral $\iint_D x \cos(2y^2)\, dA$, where D is the region in the xy-plane defined by $0 \le x \le 3$ and $x^2 \le y \le 9$; in other words, it is the region below the horizontal line $y = 9$ and above the parabola $y = x^2$, over the interval $[0, 3]$ on the x-axis. Reversing the order of integration, we get

$$\iint_D x \cos(2y^2)\, dA = \int_0^9 \left(\int_0^{\sqrt{y}} x \cos(2y^2)\, dx \right) dy$$

$$= \int_0^9 \tfrac{1}{2} x^2 \cos(2y^2) \Big|_0^{\sqrt{y}}\, dy = \tfrac{1}{2} \int_0^9 y \cos(2y^2)\, dy$$

$$= \tfrac{1}{2}\tfrac{1}{4} \sin(2y^2) \Big|_0^9 = \tfrac{1}{8} \sin 162 \approx -0.1223.$$

The integral $\int y \cos(2y^2)\, dy$ is solved using the substitution $u = 2y^2$.

21. The given integral is equal to the double integral $\iint_D 5\, dA$, where D consists of regions $1 \le y \le 2$ and $1 \le x \le \sqrt{y}$ and $2 \le y \le 4$ and $y/2 \le x \le \sqrt{y}$; see Figure 21. In other words, D is the region over the interval $[1, 2]$ on the x-axis bounded from above by the line $y = 2x$ and from below by the parabola $y = x^2$. Hence

$$\iint_D 5\, dA = \int_1^2 \left(\int_{x^2}^{2x} 5\, dy \right) dx$$

$$= 5 \int_1^2 \left(y \big|_{x^2}^{2x} \right) dx = 5 \int_1^2 (2x - x^2)\, dx$$

$$= 5 \left(x^2 - \tfrac{1}{3} x^3 \right) \Big|_1^2 = 5 \left(\tfrac{4}{3} - \tfrac{2}{3} \right) = \tfrac{10}{3}.$$

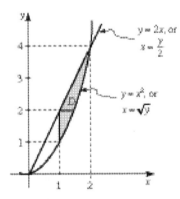

Figure 21

23. The region of integration D is the region below the parabola $y = x^2$ and over the interval $[0, 1]$ on the x-axis. Hence

$$\iint_D e^{x^3}\, dA = \int_0^1 \left(\int_0^{x^2} e^{x^3}\, dy \right) dx$$

$$= \int_0^1 y e^{x^3} \Big|_0^{x^2} dx$$

$$= \int_0^1 x^2 e^{x^3} dx = \tfrac{1}{3} e^{x^3} \Big|_0^1 = \tfrac{1}{3}(e-1).$$

The integral $\int x^2 e^{x^3} dx$ is solved by the substitution $u = x^3$, $du = 3x^2 dx$.

Note that in the reversed order of integration

$$\iint_D e^{x^3} dA = \int_0^1 \left(\int_{\sqrt{y}}^1 e^{x^3} dx \right) dy$$

we obtain the integral $\int e^{x^3} dx$, that cannot be solved as a compact formula (i.e., a numerical method or a power series method is needed).

6.4. Change of Variables in a Double Integral

1. The given integral is equal to the double integral $\iint_D e^{x^2+y^2} dA$, where D is the region defined by $0 \le x \le 2$ and $0 \le y \le \sqrt{4-x^2}$; it is the quarter-disk of radius 2 in the first quadrant. Consequently,

$$\iint_D e^{x^2+y^2} dA = \iint_{D^*} e^{r^2} r \, dA^*,$$

where D^* is defined by $0 \le r \le 2$ and $0 \le \theta \le \pi/2$. Therefore,

$$\iint_D e^{x^2+y^2} dA = \int_0^{\pi/2} \left(\int_0^2 r e^{r^2} dr \right) d\theta$$

$$= \int_0^{\pi/2} \left(\tfrac{1}{2} e^{r^2} \right) \Big|_0^2 d\theta = \tfrac{1}{2}(e^4 - 1) \int_0^{\pi/2} d\theta = \frac{\pi(e^4 - 1)}{4}.$$

3. The given integral is equal to the double integral $\iint_D \arctan(y/x) \, dA$, where D is the region defined by $-1 \le x \le 1$ and $0 \le y \le \sqrt{1-x^2}$; it is the upper half-disk of radius 1. Since $\arctan(y/x) = \theta$, it follows that

$$\iint_D \arctan(y/x) \, dA = \iint_{D^*} \theta r \, dA^*,$$

where D^* is defined by $0 \le r \le 1$ and $0 \le \theta \le \pi$. So

$$\iint_D \arctan(y/x) \, dA = \int_0^\pi \left(\int_0^1 \theta r \, dr \right) d\theta$$

$$= \int_0^\pi \left(\tfrac{1}{2} \theta r^2 \right) \Big|_0^1 d\theta = \tfrac{1}{2} \int_0^\pi \theta \, d\theta = \frac{\pi^2}{4}.$$

5. The disk $x^2 + y^2 \le 1$ is represented in polar coordinates as the region D^* given by $0 \le r \le 1$ and $0 \le \theta \le 2\pi$. It follows that

$$\iint_D \sqrt{2x^2 + 2y^2 + 3} \, dA = \iint_{D^*} r\sqrt{2r^2 + 3} \, dA^*$$

$$= \int_0^{2\pi} \left(\int_0^1 r\sqrt{2r^2 + 3} \, dr \right) d\theta$$

$$= \int_0^{2\pi} \left(\tfrac{1}{6}(2r^2 + 3)^{3/2} \right) \Big|_0^1 d\theta$$

$$= \tfrac{1}{6}(5^{3/2} - 3^{3/2}) \int_0^{2\pi} d\theta = \frac{\pi}{3}(5^{3/2} - 3^{3/2}).$$

The integral $\int r\sqrt{2r^2 + 3}\, dr$ is solved by the change of variables $u = 2r^2 + 3$.

7. Since the solid is inside the cylinder $x^2 + y^2 = 5$, the region of integration $D \subseteq \mathbb{R}^2$ is the disk $x^2 + y^2 \leq 5$; so the volume of the solid in question is $v = \iint_D (x^2 + y^2)\, dA$. Passing to polar coordinates we get

$$v = \iint_D (x^2 + y^2)\, dA = \iint_{D^*} r^2 \, r \, dA^*,$$

where D^* is defined by $0 \leq r \leq \sqrt{5}$ and $0 \leq \theta \leq 2\pi$. It follows that

$$v = \int_0^{2\pi} \left(\int_0^{\sqrt{5}} r^3 dr \right) d\theta = \left(\int_0^{2\pi} d\theta \right) \left(\int_0^{\sqrt{5}} r^3 \, dr \right) = 2\pi \left(\tfrac{1}{4}r^4 \right) \Big|_0^{\sqrt{5}} = \tfrac{25}{2}\pi.$$

9. The solid in question is bounded from above by the ellipsoid $z = \sqrt{64 - 4x^2 - 4y^2}$, on the sides by the cylinder $x^2 + y^2 = 4$ (hence the region D of integration is the disk $x^2 + y^2 \leq 4$ in the xy-plane), and from below by the xy-plane. Its volume is equal to $v = \iint_D \sqrt{64 - 4x^2 - 4y^2}\, dA$. Passing to polar coordinates, we get

$$v = \iint_D \sqrt{64 - 4x^2 - 4y^2}\, dA = \iint_{D^*} \sqrt{64 - 4r^2} \, r \, dA^*,$$

where D^* is defined by $0 \leq r \leq 2$ and $0 \leq \theta \leq 2\pi$. It follows that

$$v = \int_0^{2\pi} \left(\int_0^2 2r\sqrt{16 - r^2}\, dr \right) d\theta$$

$$= 2 \left(\int_0^{2\pi} d\theta \right) \left(\int_0^2 r\sqrt{16 - r^2}\, dr \right)$$

$$= 4\pi \left(-\tfrac{1}{3}(16 - r^2)^{3/2} \right) \Big|_0^2 = -\frac{4\pi}{3}(12^{3/2} - 16^{3/2}) \approx 93.9578.$$

11. Assume that T is a function of two variables (the proof in the case of n variables is analogous). Let

$$A = \begin{bmatrix} a_{11} & a_{12} \\ a_{21} & a_{22} \end{bmatrix} \quad \text{and} \quad \mathbf{b} = \begin{bmatrix} b_1 \\ b_2 \end{bmatrix}.$$

Then

$$T(x, y) = \begin{bmatrix} a_{11} & a_{12} \\ a_{21} & a_{22} \end{bmatrix} \begin{bmatrix} x \\ y \end{bmatrix} + \begin{bmatrix} b_1 \\ b_2 \end{bmatrix};$$

i.e., $T(x, y) = (a_{11}x + a_{12}y + b_1, a_{21}x + a_{22}y + b_2)$. The derivative of T is

$$DT = \begin{bmatrix} a_{11} & a_{12} \\ a_{21} & a_{22} \end{bmatrix} = A,$$

and therefore the Jacobian of T (which is equal to $\det(DT)$), is equal to $\det(A)$.

13. The area of the given rectangle (call it R^*) and the area of its image R are related by

$$A(R) \approx \left| \det(DT)_{at\ (2,4)} \right| A(R^*)$$

$$= \left| \det \begin{bmatrix} u(u^2 + v^2)^{-1/2} & v(u^2 + v^2)^{-1/2} \\ v & u \end{bmatrix}_{at\ (2,4)} \right| \cdot 0.005$$

$$= \left| u^2(u^2 + v^2)^{-1/2} - v^2(u^2 + v^2)^{-1/2} \right|_{at\ (2,4)} \cdot 0.005$$

$$= \left| 20^{-1/2}(-12) \right| \cdot 0.005 \approx 0.01342.$$

15. The map T is an expansion by the factor a in the u-direction and a translation by b in the v-direction. Since translation does not change areas, it follows that the area of the image of D is equal to $a \cdot \text{area}(D)$.

17. Passing to polar coordinates, we get

$$\iint_D (5x + y^2 + x^2)\, dA = \iint_{D^*} (5r \cos\theta + r^2) r\, dA^*,$$

where D^* is the region in the $r\theta$-plane defined by $1 \leq r \leq 2$ and $0 \leq \theta \leq \pi$. So

$$\iint_D (5x + y^2 + x^2)\, dA = \int_0^\pi \left(\int_1^2 (5r^2 \cos\theta + r^3)dr \right) d\theta$$

$$= \int_0^\pi \left(\tfrac{5}{3} \cos\theta\, r^3 + \tfrac{1}{4} r^4 \right) \Big|_1^2 d\theta = \int_0^\pi \left(\tfrac{35}{3} \cos\theta + \tfrac{15}{4} \right) d\theta$$

$$= \left(\tfrac{35}{3} \sin\theta + \tfrac{15}{4}\theta \right) \Big|_0^\pi = \frac{15\pi}{4}.$$

19. Let D denote the region inside $r = 1 - \sin\theta$. Its area is given by

$$A(D) = \iint_D dA = \iint_{D^*} r\, dA^*,$$

where D^* is the region in the $r\theta$-plane defined by $0 \leq \theta \leq 2\pi$ and $0 \leq r \leq 1 - \sin\theta$. So

$$A(D) = \int_0^{2\pi} \left(\int_0^{1-\sin\theta} r\, dr \right) d\theta$$

$$= \int_0^{2\pi} \left(\tfrac{1}{2} r^2 \right) \Big|_0^{1-\sin\theta} d\theta = \tfrac{1}{2} \int_0^{2\pi} (1 - \sin\theta)^2\, d\theta$$

$$= \tfrac{1}{2} \int_0^{2\pi} (1 - 2\sin\theta + \sin^2\theta)\, d\theta$$

(using $\sin^2\theta = (1 - \cos 2\theta)/2$)

$$= \tfrac{1}{2} \int_0^{2\pi} \left(\tfrac{3}{2} - 2\sin\theta - \tfrac{1}{2}\cos 2\theta \right) d\theta$$

$$= \tfrac{1}{2} \left(\tfrac{3}{2}\theta + 2\cos\theta - \tfrac{1}{4}\sin 2\theta \right) \Big|_0^{2\pi} = \frac{3\pi}{2}.$$

21. It is a good idea to sketch the region. Let us consider $r = \theta$ first. When $\theta = 0$, $r = 0$. As θ increases, the distance r from a point on the curve to the origin increases at the same rate as θ; plot a few points, for example $(\pi/6, \pi/6)$, $(\pi/4, \pi/4)$, $(\pi/3, \pi/3)$, $(\pi/2, \pi/2)$, etc. Such a curve is called a spiral. The curve $r = 2\theta$ is another spiral: this time, the distance r increases twice as fast as the angle, and the spiral "opens up" faster; see Figure 21. The area of the region D in the first quadrant bounded by $r = \theta$ and $r = 2\theta$ is

$$A(D) = \iint_D dA = \iint_{D^*} r \, dA^*,$$

where D^* is the region in the $r\theta$-plane defined by $0 \leq \theta \leq \pi/2$ and $\theta \leq r \leq 2\theta$. So

$$A(D) = \int_0^{\pi/2} \left(\int_\theta^{2\theta} r \, dr \right) d\theta$$

$$= \int_0^{\pi/2} \left(\tfrac{1}{2} r^2 \right) \Big|_\theta^{2\theta} d\theta = \tfrac{3}{2} \int_0^{\pi/2} \theta^2 \, d\theta = \tfrac{3}{2} \tfrac{1}{3} \theta^3 \Big|_0^{\pi/2} = \tfrac{1}{2} \left(\tfrac{\pi}{2} \right)^3.$$

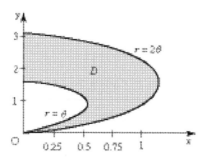

Figure 21

23. The change of variables is given by $T(u, v) = (u + v, -u + v)$. T is a linear map, and therefore the corresponding matrix $A = \begin{bmatrix} 1 & 1 \\ -1 & 1 \end{bmatrix}$ is equal to DT. In any case, the Jacobian of the change of variables T is

$$\frac{\partial(x, y)}{\partial(u, v)} = \det(A) = 2$$

and so (since $x^2 - y^2 = (u + v)^2 - (-u + v)^2 = 4uv$)

$$\iint_D (x^2 - y^2) \, dA = \iint_{D^*} 4uv \, |2| \, dA^*,$$

where D^* is the region that maps to D under T. From $y = x - 1$ it follows that $-u + v = u + v - 1$; i.e., $2u = 1$ or $u = 1/2$. Similarly, from $y = x + 1$ it follows that $-u + v = u + v + 1$; i.e., $2u = -1$ or $u = -1/2$. Finally, $xy = 1$ implies $(-u + v)(u + v) = 1$, i.e., $v^2 - u^2 = 1$; see Figure 23. It follows that (from $v^2 - u^2 = 1$ we get $v = \pm\sqrt{1 + u^2}$)

$$\iint_D (x^2 - y^2) \, dA = 8 \iint_{D^*} uv \, dA^* = 8 \int_{-1/2}^{1/2} \left(\int_{-\sqrt{1+u^2}}^{\sqrt{1+u^2}} uv \, dv \right) du$$

$$= 8 \int_{-1/2}^{1/2} \tfrac{1}{2} uv^2 \Big|_{-\sqrt{1+u^2}}^{\sqrt{1+u^2}} du$$

$$= 4 \int_{-1/2}^{1/2} \left(u(1 + u^2) - u(1 + u^2) \right) du = 0.$$

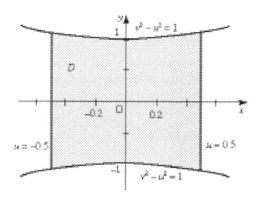

Figure 23

25. See Figure 25. The volume of the wedge is $v = \iint_D (y + 3)\, dA$, where D is the disk $x^2 + y^2 \leq 9$. Changing to polar coordinates, we get

$$v = \iint_D (y + 3)\, dA = \iint_{D^*} (r\sin\theta + 3)r\, dA^*,$$

where D^* is the region in the $r\theta$-plane defined by $0 \leq \theta \leq 2\pi$ and $0 \leq r \leq 3$. So

$$v = \int_0^{2\pi} \left(\int_0^3 (r^2 \sin\theta + 3r)dr \right) d\theta = \int_0^{2\pi} \left(\tfrac{1}{3}r^3 \sin\theta + \tfrac{3}{2}r^2 \right) \Big|_0^3 d\theta$$

$$= \int_0^{2\pi} \left(9\sin\theta + \tfrac{27}{2} \right) d\theta = \left(-9\cos\theta + \tfrac{27}{2}\theta \right) \Big|_0^{2\pi} = 27\pi.$$

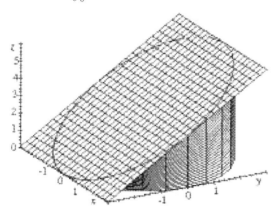

Figure 25

27. Represent the change of variables as the map $T(u, v) = (v, u/v)$. Then

$$DT(u, v) = \begin{bmatrix} 0 & 1 \\ 1/v & -u/v^2 \end{bmatrix},$$

and the Jacobian is

$$\frac{\partial(x, y)}{\partial(u, v)} = \det(DT) = -\frac{1}{v}.$$

From $xy^3 = u^3/v^2$ it follows that

$$\iint_D xy^3\, dA = \iint_{D^*} \frac{u^3}{v^2} \left| -\frac{1}{v} \right| dA^*,$$

where D^* is the region that maps to D under T. From $x = 1$ it follows that $v = 1$; from $x = 2$ it follows that $v = 2$. Replacing x and y in $xy = 1$ by u and v, we get $v(u/v) = 1$; i.e., $u = 1$. Similarly, $xy = 3$ implies that $u = 3$. In words, D^* is the rectangle $[1,3] \times [1,2]$ in the uv-plane. Therefore (since $v > 0$, replace $|v|$ by v)

$$\iint_D xy^3 \, dA = \int_1^2 \left(\int_1^3 \frac{u^3}{v^3} du \right) dv = \tfrac{1}{4} \int_1^2 \frac{1}{v^3} u^4 \Big|_1^3 \, dv$$

$$= 20 \int_1^2 \frac{1}{v^3} \, dv = 20 \left(-\tfrac{1}{2} v^{-2} \right) \Big|_1^2 = \tfrac{15}{2}.$$

29. Let $u = 3x - 2y$ and $v = x + y$. Solving for x and y we get

$$x = \tfrac{1}{5} u + \tfrac{2}{5} v \qquad \text{and} \qquad y = -\tfrac{1}{5} u + \tfrac{3}{5} v.$$

So the change of variables can be represented as a linear map

$$T(u, v) = \left(\tfrac{1}{5} u + \tfrac{2}{5} v, -\tfrac{1}{5} u + \tfrac{3}{5} v \right).$$

It follows that

$$DT = \begin{bmatrix} 1/5 & 2/5 \\ -1/5 & 3/5 \end{bmatrix}$$

and so

$$\frac{\partial(x, y)}{\partial(u, v)} = \det(DT) = \tfrac{3}{25} + \tfrac{2}{25} = \tfrac{1}{5}.$$

By the Change of Variables Theorem,

$$\iint_D 5(x + y) \, dA = \iint_{D^*} 5v \tfrac{1}{5} \, dA^*,$$

where D^* is the region that maps to D under T. The new variables u and v were chosen so that D^* is simple. Indeed, $3x - 2y = 5$ implies that $u = 5$ and $3x - 2y = -2$ gives $u = -2$. The remaining two equations imply $v = -2$ and $v = 1$. In other words, D^* is the rectangle $[-2, 5] \times [-2, 1]$ in the uv-plane. Consequently,

$$\iint_D 5(x + y) \, dA = \int_{-2}^5 \left(\int_{-2}^1 v \, dv \right) du = \int_{-2}^5 \tfrac{1}{2} v^2 \Big|_{-2}^1 \, du$$

$$= -\tfrac{3}{2} \int_{-2}^5 du = -\tfrac{3}{2} (7) = -\tfrac{21}{2}.$$

31. Let $u = x + y$ and $v = y/x$. The corresponding region D^* in the uv-plane is given by $u = 0$, $u = 2$, $v = 1$ and $v = 2$; i.e., it is the rectangle $[0, 2] \times [1, 2]$. Solving $u = x + y$ and $v = y/x$ for x and y, we get (for example, substitute $y = u - x$ into $v = y/x$ and solve for x)

$$x = \frac{u}{1 + v} \qquad \text{and} \qquad y = \frac{uv}{1 + v}.$$

Therefore, the change of variables is

$$T(u, v) = \left(\frac{u}{1 + v}, \frac{uv}{1 + v} \right);$$

its derivative is the matrix

$$DT = \begin{bmatrix} 1/(1 + v) & -u/(1 + v)^2 \\ v/(1 + v) & u/(1 + v)^2 \end{bmatrix}$$

and so

$$\frac{\partial(x,y)}{\partial(u,v)} = \det(DT) = \frac{u+uv}{(1+v)^3} = \frac{u}{(1+v)^2}.$$

By the Change of Variables Theorem,

$$\iint_D e^x \, dA = \iint_{D^*} e^{u/(1+v)} \left| \frac{u}{(1+v)^2} \right| dA^*$$

(since $0 \le u \le 2$, $|u| = u$)

$$= \int_0^2 \left(\int_1^2 e^{u/(1+v)} \frac{u}{(1+v)^2} dv \right) du.$$

Use the substitution $t = u/(1+v)$, $dt = -u \, dv/(1+v)^2$ for the inner integration:

$$\int_1^2 e^{u/(1+v)} \frac{u}{(1+v)^2} \, dv = -\int_{u/2}^{u/3} e^t \, dt = -e^t \Big|_{u/2}^{u/3} = e^{u/2} - e^{u/3}.$$

Therefore,

$$\iint_D e^x \, dA = \int_0^2 (e^{u/2} - e^{u/3}) \, du = \left(2e^{u/2} - 3e^{u/3} \right) \Big|_0^2$$

$$= (2e - 3e^{2/3}) - (2 - 3) = 1 + 2e - 3e^{2/3}.$$

33. Let $u = x + y$ and $v = x - y$. Solving for x and y we get $x = (u+v)/2$ and $y = (u-v)/2$. So this change of variables is represented by the linear map

$$T(u,v) = \left(\tfrac{1}{2}(u+v), \tfrac{1}{2}(u-v) \right).$$

Since

$$DT = \begin{bmatrix} 1/2 & 1/2 \\ 1/2 & -1/2 \end{bmatrix},$$

it follows that the Jacobian is

$$\frac{\partial(x,y)}{\partial(u,v)} = \det(DT) = -\tfrac{1}{2}.$$

By the Change of Variables Theorem,

$$\iint_D \sin\left(\frac{x+y}{x-y} \right) dA = \iint_{D^*} \sin(u/v) \tfrac{1}{2} \, dA^*,$$

where D^* is the region that maps to D under T. From $x - y = 1$ and $x - y = 5$ it follows that $v = 1$ and $v = 5$. The equation $x = 0$ implies $(u+v)/2 = 0$; i.e., $u + v = 0$. Similarly, $y = 0$ implies $(u-v)/2 = 0$; i.e., $u - v = 0$. It follows that D^* is the part of the horizontal strip in the uv-plane defined by $v = 1$ and $v = 5$, bounded by the lines $v = u$ and $v = -u$. So

$$\iint_D \sin\left(\frac{x+y}{x-y} \right) dA = \tfrac{1}{2} \iint_{D^*} \sin(u/v) \, dA^*$$

$$= \tfrac{1}{2} \int_1^5 \left(\int_{-v}^v \sin(u/v) du \right) dv = \tfrac{1}{2} \int_1^5 \left(-v \cos(u/v) \right) \Big|_{-v}^v dv$$

$$= -\tfrac{1}{2} \int_1^5 \left((v \cos 1) - (v \cos(-1)) \right) dv = 0,$$

since $\cos 1 = \cos(-1)$.

6.5. Triple Integrals

1. We evaluate the given triple integral as the following iterated integral:

$$\iiint_W (2x - y - z)\, dV = \int_0^3 \left(\int_0^2 \left(\int_0^2 (2x - y - z)dz \right) dy \right) dx$$

$$= \int_0^3 \left(\int_0^2 (2xz - zy - \tfrac{1}{2}z^2) \Big|_0^2\, dy \right) dx$$

$$= \int_0^3 \left(\int_0^2 (4x - 2y - 2)dy \right) dx$$

$$= \int_0^3 (4xy - y^2 - 2y) \Big|_0^2\, dx$$

$$= \int_0^3 (8x - 8)\, dx = (4x^2 - 8x) \Big|_0^3 = 12.$$

3. The solid W can be viewed as a type-1(3D) region: the corresponding elementary (2D) region D is the triangle bounded by $x = 0$, $y = 0$ and $x + y = 1$ (the last equation is obtained by substituting $z = 0$ into $x + y + z = 1$). The bottom and the top surfaces are $z = 0$ and $z = 1 - x - y$. It follows that

$$\iiint_W y^2\, dV = \iint_D \left(\int_0^{1-x-y} y^2 dz \right) dA$$

$$= \int_0^1 \left(\int_0^{1-y} \left(\int_0^{1-x-y} y^2 dz \right) dx \right) dy$$

$$= \int_0^1 \left(\int_0^{1-y} (y^2 z) \Big|_0^{1-x-y}\, dx \right) dy$$

$$= -\int_0^1 \left(\int_0^{1-y} y^2 (x + y - 1)dx \right) dy$$

$$= -\int_0^1 y^2 \left(\tfrac{1}{2}x^2 + xy - x \right) \Big|_0^{1-y}\, dy$$

$$= \int_0^1 \tfrac{1}{2}y^2 (1 - 2y + y^2)\, dy = \left(\tfrac{1}{10}y^5 - \tfrac{1}{4}y^4 + \tfrac{1}{6}y^3 \right) \Big|_0^1 = \tfrac{1}{60}.$$

5. Combining $z = 2y^2$ and $z = 8 - 2y^2$ we get $y^2 = 2$ and $y = \pm\sqrt{2}$. The solid W can be viewed as a type-1(3D) region: the corresponding elementary (2D) region D is the rectangle $[0, 1] \times [-\sqrt{2}, \sqrt{2}]$; the bottom surface is $z = 2y^2$ and the top surface is $z = 8 - 2y^2$. It follows that

$$\iiint_W (2z - 5)\, dV = \iint_D \left(\int_{2y^2}^{8-2y^2} (2z - 5)dz \right) dA$$

$$= \int_{-\sqrt{2}}^{\sqrt{2}} \left(\int_0^1 \left(\int_{2y^2}^{8-2y^2} (2z - 5)dz \right) dx \right) dy$$

$$= \int_{-\sqrt{2}}^{\sqrt{2}} \left(\int_0^1 (z^2 - 5z) \Big|_{2y^2}^{8-2y^2}\, dx \right) dy$$

$$= \int_{-\sqrt{2}}^{\sqrt{2}} \left(\int_0^1 (24 - 12y^2) dx \right) dy$$

$$= \int_{-\sqrt{2}}^{\sqrt{2}} (24 - 12y^2) x \Big|_0^1 \, dy$$

$$= \int_{-\sqrt{2}}^{\sqrt{2}} (24 - 12y^2) \, dy = (24y - 4y^3) \Big|_{-\sqrt{2}}^{\sqrt{2}} = 32\sqrt{2}.$$

7. Combining $z = 12 - x^2 - y^2$ and $z = 2x^2 + 2y^2$ we get $x^2 + y^2 = 4$. It follows that the solid W can be viewed as a type-1(3D) region: the corresponding elementary (2D) region D is the disk $x^2 + y^2 \leq 4$; the bottom surface is $z = 2x^2 + 2y^2$ and the top surface is $12 - x^2 - y^2$. Therefore

$$\iiint_W 4y \, dV = \iint_D \left(\int_{2x^2+2y^2}^{12-x^2-y^2} 4y \, dz \right) dA.$$

Passing to cylindrical coordinates, we get

$$\iiint_W 4y \, dV = \int_0^{2\pi} \left(\int_0^2 \left(\int_{2r^2}^{12-r^2} (4r \sin \theta) r \, dz \right) dr \right) d\theta$$

$$= 4 \left(\int_0^{2\pi} \sin \theta \, d\theta \right) \left(\int_0^2 \left(\int_{2r^2}^{12-r^2} 4r^2 dz \right) dr \right) = 0,$$

since $\int_0^{2\pi} \sin \theta d\theta = 0$.

9. The solid W can be viewed as a type-1(3D) region: the corresponding elementary (2D) region D is the disk $x^2 + y^2 \leq 1$; the bottom surface is $z = -\sqrt{2 - x^2 - y^2}$ and the top surface is $z = \sqrt{2 - x^2 - y^2}$. Therefore

$$\iiint_W xyz \, dV = \iint_D \left(\int_{-\sqrt{2-x^2-y^2}}^{\sqrt{2-x^2-y^2}} xyz \, dz \right) dA.$$

Passing to cylindrical coordinates, we get

$$\iiint_W xyz \, dV = \int_0^{2\pi} \left(\int_0^1 \left(\int_{-\sqrt{2-r^2}}^{\sqrt{2-r^2}} (r^2 z \cos \theta \sin \theta) r \, dz \right) dr \right) d\theta.$$

The product $\cos \theta \sin \theta$ can be factored out of the inner two integrations, so that

$$\iiint_W xyz \, dV = \left(\int_0^{2\pi} \cos \theta \sin \theta \, d\theta \right) \left(\int_0^1 \left(\int_{-\sqrt{2-r^2}}^{\sqrt{2-r^2}} r^3 z \cos \theta \sin \theta dz \right) dr \right).$$

Since

$$\int_0^{2\pi} \cos \theta \sin \theta \, d\theta = \frac{1}{2} \int_0^{2\pi} \sin 2\theta \, d\theta = 0,$$

it follows that $\iiint_W xyz \, dV = 0$.

11. The solid W that is the region of integration is defined by the inequalities $0 \leq y \leq 1$, $0 \leq x \leq 2 - 2y$ and $0 \leq z \leq 4 - 2x - 4y$. The corresponding elementary (2D) region D in the xy-plane is the triangular region bounded by $y = 0$, $x = 0$ and $x = 2 - 2y$ (or $y = 1 - x/2$). W is bounded from above by the plane $z = 4 - 2x - 4y$, that crosses the xy-plane when

$4 - 2x - 4y = 0$; i.e., when $x = 2 - 2y$. Consequently, the region of integration is the three-dimensional region in the first octant bounded from above by the plane $z = 4 - 2x - 4y$.

The iterated integral is computed to be

$$\int_0^1 \left(\int_0^{2-2y} \left(\int_0^{4-2x-4y} 3dz \right) dx \right) dy = \int_0^1 \left(\int_0^{2-2y} (3z) \Big|_0^{4-2x-4y} dx \right) dy$$

$$= \int_0^1 \left(\int_0^{2-2y} (12 - 6x - 12y)dx \right) dy$$

$$= \int_0^1 \left(12x - 3x^2 - 12xy \right) \Big|_0^{2-2y} dy$$

$$= \int_0^1 \left(12 - 24y + 12y^2 \right) dy$$

$$= \left(12y - 12y^2 + 4y^3 \right) \Big|_0^1 = 4.$$

13. The limits $0 \le \theta \le 2\pi$ and $0 \le r \le 2$ define the disk in the xy-plane of radius 2 centered at the origin. The equation $z = r^2$ represents the paraboloid $z = x^2 + y^2$. It follows that the region of integration is the three-dimensional solid inside the cylinder $x^2 + y^2 = 4$, above the xy-plane and below the paraboloid $z = x^2 + y^2$. The iterated integral is computed to be

$$\int_0^{2\pi} \left(\int_0^2 \left(\int_0^{r^2} r^2 z dz \right) dr \right) d\theta = \left(\int_0^{2\pi} d\theta \right) \left(\int_0^2 \tfrac{1}{2} r^2 z^2 \Big|_0^{r^2} dr \right)$$

$$= 2\pi \tfrac{1}{2} \int_0^2 r^6 dr = \tfrac{128}{7} \pi.$$

15. The inequalities $-\sqrt{8} \le x \le \sqrt{8}$ and $-\sqrt{8 - x^2} \le y \le \sqrt{8 - x^2}$ define the disk $x^2 + y^2 = 8$ in the xy-plane. The region of integration is the three-dimensional solid inside the cylinder, bounded from below by the plane $z = -3$ and from above by the paraboloid $z = 8 - x^2 - y^2$. Passing to cylindrical coordinates, we get

$$\int_{-\sqrt{8}}^{\sqrt{8}} \left(\int_{-\sqrt{8-x^2}}^{\sqrt{8-x^2}} \left(\int_{-3}^{8-x^2-y^2} 2dz \right) dy \right) dx = \int_0^{2\pi} \left(\int_0^{\sqrt{8}} \left(\int_{-3}^{8-r^2} 2rdz \right) dr \right) d\theta$$

$$= 2 \left(\int_0^{2\pi} d\theta \right) \left(\int_0^{\sqrt{8}} rz \Big|_{-3}^{8-r^2} dr \right)$$

$$= 4\pi \int_0^{\sqrt{8}} \left(11r - r^3 \right) dr$$

$$= 4\pi \left(\tfrac{11}{2} r^2 - \tfrac{1}{4} r^4 \right) \Big|_0^{\sqrt{8}} = 112\pi.$$

17. The tetrahedron in question (call it W) is the solid in the first octant bounded by the plane that goes through the points $(1, 0, 0)$, $(0, 1, 0)$, and $(0, 0, 1)$. We need the equation of that plane — and to get it, we need a normal. The vectors $\mathbf{v} = (-1, 1, 0)$ from $(1, 0, 0)$ to

$(0, 1, 0)$ and $\mathbf{w} = (-1, 0, 1)$ from $(1, 0, 0)$ to $(0, 0, 1)$ lie in the plane. So their cross product

$$\mathbf{v} \times \mathbf{w} = \begin{vmatrix} \mathbf{i} & \mathbf{j} & \mathbf{k} \\ -1 & 1 & 0 \\ -1 & 0 & 1 \end{vmatrix} = \mathbf{i} + \mathbf{j} + \mathbf{k},$$

is normal to the plane; its equation is (we need one point — take $(1, 0, 0)$):

$$1(x - 1) + 1(y - 0) + 1(z - 0) = 0;$$

i.e., $x + y + z = 1$. The region D in the xy-plane is the triangular region bounded by the coordinate axes and the line $x + y = 1$. So the volume of the tetrahedron is

$$\begin{aligned}
v = \iiint_W 1 \, dV &= \iint_D \left(\int_0^{1-x-y} 1 \, dz \right) dA \\
&= \int_0^1 \left(\int_0^{1-x} \left(\int_0^{1-x-y} dz \right) dy \right) dx \\
&= \int_0^1 \left(\int_0^{1-x} (1 - x - y) dy \right) dx \\
&= \int_0^1 \left(\tfrac{1}{2} - x + \tfrac{1}{2}x^2 \right) dx = \left. \left(\tfrac{1}{2}x - \tfrac{1}{2}x^2 + \tfrac{1}{6}x^3 \right) \right|_0^1 = \tfrac{1}{6}.
\end{aligned}$$

19. The sphere has radius 2. The semi-axes of the ellipsoid (rewrite the equation as $x^2 + y^2 + z^2/4 = 1$) are 1 (in the x-direction), 1 (in the y-direction) and 2 (in the z-direction). So the ellipsoid is inside the sphere, touching it at $(0, 0, \pm 2)$. Therefore, to get the volume of the solid in question, we will subtract the volume of the ellipsoid from $32\pi/3$ (which is the volume of the sphere).

When $z = 0$, the equation of the ellipsoid implies $x^2 + y^2 = 1$; hence, the region D in the xy-plane is the disk $x^2 + y^2 \le 1$. The solid is bounded from below by $z = -\sqrt{4 - 4x^2 - 4y^2} = -2\sqrt{1 - x^2 - y^2}$ and from above by $z = \sqrt{4 - 4x^2 - 4y^2} = 2\sqrt{1 - x^2 - y^2}$. Its volume is (we use cylindrical coordinates)

$$\begin{aligned}
v = \iiint_W 1 \, dV &= \iint_D \left(\int_{-2\sqrt{1-x^2-y^2}}^{2\sqrt{1-x^2-y^2}} dz \right) dA \\
&= \int_0^{2\pi} \left(\int_0^1 \left(\int_{-2\sqrt{1-r^2}}^{2\sqrt{1-r^2}} r \, dz \right) dr \right) d\theta \\
&= \left(\int_0^{2\pi} d\theta \right) \left(\int_0^1 \left. rz \right|_{-2\sqrt{1-r^2}}^{2\sqrt{1-r^2}} dr \right) \\
&= 2\pi \int_0^1 4r\sqrt{1 - r^2} \, dr \\
&= \left. 8\pi \left(-\tfrac{1}{3} \right) (1 - r^2)^{3/2} \right|_0^1 = \tfrac{8}{3}\pi.
\end{aligned}$$

So the volume of the solid in question is $32\pi/3 - 8\pi/3 = 8\pi$.

21. The solid in question (denote it by W) is the region below the cone $z^2 = x^2 + y^2$ and above the disk centered at $(1,0)$ of radius 1 in the xy-plane. Its volume is

$$v = \iiint_W dV = \iint_D \left(\int_0^{\sqrt{x^2+y^2}} dz \right) dA = \iint_D \sqrt{x^2 + y^2}\, dA.$$

In polar coordinates, the equation $(x - 1)^2 + y^2 = 1$ can be expressed as $r^2 - 2r\cos\theta = 0$; i.e., $r = 2\cos\theta$. Notice that $-\pi/2 \le \theta \le \pi/2$. Hence

$$\iint_D \sqrt{x^2 + y^2}\, dA = \int_{-\pi/2}^{\pi/2} \left(\int_0^{2\cos\theta} r^2 dr \right) d\theta$$

$$= \tfrac{1}{3} \int_{-\pi/2}^{\pi/2} r^3 \Big|_0^{2\cos\theta} d\theta$$

$$= \tfrac{8}{3} \int_{-\pi/2}^{\pi/2} \cos^3\theta\, d\theta$$

(use a formula from a table of integrals)

$$= \tfrac{8}{3}\tfrac{1}{3}(2 + \cos^2\theta)\sin\theta \Big|_{-\pi/2}^{\pi/2} = \tfrac{32}{9}.$$

23. The paraboloid intersects the xy-plane along the circle $x^2 + y^2 = 16$ (substitute $z = 0$ into $z = 16 - x^2 - y^2$). The solid in question (denote it by W) is the three-dimensional region below $z = 16 - x^2 - y^2$ and above the smaller of the two two-dimensional regions in the xy-plane defined by $x + y = 4$ and $x^2 + y^2 = 16$; see Figure 23. (call that region D). The volume of W is

$$v(W) = \iiint_W dV = \iint_D \left(\int_0^{16-x^2-y^2} dz \right) dA$$

$$= \iint_D (16 - x^2 - y^2)\, dA$$

$$= \int_0^4 \left(\int_{4-x}^{\sqrt{16-x^2}} (16 - x^2 - y^2)dy \right) dx.$$

In cylindrical coordinates (from $x + y = 4$ it follows that $r\sin\theta + r\cos\theta = 4$ and $r = 4/(\sin\theta + \cos\theta)$),

$$v(W) = \iint_D (16 - x^2 - y^2)\, dA = \int_0^{\pi/2} \left(\int_{4/(\sin\theta+\cos\theta)}^4 (16 - r^2)r dr \right) d\theta.$$

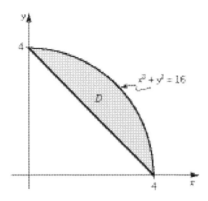

Figure 23

25. The solid W is bounded on the sides by the cylinder and from above and from below by the sphere. The region of integration D in the xy-plane is the disk $x^2 + y^2 \le b^2$. The volume of W is

$$v(W) = \iiint_W dV = \iint_D \left(\int_{-\sqrt{a^2-x^2-y^2}}^{\sqrt{a^2-x^2-y^2}} dz \right) dA = 2 \iint_D \sqrt{a^2 - x^2 - y^2}\, dA.$$

Passing to cylindrical coordinates, we get

$$v(W) = 2 \int_0^{2\pi} \left(\int_0^b r\sqrt{a^2 - r^2}\, dr \right) d\theta$$

$$= 2 \left(\int_0^{2\pi} d\theta \right) \left(-\tfrac{1}{3}(a^2 - r^2)^{3/2} \right) \Big|_0^b$$

$$= -\tfrac{4}{3}\pi((a^2 - b^2)^{3/2} - a^3) = \tfrac{4}{3}\pi(a^3 - (a^2 - b^2)^{3/2}).$$

27. The region of integration is defined by $-\sqrt{2} \le x \le \sqrt{2}$, $-\sqrt{2 - x^2} \le y \le \sqrt{2 - x^2}$ and $x^2 + y^2 - 2 \le z \le 2 - x^2 - y^2$. In words, it is the solid bounded by the paraboloids $z = x^2 + y^2 - 2$ and $z = 2 - x^2 - y^2$. The corresponding region in the xy-plane is the circle $x^2 + y^2 = 2$. Hence, in cylindrical coordinates,

$$\int_{-\sqrt{2}}^{\sqrt{2}} \int_{-\sqrt{2-x^2}}^{\sqrt{2-x^2}} \int_{x^2+y^2-2}^{2-x^2-y^2} (x^2 + y^2 - 2)dzdydx = \int_0^{2\pi} \left(\int_0^{\sqrt{2}} \left(\int_{r^2-2}^{2-r^2} (r^2 - 2)rdz \right) dr \right) d\theta.$$

29. We use the change of variables $u = x + y$, $v = y + 2z$ and $w = 2x + y + z$. The new region of integration W^* is defined by $1 \le u \le 3$, $-2 \le v \le 4$ and $-1 \le w \le 1$. To find the Jacobian, we have to solve for x, y and z.

Subtracting the equation for v from the equation for u, we get $u - v = x - 2z$; eliminating y from equations for v and w, we get $v - w = z - 2x$; now solve this new system for x and z: $x = -u/3 - v/3 + 2w/3$ and $z = -2u/3 + v/3 + w/3$; consequently, $y = u - x = 4u/3 + v/3 - 2w/3$. It follows that the change of variables can be represented as the map

$$T(u, v, w) = \left(-\tfrac{1}{3}u - \tfrac{1}{3}v + \tfrac{2}{3}w, \tfrac{4}{3}u + \tfrac{1}{3}v - \tfrac{2}{3}w, -\tfrac{2}{3}u + \tfrac{1}{3}v + \tfrac{1}{3}w \right).$$

Then

$$DT = \begin{bmatrix} -1/3 & -1/3 & 2/3 \\ 4/3 & 1/3 & -2/3 \\ -2/3 & 1/3 & 1/3 \end{bmatrix}$$

and

$$\det(DT) = \tfrac{1}{27} \begin{vmatrix} -1 & -1 & 2 \\ 4 & 1 & -2 \\ -2 & 1 & 1 \end{vmatrix} = \tfrac{1}{27}(-3 + 0 + 12) = \tfrac{1}{3}.$$

So

$$v = \iiint_W dV = \iiint_{W^*} \tfrac{1}{3} dV^*$$
$$= \int_1^3 \left(\int_{-2}^4 \left(\int_{-1}^1 \tfrac{1}{3} dw \right) dv \right) du$$
$$= \tfrac{1}{3} \left(\int_1^3 du \right) \left(\int_{-2}^4 dv \right) \left(\int_{-1}^1 dw \right)$$
$$= \tfrac{1}{3}(2)(6)(2) = 8.$$

31. Combining $z = 1$ and $z = x^2 + y^2$ we get $x^2 + y^2 = 1$. Similarly, combining $z = 1$ and $z = 2(x^2 + y^2)$ we get $x^2 + y^2 = 1/2$. So the solid W is bounded from above by $z = 2(x^2 + y^2)$ and from below by $z = x^2 + y^2$, if (x, y) belongs to the disk $x^2 + y^2 \le 1/2$. If (x, y) belongs to the region $1/2 \le x^2 + y^2 \le 1$, then W is bounded from below by $z = x^2 + y^2$ and from above by $z = 1$; see Figure 31.

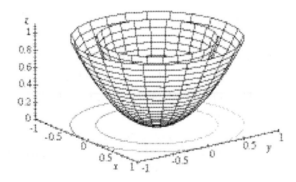

Figure 31

Consequently,

$$\iiint_W 2x \, dV = \iint_{\{x^2+y^2 \le 1/2\}} \left(\int_{x^2+y^2}^{2(x^2+y^2)} 2x \, dz \right) dA$$
$$+ \iint_{\{1/2 \le x^2+y^2 \le 1\}} \left(\int_{x^2+y^2}^1 2x \, dz \right) dA$$

Passing to cylindrical coordinates, we get

$$\iiint_W 2x\,dV = \int_0^{2\pi}\left(\int_0^{\sqrt{1/2}}\left(\int_{r^2}^{2r^2}2r^2\cos\theta dz\right)dr\right)d\theta$$

$$+\int_0^{2\pi}\left(\int_{\sqrt{1/2}}^1\left(\int_{r^2}^1 2r^2\cos\theta dz\right)dr\right)d\theta.$$

Actually it is not hard to evaluate this integral. We can factor out $\cos\theta$ from the inner two integrations in both integrals in cylindrical coordinates, thus getting $\int_0^{2\pi}\cos\theta d\theta$ as a factor. But $\int_0^{2\pi}\cos\theta d\theta = 0$, and so $\iiint_W 2x\,dV = 0$.

Chapter Review

True/false Quiz

1. False. For instance, the Jacobian of the map $T(u,v) = (u+2v, 3u-v)$ of Example 6.34 is equal to -7.

3. False. The region of integration is bounded by the parabolas $y = -\sqrt{x}$ and $y = \sqrt{x}$ (on the interval $[0,2]$).

5. False. The iterated integral on the right side does not make sense (the outer limit of integration contains a variable). Or, accepting that it makes sense - the equality still is not true, since the left side is a real number, whereas the right side is a function of x.

7. True. Define

$$\phi(y) = \begin{cases} 0 & \text{if } -\sqrt{2} \leq y \leq -1 \\ \sqrt{1-y^2} & \text{if } -1 \leq y \leq 1 \\ 0 & \text{if } 1 \leq y \leq \sqrt{2} \end{cases}$$

and $\psi(y) = \sqrt{2-y^2}$. Then D can be described as $\phi(y) \leq x \leq \psi(y)$, where $[c = -\sqrt{2}, d = \sqrt{2}]$.

9. True. Since $\sin(x^2+y^2) \leq 1$, we conclude that

$$\int_0^1\left(\int_0^x \sin(x^2+y^2)\,dy\right)dx \leq \int_0^1\left(\int_0^x 1\,dy\right)dx = \int_0^1 x\,dx = \frac{1}{2}.$$

11. False. The square $[0,1] \times [0,1]$ gets mapped under T to the rectangle $[0,2] \times [0,1/2]$.

Review Exercises and Problems

1. The region of integration (call it W) is defined by $0 \leq \theta \leq 2\pi$, $0 \leq \phi \leq \pi/4$ and $0 \leq \rho \leq 2 + \cos\theta$. From the inequalities for θ and ϕ we conclude that W is inside the circular cone whose axis of symmetry is the z-axis and whose intersection with the yz-plane (or with any plane perpendicular to the xy-plane that contains the z-axis) consists of two lines that make the angles of $\pm\pi/4$ with respect to the positive z-axis (if you rotate the part of the line

$z = y$ satisfying $y \geq 0$ about the z-axis, you will obtain that cone). The surface $\rho = 2 + \cos\theta$ bounds W from above.

The integral is computed to be

$$\int_0^{2\pi} \left(\int_0^{2+\cos\theta} \left(\int_0^{\pi/4} \rho^2 \sin\phi \, d\phi \right) d\rho \right) d\theta = \int_0^{2\pi} \left(\int_0^{2+\cos\theta} \left(-\rho^2 \cos\phi \right) \Big|_0^{\pi/4} d\rho \right) d\theta$$

$$= \left(1 - \tfrac{\sqrt{2}}{2} \right) \int_0^{2\pi} \left(\int_0^{2+\cos\theta} \rho^2 \, d\rho \right) d\theta$$

$$= \tfrac{1}{3} \left(1 - \tfrac{\sqrt{2}}{2} \right) \int_0^{2\pi} (2 + \cos\theta)^3 \, d\theta$$

$$= \tfrac{1}{3} \left(1 - \tfrac{\sqrt{2}}{2} \right) \int_0^{2\pi} 8 + 12\cos\theta + 6\cos^2\theta + \cos^3\theta \, d\theta.$$

Now $\int_0^{2\pi} \cos^3\theta \, d\theta = 0$; draw the graph of $\cos^3\theta$ and interpret the integral as "area above the x-axis $-$ area below the x-axis," or use the formula $\cos^3\theta = (2 + \cos^2\theta) \sin\theta/2$ and a substitution. Next,

$$\int_0^{2\pi} \cos^2\theta \, d\theta = \left(\tfrac{1}{2}\theta + \tfrac{1}{4}\sin 2\theta \right) \Big|_0^{2\pi} = \pi$$

and $\int_0^{2\pi} \cos\theta \, d\theta = 0$. It follows that the given integral is equal to

$$\tfrac{1}{3} \left(1 - \tfrac{\sqrt{2}}{2} \right) (16\pi + 6\pi) = \tfrac{22}{3}\pi \left(1 - \tfrac{\sqrt{2}}{2} \right).$$

3. The volume is given by

$$v = \iint_D \left(\int_0^{my+am} dz \right) dA,$$

where D is the disk $x^2 + y^2 \leq a^2$ in the xy-plane. Passing to cylindrical coordinates, we get

$$v = \int_0^{2\pi} \left(\int_0^a \left(\int_0^{mr\sin\theta + am} r \, dz \right) dr \right) d\theta$$

$$= \int_0^{2\pi} \int_0^a rz \Big|_0^{mr\sin\theta + am} dr \, d\theta$$

$$= \int_0^{2\pi} \left(\int_0^a (mr^2 \sin\theta + amr) dr \right) d\theta = \int_0^{2\pi} \left(\tfrac{1}{3}mr^3 \sin\theta + \tfrac{1}{2}amr^2 \right) \Big|_0^a d\theta$$

$$= \int_0^{2\pi} \left(\tfrac{1}{3}ma^3 \sin\theta + \tfrac{1}{2}a^3 m \right) d\theta$$

$$= \tfrac{1}{3}a^3 m \int_0^{2\pi} \sin\theta \, d\theta + \tfrac{1}{2}a^3 m \int_0^{2\pi} d\theta = a^3 m\pi,$$

since $\int_0^{2\pi} \sin\theta \, d\theta = 0$.

5. Assume that $\iint_R f \, dA = 0$ and that there is a point $(x_0, y_0) \in R$ such that $f(x_0, y_0) > 0$. We have to show that this leads to a contradiction (and thus, $f(x_0, y_0) = 0$ for all $(x_0, y_0) \in R$).

If $f(x_0, y_0) > 0$, then there must be a ball $B((x_0, y_0), \delta)$ of radius δ centered at (x_0, y_0), so that $f(x, y) > 0$ for all (x, y) in that ball. This statement follows directly from the definition of continuity. To be precise: choose $\epsilon > 0$ so that $f(x_0, y_0) - \epsilon > 0$; for that ϵ, by continuity, there must be a ball $B((x_0, y_0), \delta)$ of radius $\delta > 0$ so that, for all (x, y) in $B((x_0, y_0), \delta)$,

$$|f(x, y) - f(x_0, y_0)| < \epsilon;$$

i.e., $f(x, y)$ belongs to the interval $(f(x_0, y_0) - \epsilon, f(x_0, y_0) + \epsilon)$. However, from the way ϵ was chosen, it follows that all numbers in this interval are positive. Therefore, $f(x, y) > 0$ for all x in $B((x_0, y_0), \delta)$.

Now

$$\iint_R f \, dA = \iint_{B((x_0, y_0), \delta)} f \, dA + \iint_{R - B((x_0, y_0), \delta)} f \, dA,$$

where $R - B((x_0, y_0), \delta)$ is the part of the rectangle that does not include the ball $B((x_0, y_0), \delta)$. Since $f \geq 0$, it follows that

$$\iint_{R - B((x_0, y_0), \delta)} f \, dA \geq 0.$$

However, on $B((x_0, y_0), \delta)$, the function f satisfies $f(x, y) > 0$, and so

$$\iint_{B((x_0, y_0), \delta)} f \, dA > 0.$$

Therefore, $\iint_R f \, dA > 0$, and that contradicts the assumption.

7. In polar coordinates, the line $y = x$ has the form $\theta = \pi/4$, and the line $x = 1$ can be wtitten as $r \cos \theta = 1$, i.e., $r = 1/\cos \theta$. Thus,

$$\iint_D f \, dA = \int_0^{\pi/4} \left(\int_0^{\sec \theta} f(r \cos \theta, r \sin \theta) r \, dr \right) d\theta.$$

If we wish to reverse the order of integration: using the circular arc of radius 1 centred at the origin, we break up the region as indicated in the figure below. Thus,

$$\iint_D f \, dA = \int_0^1 \left(\int_0^{\pi/4} f(r \cos \theta, r \sin \theta) r \, d\theta \right) dr + \int_1^{\sqrt{2}} \left(\int_{\arccos(1/r)}^{\pi/4} f(r \cos \theta, r \sin \theta) r \, d\theta \right) dr.$$

Note: line $x = 1$, expressed in polar coordinates as $r \cos \theta = 1$, can be written as $\theta = \arccos(1/r)$.

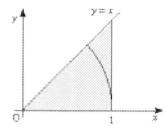

Figure 7

9. D is the quarter-circle of radius $\sqrt{2}$ in the third quadrant. Thus,

$$\iint_D f\,dA = \int_\pi^{3\pi/2} \left(\int_0^{\sqrt{2}} f(r\cos\theta, r\sin\theta)r\,dr \right) d\theta.$$

In the reversed order,

$$\iint_D f\,dA = \int_0^{\sqrt{2}} \left(\int_\pi^{3\pi/2} f(r\cos\theta, r\sin\theta)r\,d\theta \right) dr.$$

11. Let us make a sketch of the region in question (call it D). The equation $r = 2$ represents the circle $x^2 + y^2 = 4$. The equation $r = 4\cos\theta$ implies (multiply both sides by r) $r^2 = r\cos\theta$; i.e., $x^2 + y^2 = 4x$ and (after completing the square) $(x-2)^2 + y^2 = 4$. So D is the region common to both the circle centered at the origin of radius 2 and the circle centered at $(2,0)$ of radius 2; see Figure 11.

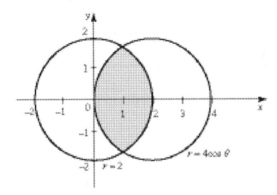

Figure 11

Let D_1 be the region inside $x^2 + y^2 = 4x$ and outside $x^2 + y^2 = 4$. We will compute the area of D_1 (since the integration is easier; see the comment at the end of this solution), and then get the area of D from $A(D) + A(D_1) = 4\pi$.

Combining $r = 2$ and $r = 4\cos\theta$ we get $\cos\theta = 1/2$; i.e., $\theta = \pm\pi/3$. Therefore, $-\pi/3 \le \theta \le \pi/3$; since $2 \le r \le 4\cos\theta$, it follows that the area of D_1 is

$$A(D_1) = \iint_{D_1} dA = \int_{-\pi/3}^{\pi/3} \left(\int_2^{4\cos\theta} r\,dr \right) d\theta = 2 \int_0^{\pi/3} \left(\int_2^{4\cos\theta} r\,dr \right) d\theta,$$

by symmetry. So

$$A(D_1) = 2 \int_0^{\pi/3} \tfrac{1}{2}r^2 \Big|_2^{4\cos\theta} d\theta$$

$$= \int_0^{\pi/3} (16\cos^2\theta - 4)\,d\theta$$

$$= \left(16\left(\tfrac{1}{2}\theta + \tfrac{1}{4}\sin 2\theta\right) - 4\theta\right)\Big|_0^{\pi/3} = \tfrac{4}{3}\pi + 2\sqrt{3}.$$

We used the identity $\cos^2\theta = \frac{1}{2}(1 + \cos 2\theta)$ in computing the integral of $\cos^2\theta$. Therefore,

$$A(D) = 4\pi - \left(\tfrac{4}{3}\pi + 2\sqrt{3}\right) = \tfrac{8}{3}\pi - 2\sqrt{3}.$$

Comment: the area of D can be computed directly, as

$$A(D) = \int_{-\pi/2}^{-\pi/3}\left(\int_0^{4\cos\theta} r\,dr\right)d\theta + \int_{-\pi/3}^{\pi/3}\left(\int_0^2 r\,dr\right)d\theta + \int_{\pi/3}^{\pi/2}\left(\int_0^{4\cos\theta} r\,dr\right)d\theta.$$

13. Enclose D into a rectangle R and divide R into n^2 equal subrectangles R_{ij}, $i,j = 1,\ldots,n$. Take only those that have a non-empty intersection with D. Since D is symmetric with respect to the x-axis, the subdivision is symmetric as well: every rectangle above the x-axis has its symmetric counterpart below the x-axis (if a rectangle crosses the x-axis, then it is symmetric to itself). Notice that from $f(x,-y) = -f(x,y)$ it follows that (read it with $y = 0$) $f(x,0) = -f(x,0)$; i.e., $f(x,0) = 0$, so the value of f along the x-axis is zero. Form a Riemann sum

$$\mathcal{R}_n = \sum_{i=1}^n \sum_{j=1}^n f(x_i^*, y_j^*)\Delta A_{ij}$$

in the following way: if a rectangle R_{ij} contains a point on the x-axis, then choose (x_i^*, y_j^*) to be that point (so, in this case, $f(x_i^*, y_j^*) = 0$). If a rectangle does not intersect the x-axis, take any point (x_i^*, y_j^*) in it (that is, of course, in D). For its symmetric counterpart, select the point $(x_i^*, -y_j^*)$. The contribution to $\mathcal{R}_n$ from this pair of rectangles is

$$f(x_i^*, y_j^*)\Delta A_{ij} + f(x_i^*, -y_j^*)\Delta A_{ij}^s$$

(where ΔA_{ij} is the area of R_{ij} and ΔA_{ij}^s is the area of the rectangle that is symmetric to R_{ij}; by construction, $\Delta A_{ij}^s = \Delta A_{ij}$). Since $f(x_i^*, -y_j^*) = -f(x_i^*, y_j^*)$, the above contribution is zero. It follows that the Riemann sum thus constructed is zero (for any n). Hence

$$\iint_R f\,dA = \lim_{n\to\infty}\mathcal{R}_n = \lim_{n\to\infty} 0 = 0.$$

15. Think of the tetrahedron as the solid below the plane through the points $(2,0,0)$, $(1,4,0)$ and $(0,0,10)$, and over the region in the xy-plane bounded by the triangle whose vertices are at $(0,0)$, $(2,0)$ and $(1,4)$.

Recall that the equation of the plane is $ax + by + cz + d = 0$. Since the plane does not go through the origin, $d \neq 0$, and we can write the equation as $Ax + By + Cz + 1 = 0$, where $A = a/d$, $B = b/d$ and $C = c/d$. Substituting the coordinates of the three given points, we get $2A + 1 = 0$ (and thus $A = -1/2$), $10C + 1 = 0$ (and thus $C = -1/10$), and $A + 4B + 1 = 0$, and thus $B = -1/8$. So, the equation of the plane is $-\frac{1}{2}x - \frac{1}{8}y - \frac{1}{10}z + 1 = 0$, i.e., $z = 10 - 5x - 5y/4$.

The triangle in the xy-plane is bounded by the lines $y = 4x$, $y = 0$ and $y = -4x + 8$. Thus, the volume is

$$\int_0^4 \left(\int_{y/4}^{-(y-8)/4} \left(\int_0^{10-5x-5y/4} 1\, dz \right) dx \right) dy$$

$$= \int_0^4 \left(\int_{y/4}^{-(y-8)/4} \left(10 - 5x - \frac{5}{4}y \right) dx \right) dy = \int_0^4 \left(10x - \frac{5}{2}x^2 - \frac{5}{4}xy \right) \Big|_{y/4}^{-(y-8)/4} dy$$

$$= \int_0^4 \left(10 - 5y + \frac{5}{8}y^2 \right) dy = 10y - \frac{5}{2}y^2 + \frac{5}{24}y^3 \Big|_0^4 = \frac{40}{3}.$$

Comments. This was an exercise in integration. We could have computed the equation of the plane using vector methods (for instance, using the cross product to get the normal vector to the plane). Also, we could have used the scalar triple product. Note that the volume could have been computed much faster: it is $(1/3)$ times the base area (4) times the height (10), i.e., $40/3$.

17. (a) Using separation of variables (as in Example 6.24 in Section 6.3), we get

$$\int_{-\infty}^{\infty} \left(\int_{-\infty}^{\infty} e^{-x^2-y^2}\, dy \right) dx = \int_{-\infty}^{\infty} \left(\int_{-\infty}^{\infty} e^{-x^2}e^{-y^2}\, dy \right) dx$$

$$= \left(\int_{-\infty}^{\infty} e^{-x^2}\, dx \right) \left(\int_{-\infty}^{\infty} e^{-y^2}\, dy \right) = \left(\int_{-\infty}^{\infty} e^{-x^2}\, dx \right)^2,$$

since $\int_{-\infty}^{\infty} e^{-y^2}\, dy = \int_{-\infty}^{\infty} e^{-x^2}\, dx$.

(b) Switching to polar coordinates, we get

$$\int_{-\infty}^{\infty} \left(\int_{-\infty}^{\infty} e^{-x^2-y^2}\, dy \right) dx = \int_0^{2\pi} \left(\int_0^{\infty} e^{-r^2} r\, dr \right) d\theta$$

$$= \int_0^{2\pi} \left(-\frac{1}{2}e^{-r^2} \right) \Big|_0^{\infty} d\theta = \int_0^{2\pi} \frac{1}{2}\, d\theta = \pi.$$

(c) Combining (a) and (b), we get that $\int_{-\infty}^{\infty} e^{-x^2}\, dx = \sqrt{\pi}$.

19. The required volume is given by $\iiint_W dV$, where W is the ellipsoid $x^2/a^2 + y^2/b^2 + z^2/c^2 = 1$. Using $x = au$, $y = bv$, $z = cw$, we get $u^2 + v^2 + w^2 = 1$ (i.e., under the given change of variables, the ellipsoid W (in the xyz-coordinate system) transforms to the sphere $u^2 + v^2 + w^2 = 1$ of radius 1 (in the uvw-coordinate system)). By the Change of Variables Theorem,

$$\iiint_W dV = \iiint_{W^*} \left| \frac{\partial(x,y,z)}{\partial(u,v,w)} \right| dV^*,$$

where

$$\frac{\partial(x,y,z)}{\partial(u,v,w)} = \begin{vmatrix} a & 0 & 0 \\ 0 & b & 0 \\ 0 & 0 & c \end{vmatrix} = abc.$$

(Keep in mind that $a, b, c > 0$.) Thus,

$$\iiint_W dV = \iiint_{W^*} abc\, dV^* = abc \iiint_{W^*} dV^* = \frac{4abc\pi}{3}.$$

Note that $\iiint_{W^*} dV^* = 4\pi/3$, since it is the volume of the sphere of radius 1.

21. Divide D as shown in figure below. Thus

$$\iint_D f\, dA = \int_0^1 \left(\int_0^{\pi/2} f(r\cos\theta, r\sin\theta) r\, d\theta \right) dr$$

$$+ \int_1^{\sqrt{2}} \left(\int_{\arccos(1/r)}^{\arcsin(1/r)} f(r\cos\theta, r\sin\theta) r\, d\theta \right) dr.$$

Reversing the order, we get

$$\iint_D f\, dA = \int_0^{\pi/4} \left(\int_0^{\sec\theta} f(r\cos\theta, r\sin\theta) r\, dr \right) d\theta$$

$$+ \int_{\pi/4}^{\pi/2} \left(\int_0^{\csc\theta} f(r\cos\theta, r\sin\theta) r\, dr \right) d\theta.$$

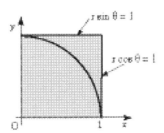

Figure 21

7. INTEGRATION OVER SURFACES.
PROPERTIES AND APPLICATIONS OF INTEGRALS

7.1. Parametrized Surfaces

1. Let us consider the sphere first. The lines $u = 0$ and $u = 2\pi$ map to the same meridian (i.e., the semi-circle) $(a \cos v, 0, a \sin v)$, $-\pi/2 \le v \le \pi/2$, that is the intersection of the sphere and the xz-plane, with $x \ge 0$. The line $v = \pi/2$ maps to $(0,0,a)$ and the line $v = -\pi/2$ maps to $(0,0,-a)$. So the top and the bottom line segments map to the north and the south poles respectively, and the vertical boundary line segments map to the meridian that is the intersection of the sphere and the xz-plane with $x \ge 0$.

Consider the cylinder $\mathbf{r}(u,v) = (a \cos u, a \sin u, v)$, $0 \le v \le b$, $0 \le u \le 2\pi$. Both vertical lines $u = 0$ and $u = 2\pi$ map to the line segment $(a, 0, v)$, $0 \le v \le b$, that is the intersection of the cylinder and the xz-plane with $x \ge 0$. The horizontal lines $v = 0$ and $v = b$ map to the bottom and the top boundary circles $\mathbf{r}(u,0) = (a \cos u, a \sin u, 0)$, $0 \le u \le 2\pi$ (in the xy-plane) and $\mathbf{r}(u,b) = (a \cos u, a \sin u, b)$, $0 \le u \le 2\pi$ (in the plane $y = b$).

3. We take a parametrization of the sphere $x^2 + y^2 + z^2 = a^2$ (for example, from Example 7.4) and adjust the range of the parameter(s) so that the z-coordinate is positive. So the required parametrization is

$$\mathbf{r}(u,v) = (a \cos v \cos u, a \cos v \sin u, a \sin v), \quad 0 \le u \le 2\pi, \quad 0 \le v \le \pi/2.$$

Notice that $0 \le v \le \pi/2$ implies that the z-coordinate $a \sin v$ is positive, which is exactly what is needed.

There are other parametrizations; for example,

$$\mathbf{r}(u,v) = (a \cos u \sin v, a \sin u \sin v, a \cos v),$$

where $0 \le u \le 2\pi$ and $0 \le v \le \pi/2$.

5. We adjust any parametrization of the sphere. For instance, from Example 7.4 we get

$$\mathbf{r}(u,v) = (2 \cos v \cos u - 2, 2 \cos v \sin u + 3, 2 \sin v + 7),$$

where $0 \le u \le 2\pi$, and $-\pi/2 \le v \le \pi/2$.

7. Parametrize the plane as the graph of $z = 2 + 3y - x$; hence $\mathbf{r}(u,v) = (u, v, 2 + 3v - u)$. We need only those points on the plane that are inside the cylinder; i.e., that satisfy $u^2 + v^2 \le 4$. The required parametrization $\mathbf{r} \colon D \subseteq \mathbb{R}^2 \to \mathbb{R}^3$ is given by

$$\mathbf{r}(u,v) = (u, v, 2 + 3v - u), \qquad (u,v) \in D,$$

where D is the disk $u^2 + v^2 \le 4$.

9. Parametrize the given plane as the graph of the function $z = 6 - x - 2y$; the domain $D \subseteq \mathbb{R}^2$ of the parametrization is defined by the requirement that only the part in the first octant is needed. So $x \geq 0$, $y \geq 0$ and $6 - x - 2y \geq 0$, or $x + 2y \leq 6$. It follows that D is the triangular region bounded by the coordinate axes and the line $x + 2y = 6$; i.e., the boundary of D is the triangle with vertices at $(0,0)$, $(6,0)$ and $(0,3)$. So the required parametrization is

$$\mathbf{r}(u, v) = (u, v, 6 - u - 2v), \qquad (u, v) \in D,$$

where D is the triangular region bounded by the coordinate lines and the line $u + 2v = 6$.

11. Parametrize the paraboloid as the graph of the surface $z = x^2 + y^2$; i.e., $\mathbf{r}(u, v) = (u, v, u^2 + v^2)$. To make sure that $\mathbf{r}(u, v)$ belongs to the first octant, we must take $u \geq 0$ and $v \geq 0$. Therefore

$$\mathbf{r}(u, v) = (u, v, u^2 + v^2), \qquad u \geq 0 \quad v \geq 0$$

is a required parametrization.

13. (a) From $\mathbf{r}(u, v) = (2u, u^2 + v, v^2)$ it follows that $\mathbf{T}_u = (2, 2u, 0)$, $\mathbf{T}_v = (0, 1, 2v)$ and

$$\mathbf{N} = \mathbf{T}_u \times \mathbf{T}_v = \begin{vmatrix} \mathbf{i} & \mathbf{j} & \mathbf{k} \\ 2 & 2u & 0 \\ 0 & 1 & 2v \end{vmatrix} = (4uv, -4v, 2).$$

(b) Since the z-component of $\mathbf{N}$ is 2, $\mathbf{N} \neq \mathbf{0}$, no matter what u and v are; so S is smooth for all $u, v \geq 0$.

15. (a) From $\mathbf{r}(u, v) = (\sin u \cos v, \sin u \sin v, 2 \cos u)$ it follows that

$$\mathbf{T}_u = (\cos u \cos v, \cos u \sin v, -2 \sin u),$$

$$\mathbf{T}_v = (-\sin u \sin v, \sin u \cos v, 0)$$

and

$$\mathbf{N} = \mathbf{T}_u \times \mathbf{T}_v = \begin{vmatrix} \mathbf{i} & \mathbf{j} & \mathbf{k} \\ \cos u \cos v & \cos u \sin v & -2 \sin u \\ -\sin u \sin v & \sin u \cos v & 0 \end{vmatrix}$$

$$= (2 \sin^2 u \cos v, 2 \sin^2 u \sin v, \sin u \cos u \cos^2 v + \sin u \cos u \sin^2 v)$$

$$= \sin u (2 \sin u \cos v, 2 \sin u \sin v, \cos u).$$

(b) To find the points where S is not smooth, we compute the magnitude of the normal $\mathbf{N}$:

$$\|\mathbf{N}\| = |\sin u| \sqrt{4 \sin^2 u \cos^2 v + 4 \sin^2 u \sin^2 v + \cos^2 u}$$

$$= |\sin u| \sqrt{4 \sin^2 u + \cos^2 u}.$$

Now $\|\mathbf{N}\| = 0$ implies that $|\sin u| = 0$; i.e., $u = 0, \pi, 2\pi$. Notice that $4 \sin^2 u + \cos^2 u > 0$ (since there is no u such that $\sin u = \cos u = 0$). It follows that S is not smooth at $\mathbf{r}(0, v) = (0, 0, 2)$, $\mathbf{r}(\pi, v) = (0, 0, -2)$, and $\mathbf{r}(2\pi, v) = \mathbf{r}(0, v) = (0, 0, 2)$.

17. (a) From $\mathbf{r}(u, v) = ((1 + \cos v) \cos u, (1 + \cos v) \sin u, \sin v)$ we get

$$\mathbf{T}_u = (-(1 + \cos v) \sin u, (1 + \cos v) \cos u, 0),$$

$$\mathbf{T}_v = (-\sin v \cos u, -\sin v \sin u, \cos v)$$

and

$$\mathbf{N} = \mathbf{T}_u \times \mathbf{T}_v = \begin{vmatrix} \mathbf{i} & \mathbf{j} & \mathbf{k} \\ -(1+\cos v)\sin u & (1+\cos v)\cos u & 0 \\ -\sin v \cos u & -\sin v \sin u & \cos v \end{vmatrix}$$

$$= ((1+\cos v)\cos v \cos u, (1+\cos v)\sin u \cos v,$$

$$(1+\cos v)\sin^2 u \sin v + (1+\cos v)\sin v \cos^2 u)$$

$$= (1+\cos v)(\cos u \cos v, \sin u \cos v, \sin v).$$

(b) To find the points where S is not smooth, we compute the magnitude of the surface normal $\mathbf{N}$:

$$\|\mathbf{N}\| = |1+\cos v|\sqrt{\cos^2 u \cos^2 v + \sin^2 u \cos^2 v + \sin^2 v}$$

$$= |1+\cos v|\sqrt{\cos^2 v + \sin^2 v} = |1+\cos v|.$$

Since $|1+\cos v| = 0$ only when $v = \pi$, it follows that S is smooth at all points except at $\mathbf{r}(u,\pi) = (0,0,0)$.

19. (a) From $\mathbf{r}(u,v) = (u, v, 1-(u^2+v^2))$ we get $\mathbf{T}_u = (1, 0, -2u)$, $\mathbf{T}_v = (0, 1, -2v)$ and

$$\mathbf{N} = \mathbf{T}_u \times \mathbf{T}_v = \begin{vmatrix} \mathbf{i} & \mathbf{j} & \mathbf{k} \\ 1 & 0 & -2u \\ 0 & 1 & -2v \end{vmatrix} = (2u, 2v, 1).$$

(b) Since the z-coordinate of $\mathbf{N}$ is 1, it follows that $\mathbf{N}(u,v) \neq 0$ for all (u,v); so S is smooth for all $0 \leq u, v < \infty$.

21. A normal to the surface is given by

$$\mathbf{N} = \mathbf{T}_u \times \mathbf{T}_v = \begin{vmatrix} \mathbf{i} & \mathbf{j} & \mathbf{k} \\ e^u & 0 & v \\ 0 & e^v & u \end{vmatrix} = (-ve^v, -ue^u, e^u e^v).$$

The point $(1,1,0)$ is obtained for $u = 0$ and $v = 0$; so $\mathbf{N}(0,0) = (0,0,1)$ is a normal to the surface at $\mathbf{r}(0,0) = (1,1,0)$. Consequently, the tangent plane has the equation

$$0(x-1) + 0(y-1) + 1(z-0) = 0;$$

i.e., $z = 0$.

23. Parametrize the given surface as $\mathbf{r}(u,v) = (u, u^2 + 2uv, v)$. In that case, the point of tangency is $\mathbf{r}(1,1) = (1,3,1)$. A normal to the surface is given by

$$\mathbf{N}(u,v) = \mathbf{T}_u \times \mathbf{T}_v = \begin{vmatrix} \mathbf{i} & \mathbf{j} & \mathbf{k} \\ 1 & 2u+2v & 0 \\ 0 & 2u & 1 \end{vmatrix} = (2u+2v, -1, 2u),$$

and hence $\mathbf{N}(1,1) = (4,-1,2)$. The equation of the tangent plane is

$$4(x-1) - 1(y-3) + 2(z-1) = 0;$$

i.e., $4x - y + 2z - 3 = 0$.

25. Parametrize the cone as in Example 7.3 as

$$\mathbf{r}(u, v) = (v \cos u, v \sin u, v), \qquad 0 \le u \le 2\pi;$$

but this time, $v \in \mathbb{R}$. The normal vector $\mathbf{N}$ is

$$\mathbf{N} = \mathbf{T}_u \times \mathbf{T}_v = \begin{vmatrix} \mathbf{i} & \mathbf{j} & \mathbf{k} \\ -v \sin u & v \cos u & 0 \\ \cos u & \sin u & 1 \end{vmatrix} = (v \cos u, v \sin u, -v).$$

Since

$$\|\mathbf{N}\| = \sqrt{v^2 \cos^2 u + v^2 \sin^2 u + v^2} = |v| \sqrt{2},$$

it follows that the cone is not smooth only when $v = 0$; i.e., at the point $\mathbf{r}(u, 0) = (0, 0, 0)$. At any other point $\mathbf{r}(u_0, v_0) = (v_0 \cos u_0, v_0 \sin u_0, v_0)$, $0 \le u_0 \le 2\pi$, $v \ne 0$, the tangent plane has the equation

$$v_0 \cos u_0 (x - v_0 \cos u_0) + v_0 \sin u_0 (y - v_0 \sin u_0) - v_0 (z - v_0) = 0;$$

i.e.,

$$(v_0 \cos u_0) x + (v_0 \sin u_0) y - v_0 z = 0.$$

Clearly, this plane goes through the origin.

27. An ellipsoid can be obtained by stretching or compressing a sphere in the directions of the coordinate axes. So we take a parametrization of a sphere (say, from Example 7.4) and adjust it (i.e., change the "radii" in the coordinate directions); thus

$$\mathbf{r}(u, v) = (a \cos v \cos u, b \cos v \sin u, c \sin v), \qquad 0 \le u \le 2\pi, \quad -\pi/2 \le v \le \pi/2.$$

It follows that

$$\frac{x^2}{a^2} + \frac{y^2}{b^2} + \frac{z^2}{c^2} = \cos^2 v \cos^2 u + \cos^2 v \sin^2 u + \sin^2 v = 1.$$

Here is another parametrization (and there are many possibilities):

$$\mathbf{r}(u, v) = (a \cos u \sin v, b \sin u \sin v, c \cos u), \qquad 0 \le u \le 2\pi, \quad 0 \le v \le \pi.$$

29. We have to find all points on the given surface where the surface normal $\mathbf{N}$ is parallel to $(-1, 1, 1)$ (which is a normal vector to the plane $-x + y + z = 0$). Now

$$\mathbf{N} = \mathbf{T}_u \times \mathbf{T}_v = \begin{vmatrix} \mathbf{i} & \mathbf{j} & \mathbf{k} \\ 2uv & v^2 & 0 \\ u^2 & 2uv & 0 \end{vmatrix} = (0, 0, 3u^2 v^2).$$

It follows that there are no points where $(0, 0, 3u^2 v^2)$ is parallel to $(-1, 1, 1)$.

31. We are asked to find the matrix A such that $D\mathbf{r}(u_0, v_0)\begin{bmatrix} u \\ v \end{bmatrix} = A\begin{bmatrix} u \\ v \end{bmatrix}$, where $\mathbf{r}(u, v) = (x(u, v), y(u, v), z(u, v))$. By (2.14) in Section 2.4, we get

$$D\mathbf{r} = \begin{bmatrix} \partial x/\partial u & \partial x/\partial v \\ \partial y/\partial u & \partial y/\partial v \\ \partial z/\partial u & \partial z/\partial v \end{bmatrix}.$$

Thus, $D\mathbf{r}(u_0, v_0)\begin{bmatrix} u \\ v \end{bmatrix} = A\begin{bmatrix} u \\ v \end{bmatrix}$, where

$$A = \begin{bmatrix} \partial x/\partial u(u_0, v_0) & \partial x/\partial v(u_0, v_0) \\ \partial y/\partial u(u_0, v_0) & \partial y/\partial v(u_0, v_0) \\ \partial z/\partial u(u_0, v_0) & \partial z/\partial v(u_0, v_0) \end{bmatrix}.$$

From

$$D\mathbf{r}(u_0, v_0)\begin{bmatrix} 1 \\ 0 \end{bmatrix} = \begin{bmatrix} \partial x/\partial u(u_0, v_0) \\ \partial y/\partial u(u_0, v_0) \\ \partial z/\partial u(u_0, v_0) \end{bmatrix} = \mathbf{T}_u(u_0, v_0)$$

and

$$D\mathbf{r}(u_0, v_0)\begin{bmatrix} 0 \\ 1 \end{bmatrix} = \begin{bmatrix} \partial x/\partial v(u_0, v_0) \\ \partial y/\partial v(u_0, v_0) \\ \partial z/\partial v(u_0, v_0) \end{bmatrix} = \mathbf{T}_v(u_0, v_0),$$

it follows that the range of $D\mathbf{r}(u_0, v_0)$ is the plane spanned by $\mathbf{T}_u(u_0, v_0)$ and $\mathbf{T}_v(u_0, v_0)$.

(b) Let P be a point in the tangent plane. Then $\overrightarrow{OP} - \mathbf{r}(u_0, v_0)$ is a vector in the tangent plane at $\mathbf{r}(u_0, v_0)$ spanned by $\mathbf{T}_u(u_0, v_0)$ and $\mathbf{T}_v(u_0, v_0)$. In (a) we showed that thus plane is the range of $D\mathbf{r}(u_0, v_0)$. Thus,

$$\overrightarrow{OP} - \mathbf{r}(u_0, v_0) = D\mathbf{r}(u_0, v_0)\begin{bmatrix} u - u_0 \\ v - v_0 \end{bmatrix},$$

for some vector $(u - u_0, v - v_0)$ (written as a column vector). We need $(u - u_0, v - v_0)$, rather than (u, v), since $D\mathbf{r}(u_0, v_0)$ acts on tangent vectors based at (u_0, v_0); for the record, its range consists of tangent vectors based at $\mathbf{r}(u_0, v_0)$.

7.2. World of Surfaces

1. From

$$\mathbf{c}_{\text{horiz}}(u) = \mathbf{r}(u, v_0) = (v_0, f(v_0)\cos u, f(v_0)\sin u), \quad 0 \le u \le 2\pi,$$

we conclude that $\mathbf{r}$ maps a horizontal line $v = v_0$ onto a circle in the plane $x = v_0$, centered at $(v_0, 0, 0)$ of radius $|f(v_0)|$. From

$$\mathbf{c}_{\text{vert}}(v) = \mathbf{r}(u_0, v) = (v, f(v)\cos u_0, f(v)\sin u_0), \quad a \le v \le b,$$

we conclude that $\mathbf{r}$ maps a vertical line $u = u_0$ onto a curve that is obtained by rotating the graph of $y = f(x)$ through the angle of u_0 radians.

Computing

$$\mathbf{c}'_{\text{horiz}}(u) = (0, -f(v_0)\sin u, f(v_0)\cos u)$$

and

$$\mathbf{c}'_{\text{vert}}(v) = (1, f'(v) \cos u_0, f'(v) \sin u_0),$$

we get that $\mathbf{c}'_{\text{horiz}}(u_0) \cdot \mathbf{c}'_{\text{vert}}(v_0) = 0$ at the point $\mathbf{r}(u_0, v_0)$. Thus, the images of $u = u_0$ and $v = v_0$ under $\mathbf{r}$ are perpendicular to each other.

3. Let $v = t$. Since the rotation is about the y-axis, $y = v^3$, $1 \leq v \leq 2$. The rotation produces circles of radius v^2 in the xz-plane, and thus $x = v^2 \cos u$ and $y = v^2 \sin u$. Thus,

$$\mathbf{r}(u, v) = (v^2 \cos u, v^3, v^2 \sin u), \qquad 0 \leq u \leq 2\pi, \quad 1 \leq v \leq 2,$$

is the desired parametrization.

5. From $\mathbf{r}(u_1, v_1) = \mathbf{r}(u_2, v_2)$ we get that $u_1 v_1 = u_2 v_2$, $u_1 = u_2$ and $v_1^2 = v_2^2$. Dividing $u_1 v_1 = u_2 v_2$ by $u_1 = u_2$ we get $v_1 = v_2$; i.e., as long as $u_1 = u_2 \neq 0$, $\mathbf{r}$ is one-to-one. If $u_1 = u_2 = 0$, then $\mathbf{r}(0, v) = (0, 0, v^2)$ is not one-to-one, since, for instance, $\mathbf{r}(0, \pm 1/2) = (0, 0, 1/4)$. in conclusion, if $u \neq 0$, then the map $\mathbf{r}$ is one-to-one.

7. We need to prove that the parametrization $\mathbf{r}(u, v) = (u \cos v, u \sin v, v)$ is one-to-one for all $(u, v) \in [0, 1] \times [0, 2\pi]$, except possibly on the boundary of D. By comparing the z-coordinates in $\mathbf{r}(u_1, v_1) = \mathbf{r}(u_2, v_2)$ we conlcude that $v_1 = v_2$. Consequently, $u_1 \cos v_1 = u_2 \cos v_1$ and $u_1 \sin v_1 = u_2 \sin v_1$. Since there is no v_1 for which both $\cos v_1$ and $\sin v_1$ are zero, it follows that $u_1 = u_2$.

9. Parametrize the given circle by $\mathbf{c}(t) = (2 \cos t + 3, 2 \sin t + 4)$, $0 \leq t \leq 2\pi$. Imitating the parametrization constructed in Exercise 2, or from the text preceeding the formula (7.7), we get, for the rotation about the x-axis,

$$\mathbf{r}(t, u) = (2 \cos t + 3, (2 \sin t + 4) \cos u, (2 \sin t + 4) \sin u),$$

where $0 \leq u, t \leq 2\pi$. The torus, obtained by rotating the same circle about the y-axis, can be parametrized as $\mathbf{r}(t, u) = ((2 \cos t + 3) \cos u, 2 \sin t + 4, (2 \cos t + 3) \sin u)$, where $0 \leq u, t \leq 2\pi$.

11. (a) One way to do this is to substitute the coordinates of B, B', B'' and B''' into $\mathbf{r}(u, v)$ and show that $\mathbf{r}(u, v) = (R + \rho, 0, 0)$ for all of them.

Alternatively, we think of how $D = [0, 2\pi] \times [0, 2\pi]$ is deformed to obtain the torus (see end of Example 7.18). Because the left side of D gets glued to the right side, B and B' will get mapped (glued) to the same point (call it B_1), and so will B'' and B''' (let B_2 be the point they get mapped (glued) to). Since B_1 and B_2 lie on the vertical line (B_2 directly above B_1), when the cylinder is deformed into the torus, they will get mapped (glued) to the same point (like points B and B' in Figure 7.26).

(b) After the right and the left sides of D are glued together to form the cylinder, the two points A and A' will become the same point. The curve is closed because after the cylinder is deformed into the torus, the points B and B'' get glued together (i.e., map to the same point; we know this from (a)).

(c) $\mathbf{r}$ will map the four segments into one continuous curve (we argue as in (b)) that is closed (again, we argue as in (b)) and that wraps around the torus four times.

13. Initially, the segment lies on the x-axis, between $\mathbf{r}(0, -1/2) = (1/2, 0, 0)$ and $\mathbf{r}(0, 1/2) = (3/2, 0, 0)$. After completing one half of the revolution, the end of the segment that was initially at $(1/2, 0, 0)$ is now at $\mathbf{r}(\pi, -1/2) = (-3/2, 0, 0)$, and the end that was initially at $(3/2, 0, 0)$ is now at $\mathbf{r}(\pi, 1/2) = (-1/2, 0, 0)$. Note that when $t = \pi$, the line segment is again horizontal. The end that was initially closer to the origin is now, half a revolution later, farther away from it (and vice versa). Thus, the segment completed one twist from $t = 0$ to $t = \pi$. From $t = \pi$ to $t = 2\pi$, it completes another twist, ending at its initial poistion when $t = 2\pi$.

If we wish to obtain a parametrization of this surface, all we need to do is to replace $t/2$ by t in the parametrization given in Example 7.19.

The surface normal comes back to its initial position as we complete our walk around the surface (since there are two twists of π radians each). Thus, the surface is two-sided (i.e., orientable).

15. (a) From $\mathbf{r}(u, v) = (u, v, uv)$ we get that $x = u$, $y = v$ and $z = uv = xy$. The level curves $z = c$ (i.e., the intersections of the surface with planes parallel to the xy-plane) are hyperbolas $y = c/x$, when $c \neq 0$. When $c = 0$, the level set consists of the x-axis and the y-axis. The intersection of $z = xy$ with the planes $y = mx$, $m \in \mathbb{R}$ (i.e., planes perpendicular to the xy-plane) are the parabolas $z = mx^2$.

(b) When $u = u_0$, we get

$$\mathbf{r}(u_0, v) = (u_0, v, u_0 v) = (u_0, 0, 0) + v(0, 1, u_0),$$

which is a line segment going through $(u_0, 0, 0)$ in the direction of the vector $(0, 1, u_0)$. As u_0 changes from -1 to 1, i.e., as a point travels along one edge of the square, the direction vector changes from $(0, 1, -1)$ to $(0, 1, 1)$; that motion produces the twist shown in Figure 7.42 in the text.

(c) The curve

$$\mathbf{r}(u, v_0) = (u, v_0, u v_0) = (0, v_0, 0) + v(1, 0, v_0), \qquad -1 \leq u \leq 1,$$

is a line segment through $(0, v_0, 0)$ in the direction of $(1, 0, v_0)$. Continuing as in (b), we realize that the change in v_0 produces a twist in the the other pair of parallel sides.

17. A straightforward calculation shows that

$$\frac{x^2}{a^2} + \frac{y^2}{b^2} + \frac{z^2}{c^2} = \frac{a^2 \cos^2 v \cos^2 u}{a^2} + \frac{b^2 \cos^2 v \sin^2 u}{b^2} + \frac{c^2 \sin^2 v}{c^2} = \cos^2 v + \sin^2 v = 1.$$

The domain could be $0 \leq u \leq 2\pi$, $-\pi/2 \leq v \leq \pi/2$ (as for the sphere in Example 7.4). There are many possible choices for the domain.

19. Imitating the parametrization of Example 7.3, we get $\mathbf{r}(u, v) = (av \cos u, bv \sin u, cv)$, where $0 \leq u \leq 2\pi$ and $v \in \mathbb{R}$. Clearly,

$$\frac{x^2}{a^2} + \frac{y^2}{b^2} - \frac{z^2}{c^2} = \frac{a^2 v^2 \cos^2 u}{a^2} + \frac{b^2 v^2 \sin^2 u}{b^2} - \frac{c^2 v^2}{c^2} = v^2 - v^2 = 0.$$

The tangent vectors are $\mathbf{T}_u = (-av \sin u, bv \cos u, 0)$ and $\mathbf{T}_v = (a \cos u, b \sin u, c)$, and thus $\mathbf{N}(u, v) = \mathbf{T}_u \times \mathbf{T}_v = (bcv \cos u, acv \sin u, -abv)$. Clearly, $\mathbf{r}(u, v)$ is differentiable at all (u, v). When $v = 0$, the normal $\mathbf{N}(u, 0) = \mathbf{0}$, and so the parametrization is not smooth at $\mathbf{r}(u, 0) = (0, 0, 0)$. It is smooth at other points.

21. Setting $z = z_0$, we get $x^2/a^2 + y^2/b^2 = 1$. Thus, the level curve of value z_0 (for all z_0) is the ellipse with semi-axes a and b and center at $(z_0, 0, 0)$. In other words, the cylinder is obtained by moving the ellipse $x^2/a^2 + y^2/b^2 = 1$ in the xy-plane up and down (so that its center always lies on the z-axis, and the ellipse is always parallel to the xy-plane).

Combining $x^2/a^2 + y^2/b^2 = 1$ and $y = 0$ we conclude that the cylinder intersects the xz-plane along $x = \pm a$. In the same way we show that the cylinder intersects and the yz-plane along vertical lines $y = \pm b$.

Imitating the parametrization of the cylinder in Example 7.2, we write

$$\mathbf{r}(u, v) = (a \cos u, b \sin u, v),$$

where $0 \leq u \leq 2\pi$ and $v \in \mathbb{R}$. The parametrization $\mathbf{r}$ is differentiable at all points. From $\mathbf{T}_u = (-a \sin u, b \cos u, 0)$ and $\mathbf{T}_v = (0, 0, 1)$ we get $\mathbf{N}(u, v) = (b \cos u, a \sin u, 0)$. Since $\mathbf{N}(u, v) \neq 0$ for all (u, v) (of course, assuming that $a, b \neq 0$) it follows that the above parametrization is smooth.

23. Set $u = x$ and $v = y$. Then $\mathbf{r}(u, v) = (u, v, c(u^2/a^2 + v^2/b^2))$, where $u, v \in \mathbb{R}$, parametrizes the given paraboloid. We compute $\mathbf{T}_u = (1, 0, 2cu/a)$, $\mathbf{T}_v = (0, 1, 2cv/b)$, and thus $\mathbf{N}(u, v) = (-2cu/a, -2cv/b, 1)$. Since $\mathbf{r}(u, v)$ is differentiable and $\mathbf{N}(u, v) \neq 0$ for all (u, v), we conclude that $\mathbf{r}(u, v)$ is a smooth parametrization.

The surface is called a paraboloid because it is built of parabolas: the cross-section of the surface with the plane $y = mx$ (m is a real number) is given by $z/c = x^2/a^2 + m^2 x^2/b^2$, i.e., it is the parabola $z = (c/a^2 + cm^2/b^2)x^2$.

7.3. Surface Integrals of Real-Valued Functions

1. (a) Recall that, by definition,

$$\frac{\partial x}{\partial u} = \lim_{\Delta u \to 0} \frac{x(u + \Delta u, v) - x(u, v)}{\Delta u}.$$

Since Δu is small, we write

$$\frac{\partial x}{\partial u} \approx \frac{x(u + \Delta u, v) - x(u, v)}{\Delta u},$$

i.e., $x(u + \Delta u, v) - x(u, v) \approx (\partial x/\partial u)(u, v)\Delta u$.

(b) Analogously, we obtain $y(u+\Delta u, v) - y(u, v) \approx (\partial y/\partial u)(u, v)\Delta u$, and a similar expression for z. It follows that

$$\mathbf{r}(u+\Delta u, v) - \mathbf{r}(u, v) \approx \begin{bmatrix} \partial x/\partial u(u, v) \\ \partial y/\partial u(u, v) \\ \partial z/\partial u(u, v) \end{bmatrix} \Delta u = \frac{\partial \mathbf{r}}{\partial u}(u, v)\Delta u.$$

3. We could use the result of Exercise 2: to compute the surface area we need to add up the lengths of all circles as we walk along the curve. At a location $(x, f(x))$, the radius of the circle (rotation is about the y-axis) is $|x|$. The curve $y = f(x)$ can be parametrized as $\mathbf{c}(t) = (t, f(t))$, $t \in [a, b]$. We conclude that the surface area is

$$\int_{\mathbf{c}} 2\pi|x|\,ds = \int_a^b 2\pi|t|\sqrt{1+(f'(t))^2}\,dt,$$

since $ds = \|\mathbf{c}'(t)\|dt = \sqrt{1+(f'(t))^2}\,dt$.

Alternatively, parametrize S by $\mathbf{r}(u, v) = (u\cos v, f(u), u\sin v)$, where $a \leq u \leq b$, $0 \leq v \leq 2\pi$. Then $\mathbf{T}_u = (\cos v, f'(u), \sin v)$, $\mathbf{T}_v = (-u\sin v, 0, u\cos v)$, $\mathbf{N} = \mathbf{T}_u \times \mathbf{T}_v = (uf'(u)\cos v, u, uf'(u)\sin v)$, and $\|\mathbf{N}\| = \sqrt{u^2(f'(u))^2 + u^2} = |u|\sqrt{1+(f'(u))^2}$. Thus,

$$A(S) = \iint_{[a,b]\times[0,2\pi]} \|\mathbf{N}(u, v)\|\,dA$$

$$= \int_0^{2\pi} \left(\int_a^b |u|\sqrt{1+(f'(u))^2}du \right) dv = 2\pi \int_a^b |u|\sqrt{1+(f'(u))^2}\,du.$$

5. Parametrize S by

$$\mathbf{r}(u, v) = (u, v, u^2 + v^2), \qquad (u, v) \in D,$$

where D is the region in the uv-plane defined by $u^2 + v^2 \leq 4$. Then $\mathbf{T}_u = (1, 0, 2u)$, $\mathbf{T}_v = (0, 1, 2v)$ and

$$\mathbf{N} = \mathbf{T}_u \times \mathbf{T}_v = \begin{vmatrix} \mathbf{i} & \mathbf{j} & \mathbf{k} \\ 1 & 0 & 2u \\ 0 & 1 & 2v \end{vmatrix} = (-2u, -2v, 1),$$

and so $\|\mathbf{N}\| = \sqrt{1 + 4(u^2 + v^2)}$. The values of f at points on the surface are $f(\mathbf{r}(u, v)) = f(u, v, u^2 + v^2) = uv$; it follows that

$$\iint_S xy\,dS = \iint_D uv\sqrt{1 + 4(u^2 + v^2)}\,dA.$$

Passing to polar coordinates, we get

$$\iint_S xy\,dS = \int_0^{2\pi} \left(\int_0^2 r^2\sin\theta\cos\theta\sqrt{1+4r^2}\,rdr \right) d\theta$$

$$= \left(\int_0^{2\pi} \sin\theta\cos\theta\,d\theta \right) \left(\int_0^2 r^3\sqrt{1+4r^2}\,dr \right).$$

Since

$$\int_0^{2\pi} \sin\theta\cos\theta\,d\theta = \tfrac{1}{2}\int_0^{2\pi} \sin 2\theta\,d\theta = 0,$$

it follows that $\iint_S xy\,dS = 0$.

7. S is a piecewise smooth surface: it consists of four smooth surfaces, that are the faces of the tetrahedron. We will integrate f along each smooth surface and then add the results.

Let S_1 be the face containing the points $(0,0,0)$, $(2,0,0)$ and $(0,2,0)$. It is a part of the xy-plane, so parametrize it by $\mathbf{r}_1(u,v) = (u,v,0)$, $(u,v) \in D_1$, where D_1 is the triangular region in the uv-plane defined by $u = 0$, $v = 0$ and $u + v = 2$ (that is the line joining $(2,0)$ and $(0,2)$). The normal is

$$\mathbf{N} = \begin{vmatrix} \mathbf{i} & \mathbf{j} & \mathbf{k} \\ 1 & 0 & 0 \\ 0 & 1 & 0 \end{vmatrix} = \mathbf{k}$$

(as expected) and so

$$\iint_{S_1} (y + x)\,dS = \iint_{D_1} (u + v)\,1\,dA = \int_0^2 \left(\int_0^{2-v} (u + v)du \right) dv$$

$$= \int_0^2 \left(\tfrac{1}{2}u^2 + uv \right) \Big|_0^{2-v} dv = \int_0^2 \left(2 - \tfrac{1}{2}v^2 \right) dv = \left(2v - \tfrac{1}{6}v^3 \right) \Big|_0^2 = \tfrac{8}{3}.$$

Let S_2 be the face containing the points $(0,0,0)$, $(0,2,0)$ and $(0,0,2)$. It is a part of the yz-plane, so parametrize it by $\mathbf{r}_2(u,v) = (0,u,v)$, $(u,v) \in D_2$; D_2 is the triangular region in the uv-plane (that, in this case, coincides with the yz-plane) defined by $u = 0$, $v = 0$ and $u + v = 2$. The normal is

$$\mathbf{N} = \begin{vmatrix} \mathbf{i} & \mathbf{j} & \mathbf{k} \\ 0 & 1 & 0 \\ 0 & 0 & 1 \end{vmatrix} = \mathbf{i}$$

and so

$$\iint_{S_2} (y + x)\,dS = \iint_{D_2} u\,dA = \int_0^2 \left(\int_0^{2-v} u\,du \right) dv$$

$$= \int_0^2 \left(\tfrac{1}{2}u^2 \right) \Big|_0^{2-v} dv = \tfrac{1}{2} \int_0^2 (-2 + v)^2\,dv = \tfrac{1}{2}\tfrac{1}{3}(v - 2)^3 \Big|_0^2 = \tfrac{4}{3}.$$

Similarly, let S_3 be the face containing the points $(0,0,0)$, $(2,0,0)$ and $(0,0,2)$. It is a part of the xz-plane, so parametrize it by $\mathbf{r}_3(u,v) = (u,0,v)$, $(u,v) \in D_3$; D_3 is the triangular region defined by $u = 0$, $v = 0$ and $u + v = 2$. The normal is

$$\mathbf{N} = \begin{vmatrix} \mathbf{i} & \mathbf{j} & \mathbf{k} \\ 1 & 0 & 0 \\ 0 & 0 & 1 \end{vmatrix} = -\mathbf{j}$$

and so

$$\iint_{S_3} (y + x)\,dS = \iint_{D_3} u\,dA = \int_0^2 \left(\int_0^{2-v} u\,du \right) dv = \tfrac{4}{3};$$

the result is borrowed from the integration over S_2.

Finally, parametrize the face S_4 containing the points $(2,0,0)$, $(0,2,0)$ and $(0,0,2)$ (the equation of the plane containing these three points is $x+y+z = 2$) as $\mathbf{r}_4(u,v) = (u,v,2-u-v)$, $(u,v) \in D_4 = D_1$. The surface normal $\mathbf{N}$ is

$$\mathbf{N} = \begin{vmatrix} \mathbf{i} & \mathbf{j} & \mathbf{k} \\ 1 & 0 & -1 \\ 0 & 1 & -1 \end{vmatrix} = \mathbf{i} + \mathbf{j} + \mathbf{k};$$

so $\|\mathbf{N}\| = \sqrt{3}$, and thus

$$\iint_{S_4} (y+x)\,dS = \iint_{D_4} (u+v)\sqrt{3}\,dA = \tfrac{8}{3}\sqrt{3},$$

(borrowing the result of $\iint_{S_1} (y+x)\,dS$).

Hence

$$\iint_{S} (y+x)\,dS = \tfrac{8}{3} + \tfrac{4}{3} + \tfrac{4}{3} + \tfrac{8}{3}\sqrt{3} = \tfrac{16}{3} + \tfrac{8}{3}\sqrt{3}.$$

9. Imitating the parametrization used in Example 7.3 in Section 7.1, we get $\mathbf{r}(u,v) = (v, v\cos u, v\sin u)$. The requirement that $\mathbf{r}(u,v)$ be in the first octant implies $v \geq 0$ and $0 \leq u \leq \pi/2$. It is also given that $v \leq 1$. Therefore,

$$\mathbf{r}(u,v) = (v, v\cos u, v\sin u), \qquad 0 \leq u \leq \pi/2, \quad 0 \leq v \leq 1$$

is a desired parametrization. The normal vector $\mathbf{N}$ is

$$\mathbf{N} = \begin{vmatrix} \mathbf{i} & \mathbf{j} & \mathbf{k} \\ 0 & -v\sin u & v\cos u \\ 1 & \cos u & \sin u \end{vmatrix} = (-v, v\cos u, v\sin u),$$

and its magnitude is

$$\|\mathbf{N}\| = \sqrt{v^2 + v^2\cos^2 u + v^2\sin^2 u} = |v|\sqrt{2} = v\sqrt{2}.$$

The value of f at points on S is $f(\mathbf{r}(u,v)) = 2v\cos u - v$, and therefore

$$\iint_{S} (2y-x)\,dS = \iint_{D} (2v\cos u - v)v\sqrt{2}\,dA,$$

where D is defined by $0 \leq u \leq \pi/2$, $0 \leq v \leq 1$. Hence

$$\iint_{S} (2y-x)\,dS = \sqrt{2}\int_{0}^{\pi/2} \left(\int_{0}^{1} v^2(2\cos u - 1)dv \right) du$$

$$= \sqrt{2}\int_{0}^{\pi/2} \tfrac{1}{3}(2\cos u - 1)\,du$$

$$= \frac{\sqrt{2}}{3}(2\sin u - u)\Big|_{0}^{\pi/2} = \frac{\sqrt{2}}{3}\left(2 - \tfrac{1}{2}\pi\right) = \tfrac{1}{6}(4-\pi)\sqrt{2}.$$

11. The intersection of the paraboloid $z = 4-x^2-y^2$ and the xy-plane is the circle $x^2+y^2 = 4$. Denote the disk $\{(x,y) \mid x^2+y^2 \leq 4\}$ by D, and parametrize S as

$$\mathbf{r}(u,v) = (u,v,4-u^2-v^2), \qquad (u,v) \in D.$$

The values of f on S are $f(\mathbf{r}(u,v)) = (4u^2 + 4v^2 + 1)^{-1/2}$; the normal vector $\mathbf{N}$ is

$$\mathbf{N} = \begin{vmatrix} \mathbf{i} & \mathbf{j} & \mathbf{k} \\ 1 & 0 & -2u \\ 0 & 1 & -2v \end{vmatrix} = (2u, 2v, 1),$$

so $\|\mathbf{N}\| = \sqrt{4u^2 + 4v^2 + 1}$ and therefore

$$\iint_S (4x^2 + 4y^2 + 1)^{-1/2}\, dS = \iint_D (4u^2 + 4v^2 + 1)^{-1/2}\,(4u^2 + 4v^2 + 1)^{1/2}\, dA$$

$$= \iint_D 1\, dA = \text{area}(D) = 4\pi.$$

13. From $\mathbf{r}(u,v) = (2u\cos v, 2u\sin v, v)$ it follows that $\mathbf{T}_u = (2\cos v, 2\sin v, 0)$ and $\mathbf{T}_v = (-2u\sin v, 2u\cos v, 1)$; so the surface normal $\mathbf{N}$ is

$$\mathbf{N} = \begin{vmatrix} \mathbf{i} & \mathbf{j} & \mathbf{k} \\ 2\cos v & 2\sin v & 0 \\ -2u\sin v & 2u\cos v & 1 \end{vmatrix} = (2\sin v, -2\cos v, 4u).$$

Hence $\|\mathbf{N}\| = \sqrt{4 + 16u^2} = 2\sqrt{1 + 4u^2}$ and the surface area is

$$\iint_S dS = \iint_D 2\sqrt{1 + 4u^2}\, dA,$$

where D is the region in the uv-plane defined by $0 \le u \le 2$ and $0 \le v \le \pi$. So

$$\iint_S dS = 2 \int_0^\pi \left(\int_0^2 \sqrt{1 + 4u^2}\, du \right) dv = 2 \left(\int_0^\pi dv \right) \left(\int_0^2 \sqrt{1 + 4u^2}\, du \right).$$

Using integration tables (or a trigonometric substitution) we get

$$\int_0^2 \sqrt{1 + 4u^2}\, du = 2 \int_0^2 \sqrt{\tfrac{1}{4} + u^2}\, du$$

$$= 2 \left(\tfrac{1}{2} u \sqrt{\tfrac{1}{4} + u^2} + \tfrac{1}{8} \ln \left| u + \sqrt{\tfrac{1}{4} + u^2} \right| \right) \Bigg|_0^2$$

$$= 2 \left(\tfrac{1}{2}\sqrt{17} + \tfrac{1}{8} \ln \left| 2 + \tfrac{1}{2}\sqrt{17} - \tfrac{1}{8} \ln \left(\tfrac{1}{2} \right) \right| \right).$$

It follows that the surface area of the surface in question is

$$4\pi \left(\tfrac{1}{2}\sqrt{17} + \tfrac{1}{8} \ln \left| 2 + \tfrac{1}{2}\sqrt{17} + \tfrac{1}{8} \ln 2 \right| \right) \approx 29.1966.$$

15. Recall the parametrization of the cone from Example 7.3 in Section 7.1:

$$\mathbf{r}(u,v) = (v\cos u, v\sin u, v), \qquad 0 \le u \le 2\pi, \quad 0 \le v \le h.$$

This is not what we are looking for: when $v = h$, $\mathbf{r}(u,h) = (h\cos u, h\sin u, h)$; i.e., we get a circle of radius h in the plane $z = h$. What we need is a circle of radius r — so, we adjust the above parametrization:

$$\mathbf{r}(u,v) = \left(\frac{rv}{h} \cos u, \frac{rv}{h} \sin u, v \right), \qquad 0 \le u \le 2\pi, \quad 0 \le v \le h.$$

This time, $\mathbf{r}(u,h) = (r\cos u, r\sin u, h)$, exactly as needed.

Now $\mathbf{T}_u = (-(rv/h)\sin u, (rv/h)\cos u, 0)$ and $\mathbf{T}_v = ((r/h)\cos u, (r/h)\sin u, 1)$; so the surface normal $\mathbf{N}$ is

$$\mathbf{N} = \begin{vmatrix} \mathbf{i} & \mathbf{j} & \mathbf{k} \\ -(rv/h)\sin u & (rv/h)\cos u & 0 \\ (r/h)\cos u & (r/h)\sin u & 1 \end{vmatrix} = \left(\frac{rv}{h}\cos u, \frac{rv}{h}\sin u, -\frac{r^2 v}{h^2} \right).$$

Hence

$$\|\mathbf{N}\| = \sqrt{\frac{r^2 v^2}{h^2} + \frac{r^4 v^2}{h^4}} = |v|\sqrt{\frac{r^2}{h^2}\left(1 + \frac{r^2}{h^2}\right)} = \frac{vr}{h}\sqrt{1 + \frac{r^2}{h^2}}$$

and the surface area of the cone is

$$A(S) = \iint_S dS = \iint_D \|\mathbf{N}\|\, dA,$$

where D is the region in the uv-plane defined by $0 \le u \le 2\pi$ and $0 \le v \le h$. Hence

$$A(S) = \int_0^{2\pi} \left(\int_0^h \frac{vr}{h}\sqrt{1 + \frac{r^2}{h^2}}\, dv \right) du$$

$$= \left(\int_0^{2\pi} du \right) \frac{r}{h}\sqrt{1 + \frac{r^2}{h^2}} \left(\frac{v^2}{2}\Big|_0^h \right)$$

$$= \frac{2\pi r}{h}\sqrt{1 + \frac{r^2}{h^2}}\frac{h^2}{2} = \pi r h\sqrt{1 + \frac{r^2}{h^2}} = \pi r\sqrt{h^2 + r^2}.$$

17. Recall that a surface integral $\iint_S f\, dS$ is the limit as $n \to \infty$ of Riemann sums of the form

$$\mathcal{R}_n = \sum (\text{value of } f) \cdot (\text{area of a small patch of } S),$$

where the summation goes over all small patches that cover S. Take a subdivision where all small patches have the same area. Then

$$\mathcal{R}_n = (\text{area of a small patch}) \sum (\text{value of } f).$$

At a point (x, y, z), the value of the function $f(x, y, z) = x$ is x. At a symmetric point $(-x, y, z)$ on S, the value is $-x$, and the two contributions to $\mathcal{R}_n$ will cancel each other. That will happen to every such pair, and consequently, $\mathcal{R}_n = 0$ for all n and so $\iint_S x\, dS = 0$. An analogous argument proves that $\iint_S x^3\, dS = 0$.

By symmetry,

$$\iint_S x^2\, dS = \iint_S y^2\, dS = \iint_S z^2\, dS,$$

and so

$$3\iint_S x^2\, dS = \iint_S x^2\, dS + \iint_S y^2\, dS + \iint_S z^2\, dS = \iint_S (x^2 + y^2 + z^2)\, dS$$

$$= \iint_S a^2\, dS = 4\pi a^2\, a^2 = 4\pi a^4,$$

since $\iint_S dS = \text{area}(S) = 4\pi a^2$. It follows that $\iint_S x^2\, dS = 4\pi a^4/3$.

19. See Figure 19. One side of S is of length a, and the other one is the hypotenuse of the right triangle with sides b and bm (the slope of $z = my$ is m). So the area of S is

$$\text{area}(S) = a\sqrt{b^2 + b^2m^2} = ab\sqrt{1 + m^2} = \text{area}(R)\sqrt{1 + m^2}.$$

Notice that the angle between $\mathbf{k}$ and the upward normal $\mathbf{N}$ is equal to α. Since

$$\cos\alpha = \frac{b}{\sqrt{b^2 + b^2m^2}} = \frac{1}{\sqrt{1 + m^2}},$$

it follows that $\sec\alpha = \sqrt{1 + m^2}$ and (area of S) $= \sec\alpha$ (area of R).

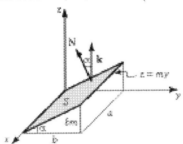

Figure 19

21. Parametrize S by $\mathbf{r}(u, v) = (u, v, 1 - v^2)$, $(u, v) \in D$, where D is defined by $0 \leq u \leq 2$ and $0 \leq v \leq 1$ (obtained from $z = 1 - y^2 = 0$ and the fact that S is in the first octant). The values of f on S are $f(\mathbf{r}(u, v)) = 4uv$; the normal is

$$\mathbf{N} = \begin{vmatrix} \mathbf{i} & \mathbf{j} & \mathbf{k} \\ 1 & 0 & 0 \\ 0 & 1 & -2v \end{vmatrix} = 2v\mathbf{j} + \mathbf{k};$$

hence $\|\mathbf{N}\| = \sqrt{1 + 4v^2}$ and therefore

$$\iint_S 4xy\, dS = \iint_D 4uv\sqrt{1 + 4v^2}\, dA$$
$$= \int_0^2 \left(\int_0^1 4uv\sqrt{1 + 4v^2}\, dv \right) du$$
$$= \left(\int_0^2 u\, du \right) \left(2\tfrac{1}{3}(1 + 4v^2)^{3/2} \right) \Big|_0^1$$
$$= \tfrac{4}{3}(5^{3/2} - 1).$$

Let us recompute this integral using a projection onto the xy-plane; that projection is the rectangle (call it D) $0 \leq x, y \leq 1$. Since the unit normal is $\mathbf{n} = (2y\mathbf{j} + \mathbf{k})/\sqrt{(1 + 4y^2)}$, it follows that

$$\iint_S f\, dS = \iint_D 4xy\, \frac{dA}{|\mathbf{n} \cdot \mathbf{k}|} = \iint_D 4xy\sqrt{1 + 4y^2}dA.$$

We proceed as above.

23. Let us project onto the xy-plane. In that case (take $z = 0$) $x + y^2 = 4$, and the projection D is defined by $x = 0$, $y = 0$ and $x + y^2 = 4$. A normal vector to S is (view it as the level surface of $g(x, y, z) = x + y^2 + z$ of value 4) $\nabla g = (1, 2y, 1)$, and a unit normal is $\mathbf{n} = \nabla g/\|\nabla g\| = (1, 2y, 1)/\sqrt{2 + 4y^2}$. Thus

$$\iint_S y \, dS = \iint_D y \, \frac{dA}{|\mathbf{n} \cdot \mathbf{k}|}$$

$$= \iint_D y\sqrt{2 + 4y^2} \, dA$$

$$= \int_0^4 \left(\int_0^{\sqrt{4-x}} y\sqrt{2 + 4y^2} \, dy \right) dx$$

$$= \int_0^4 \left(\tfrac{1}{12}(2 + 4y^2)^{3/2} \right) \Big|_0^{\sqrt{4-x}} dx$$

$$= \tfrac{1}{12} \int_0^4 \left((18 - 4x)^{3/2} - 2^{3/2} \right) dx$$

$$= \tfrac{1}{12} \left(-\tfrac{2}{20}(18 - 4x)^{5/2} - 2^{3/2}x \right) \Big|_0^4$$

$$= \tfrac{1}{12} \left(-\tfrac{1}{10}2^{5/2} - 2^{3/2}\,4 + \tfrac{1}{10}18^{5/2} \right) \approx 10.4652.$$

25. Let $g(x, y, z) = x^2 + y^2 + z^2$; then S is the level surface of g of value a^2, and so

$$\mathbf{n} = \frac{\nabla g}{\|\nabla g\|} = \frac{(2x, 2y, 2z)}{\sqrt{4x^2 + 4y^2 + 4z^2}} = \frac{(x, y, z)}{a}$$

is its unit normal. Project S onto the xz-plane; the projection D is the disk $x^2 + z^2 \leq a^2$. It follows that the surface area is

$$A(S) = \iint_D \frac{dA}{|\mathbf{n} \cdot \mathbf{j}|}$$

$$= \iint_D \frac{dA}{y/a} = a \iint_D (a^2 - x^2 - z^2)^{-1/2}.$$

Passing to polar coordinates $x = r\cos\theta$, $z = r\sin\theta$, we get

$$A(S) = a \int_0^{2\pi} \left(\int_0^a (a^2 - r^2)^{-1/2} \, r dr \right) d\theta$$

$$= a \left(\int_0^{2\pi} d\theta \right) \left(-(a^2 - r^2)^{1/2} \Big|_0^a \right)$$

$$= 2\pi a(0 - (-(a^2)^{1/2})) = 2\pi a^2.$$

7.4. Surface Integrals of Vector Fields

1. S is part of the cylinder of radius 1 (whose axis of symmetry is the z-axis) between $z = 1$ and $z = 2$. At all points, its outward normal $\mathbf{N}$ is parallel to the xy-plane. Thus, $\iint_S \mathbf{k} \cdot d\mathbf{S} = 0$.

The flow of $\mathbf{F}_1(x, y, z) = \mathbf{i}$ into the cylinder is the same as the flow out of it, since $\mathbf{F}_1$ is constant and the areas along which $\mathbf{F}_1$ flows in and out are equal. Thus, the flux of $\mathbf{F}_1$ is zero. The flux of $\mathbf{F}_2(x, y, z) = x\mathbf{i} + y\mathbf{j}$ is positive, since it flows out along the whole surface.

3. The flux $\iint_S \mathbf{F} \cdot d\mathbf{S}$ is equal to the volume that flows through S in a unit of time. Since $\|\mathbf{F}\|$ is constant, the flux is equal to $\|\mathbf{F}\| A(S) \cos \theta$, where $A(S)$ is the area of the surface and θ is the angle between the surface normal $\mathbf{N}$ and the vector field $\mathbf{F}$. Thus, the flux is

$$\|\mathbf{F}\| A(S) \cos \theta = \|\mathbf{F}\| A(S) \frac{\mathbf{F} \cdot \mathbf{N}}{\|\mathbf{F}\| \|\mathbf{N}\|} = A(S) \frac{\mathbf{F} \cdot \mathbf{N}}{\|\mathbf{N}\|}.$$

Since $A(S) = 12$ and $\mathbf{N} = \mathbf{k}$ (so that $\mathbf{F} \cdot \mathbf{N} = 4$), we get that the flux is equal to $12 \cdot 4 = 48$.

5. The equation of the plane through $(1,0,0)$, $(0,2,0)$ and $(0,0,4)$ is $x + y/2 + z/4 = 1$, i.e., $4x + 2y + z = 4$. So, the surface normal is $\mathbf{N} = (4,2,1)$.

The area of the triangle can be computed using the cross product. Let $\mathbf{v} = (-1,2,0)$ (vector from $(1,0,0)$ to $(0,2,0)$) and $\mathbf{w} = (-1,0,4)$ (vector from $(1,0,0)$ to $(0,0,4)$). Then $A(S) = \|\mathbf{v} \times \mathbf{w}\|/2 = \|(8,4,2)\|/2 = \sqrt{21}$.

We compute the flux ($\|\mathbf{N}\| = \sqrt{21}$ and $\mathbf{F} \cdot \mathbf{N} = (1,-2,4) \cdot (4,2,1) = 4$) to be $A(S)\,\mathbf{F} \cdot \mathbf{N}/\|\mathbf{N}\| = \sqrt{21}(4/\sqrt{21}) = 4$.

7. Parametrize S by $\mathbf{r}(u,v) = (u, v, 3u + v - 1)$, where $u^2 + v^2 \le 4$ (so that only the part of the plane inside the cylinder is considered). Then $\mathbf{T}_u = (1,0,3)$, $\mathbf{T}_v = (0,1,1)$ and

$$\mathbf{N} = \mathbf{T}_u \times \mathbf{T}_v = \begin{vmatrix} \mathbf{i} & \mathbf{j} & \mathbf{k} \\ 1 & 0 & 3 \\ 0 & 1 & 1 \end{vmatrix} = -3\mathbf{i} - \mathbf{j} + \mathbf{k}$$

(since the $\mathbf{k}$ component of $\mathbf{N}$ is positive, $\mathbf{N}$ is an upward normal). It follows that

$$\iint_S (4y\mathbf{i} + (3x-1)\mathbf{j} + z\mathbf{k}) \cdot d\mathbf{S} = \iint_D (4y\mathbf{i} + (3x-1)\mathbf{j} + z\mathbf{k}) \cdot (-3\mathbf{i} - \mathbf{j} + \mathbf{k})\, dA$$

$$= \iint_D (-12y - 3x + 1 + z)\, dA$$

where D is the region in the uv-plane defined by $u^2 + v^2 \le 4$. Passing to polar coordinates $u = r \cos \theta$, $v = r \sin \theta$, we get

$$\iint_S (4y\mathbf{i} + (3x-1)\mathbf{j} + z\mathbf{k}) \cdot d\mathbf{S} = \iint_D (-12v - 3u + 1 + 3u + v - 1)\, dA$$

$$= \iint_D (-11v)\, dA$$

$$= -11 \int_0^{2\pi} \left(\int_0^2 r^2 \sin\theta\, dr \right) d\theta$$

$$= -11 \left(\int_0^{2\pi} \sin\theta\, d\theta \right) \left(\int_0^2 \tfrac{1}{3} r^3\, dr \right) = 0,$$

since $\int_0^{2\pi} \sin\theta\, d\theta = 0$.

9. Parametrize S by

$$\mathbf{r}(u,v) = 3\cos v \cos u\, \mathbf{i} + 3\cos v \sin u\, \mathbf{j} + 3\sin v\, \mathbf{k}, \qquad 0 \le u \le 2\pi, \quad 0 \le v \le \pi/2$$

(if $0 \le v \le \pi/2$, then $3 \sin v \ge 0$, as needed). Let D be the rectangle $D = [0, 2\pi] \times [0, \pi/2]$ in the uv-plane. The parametrization we use has been introduced in Example 7.4; the surface normal $\mathbf{N}$ was computed in Example 7.7 to be $\mathbf{N} = 3 \cos v \, \mathbf{r}(u, v)$ (and this is an outward normal, since it points away from the origin). Therefore,

$$\iint_S (x^2 \mathbf{i} + 2z\mathbf{k}) \cdot d\mathbf{S}$$

$$= \iint_D (9 \cos^2 v \cos^2 u \, \mathbf{i} + 6 \sin v \, \mathbf{k}) \cdot 3 \cos v \, (3 \cos v \cos u \, \mathbf{i} + 3 \cos v \sin u \, \mathbf{j} + 3 \sin v \, \mathbf{k}) \, dA$$

$$= \iint_D (81 \cos^4 v \cos^3 u + 54 \sin^2 v \cos v) \, dA$$

$$= 81 \int_0^{2\pi} \left(\int_0^{\pi/2} \cos^4 v \cos^3 u \, dv \right) du + 54 \int_0^{2\pi} \left(\int_0^{\pi/2} \sin^2 v \cos v \, dv \right) du$$

$$= 81 \left(\int_0^{2\pi} \cos^3 u \, du \right) \left(\int_0^{\pi/2} \cos^4 v \, dv \right) + 54 \left(\int_0^{2\pi} du \right) \left(\tfrac{1}{3} \sin^3 v \Big|_0^{\pi/2} \right).$$

Since $\int_0^{2\pi} \cos^3 u \, du = 0$ (no computation is needed — sketch the graph of $\cos^3 u$ and think of a definite integral as a "signed area"; i.e., "area above the x-axis" − "area below the x-axis"), it follows that

$$\iint_S (x^2 \mathbf{i} + 2z\mathbf{k}) \cdot d\mathbf{S} = 54(2\pi)\tfrac{1}{3} = 36\pi.$$

11. Combining $z = x^2 + y^2$ with $z = 1$ and with $z = 2$, we get $1 \le x^2 + y^2 \le 2$. So we can parametrize S by $\mathbf{r}(u, v) = (u, v, u^2 + v^2)$, where $(u, v) \in D$, and D is the annulus $1 \le u^2 + v^2 \le 2$ in the uv-plane. The surface normal $\mathbf{N}$ is computed to be

$$\mathbf{N} = \mathbf{T}_u \times \mathbf{T}_v = \begin{vmatrix} \mathbf{i} & \mathbf{j} & \mathbf{k} \\ 1 & 0 & 2u \\ 0 & 1 & 2v \end{vmatrix} = -2u\mathbf{i} - 2v\mathbf{j} + \mathbf{k}.$$

Since the $\mathbf{k}$ component of $\mathbf{N}$ is positive, $\mathbf{N}$ points inward, into the paraboloid. We can do one of two things: use $\mathbf{N}$ to compute the surface integral and then change the sign of the result, or change the sign of the normal $\mathbf{N}$ right away. We will change it right away:

$$\iint_S z\mathbf{k} \cdot d\mathbf{S} = \iint_D z\mathbf{k} \cdot (-\mathbf{N}) \, dA$$

$$= \iint_D (u^2 + v^2)\mathbf{k} \cdot (2u\mathbf{i} + 2v\mathbf{j} - \mathbf{k}) \, dA$$

$$= -\iint_D (u^2 + v^2) \, dA$$

(passing to polar coordinates $u = r \cos \theta$, $v = r \sin \theta$)

$$= -\int_0^{2\pi} \left(\int_1^{\sqrt{2}} r^3 dr \right) d\theta$$

$$= -\left(\int_0^{2\pi} d\theta \right) \left(\tfrac{1}{4} r^4 \Big|_1^{\sqrt{2}} \right) = -2\pi \left(1 - \tfrac{1}{4} \right) = -\tfrac{3}{2}\pi.$$

13. From $\mathbf{T}_u = (e^u \cos v, e^u \sin v, 0)$ and $\mathbf{T}_v = (-e^u \sin v, e^u \cos v, 1)$ it follows that

$$\mathbf{N} = \begin{vmatrix} \mathbf{i} & \mathbf{j} & \mathbf{k} \\ e^u \cos v & e^u \sin v & 0 \\ -e^u \sin v & e^u \cos v & 1 \end{vmatrix} = (e^u \sin v, -e^u \cos v, e^{2u})$$

(this is an upward pointing normal, as needed). The value of $\mathbf{F}$ along S is $\mathbf{F}(\mathbf{r}(u, v)) = (e^u \cos v, e^u \sin v, v)$; so

$$\mathbf{F} \cdot \mathbf{N} = (e^u \cos v, e^u \sin v, v) \cdot (e^u \sin v, -e^u \cos v, e^{2u}) = e^{2u} v,$$

and (let D be the rectangle $[0, \ln 2] \times [0, \pi]$ in the uv-plane)

$$\iint_S (x\mathbf{i} + y\mathbf{j} + z\mathbf{k}) \cdot d\mathbf{S} = \iint_D e^{2u} v \, dA$$

$$= \int_0^{\ln 2} \left(\int_0^\pi e^{2u} v \, dv \right) du$$

$$= \left(\int_0^{\ln 2} e^{2u} \, du \right) \left(\int_0^\pi v \, dv \right)$$

$$= \left(\frac{1}{2} e^{2u} \Big|_0^{\ln 2} \right) \left(\frac{1}{2} v^2 \Big|_0^\pi \right) = \frac{3}{4} \pi^2.$$

15. From $\mathbf{F} = -\nabla T$ we get $\mathbf{F}(x, y, z) = -(2x, 2y, 6z)$. Parametrize the cylinder by $\mathbf{r}(u, v) = (\cos u, \sin u, v)$, where $(u, v) \in D = [0, 2\pi] \times [-1, 1]$. The surface normal is

$$\mathbf{N} = \mathbf{T}_u \times \mathbf{T}_v = \begin{vmatrix} \mathbf{i} & \mathbf{j} & \mathbf{k} \\ -\sin u & \cos u & 0 \\ 0 & 0 & 1 \end{vmatrix} = (\cos u, \sin u, 0).$$

Since $\mathbf{T}_u \times \mathbf{T}_v$ points away from the origin (hence away from the surface of the cylinder), it follows that $\mathbf{T}_u \times \mathbf{T}_v$ is an outward normal. The flux of $\mathbf{F}$ is

$$-\iint_S (2x, 2y, 6z) \cdot d\mathbf{S} = -\iint_D (2\cos u, 2\sin u, 6v) \cdot (\cos u, \sin u, 0) \, dA$$

$$= -\iint_D 2 \, dA = -2\,\text{area}(D) = -8\pi.$$

17. (a) View S as the level surface of $f(x, y, z) = x^2 + y^2 + z^2$ of value a^2. Then

$$\frac{\nabla f}{\|\nabla f\|} = \frac{(2x, 2y, 2z)}{\sqrt{4x^2 + 4y^2 + 4z^2}} = \frac{(x, y, z)}{a}$$

is a normal vector. Since (x, y, z) points away from the origin, $\nabla f / \|\nabla f\|$ is an outward unit normal. So

$$\iint_S \mathbf{F} \cdot \mathbf{n} \, dS = \iint_S c\mathbf{k} \cdot \frac{1}{a} (x\mathbf{i} + y\mathbf{j} + z\mathbf{k}) \, dS = \iint_S \frac{c}{a} z \, dS = \frac{c}{a} \iint_S z \, dS.$$

The projection D of S onto the xy-plane is the disk $x^2 + y^2 \le a^2$. Therefore, with $\mathbf{n} = \nabla f / \|\nabla f\|$,

$$\iint_S z \, dS = \iint_D z \frac{dA}{|\mathbf{n} \cdot \mathbf{k}|} = \iint_D z \frac{dA}{z/a} = a \iint_D dA = a(\pi a^2) = a^3 \pi.$$

It follows that

$$\iint_S \mathbf{F} \cdot \mathbf{n} \, dS = \frac{c}{a} a^3 \pi = a^2 c \pi.$$

(b) Parametrize the disk $x^2 + y^2 \leq a^2$ in the xy-plane as (think of polar coordinates!)

$$\mathbf{r}(u, v) = (u \cos v, u \sin v, 0), \qquad (u, v) \in D = [0, a] \times [0, 2\pi].$$

Then

$$\mathbf{N} = \begin{vmatrix} \mathbf{i} & \mathbf{j} & \mathbf{k} \\ \cos v & \sin v & 0 \\ -u \sin v & u \cos v & 0 \end{vmatrix} = u\mathbf{k},$$

and therefore (recall that $\mathbf{F} = c\mathbf{k}$)

$$\iint_S \mathbf{F} \cdot \mathbf{n} \, dS = \iint_D c\mathbf{k} \cdot u\mathbf{k} \, dA = \iint_D cu \, dA$$
$$= c \int_0^{2\pi} \left(\int_0^a u \, du \right) dv$$
$$= c(2\pi)\tfrac{1}{2}a^2 = a^2 c \pi.$$

(c) The flow is constant and occurs in the vertical direction. The volume of fluid that flows across the surface of the disk is the same as the volume of fluid that flows across the hemisphere, since their projections onto the xy-plane are the same.

19. Let S_1 denote the paraboloid $z = x^2 + y^2$ and let S_2 denote the paraboloid $z = 12 - x^2 - y^2$. Combining the two equations we get $x^2 + y^2 = 6$; it follows that the region of integration D is the disk $x^2 + y^2 \leq 6$ (for both surfaces). Parametrize S_1 by $\mathbf{r}_1(u, v) = (u, v, u^2 + v^2)$. Its surface normal

$$\mathbf{N}_1 = \begin{vmatrix} \mathbf{i} & \mathbf{j} & \mathbf{k} \\ 1 & 0 & 2u \\ 0 & 1 & 2v \end{vmatrix} = (-2u, -2v, 1)$$

points inwards, into the paraboloid. We will compute the surface integral anyway, and will change the sign of the result.

$$\iint_{S_1} \mathbf{F} \cdot d\mathbf{S} = \iint_D \mathbf{F}(\mathbf{r}_1(u, v)) \cdot \mathbf{N}_1 \, dA$$
$$= \iint_D u\mathbf{i} \cdot (-2u\mathbf{i} - 2v\mathbf{j} + \mathbf{k}) \, dA$$
$$= -2 \iint_D u^2 \, dA$$

(change to polar coordinates $u = r \cos \theta$, $v = r \sin \theta$)

$$= -2 \int_0^{2\pi} \left(\int_0^{\sqrt{6}} r^3 \cos^2 \theta \, dr \right) d\theta$$
$$= -2 \left(\int_0^{2\pi} \cos^2 \theta \, d\theta \right) \left(\int_0^{\sqrt{6}} r^3 \, dr \right)$$
$$= (-2) \left(\tfrac{1}{2}\theta + \tfrac{1}{4} \sin(2\theta) \right) \Big|_0^{2\pi} \left(\tfrac{1}{4}r^4 \Big|_0^{\sqrt{6}} \right) = (-2)(\pi)(9) = -18\pi.$$

It follows that the outward flow is 18π. Now parametrize S_2 by $\mathbf{r}_2(u, v) = (u, v, 12 - u^2 - v^2)$. Its surface normal

$$\mathbf{N}_2 = \begin{vmatrix} \mathbf{i} & \mathbf{j} & \mathbf{k} \\ 1 & 0 & -2u \\ 0 & 1 & -2v \end{vmatrix} = (2u, 2v, 1)$$

points outward, and

$$\iint_{S_2} \mathbf{F} \cdot d\mathbf{S} = \iint_D u\mathbf{i} \cdot (2u\mathbf{i} + 2v\mathbf{j} + \mathbf{k})\, dA = 2 \iint_D u^2\, dA = 18\pi,$$

as above. So the total flux across S is $18\pi + 18\pi = 36\pi$.

21. The given plane intersects the xy-plane along the line $2x + y = 16$. Therefore, its projection onto the xy-plane is the triangular region D bounded by $x = 0$, $y = 0$ and $2x + y = 16$. Parametrize the given plane by $\mathbf{r}(u, v) = (u, v, 16 - 2u - v)$, $(u, v) \in D$. Then

$$\mathbf{N} = \mathbf{T}_u \times \mathbf{T}_v = \begin{vmatrix} \mathbf{i} & \mathbf{j} & \mathbf{k} \\ 1 & 0 & -2 \\ 0 & 1 & -1 \end{vmatrix} = 2\mathbf{i} + \mathbf{j} + \mathbf{k}$$

points outward; i.e., away from the origin. The required flux is

$$\iint_S y^3(\mathbf{j} - \mathbf{k}) \cdot d\mathbf{S} = \iint_D v^3(\mathbf{j} - \mathbf{k}) \cdot (2\mathbf{i} + \mathbf{j} + \mathbf{k})\, dA$$
$$= \iint_D 0\, dA = 0.$$

23. Parametrize the given surface by $\mathbf{r}(u, v) = (u, v, 2u^2 v)$, $0 \le u \le 1$, $0 \le v \le 2$. The surface normal

$$\mathbf{N} = \mathbf{T}_u \times \mathbf{T}_v = \begin{vmatrix} \mathbf{i} & \mathbf{j} & \mathbf{k} \\ 1 & 0 & 4uv \\ 0 & 1 & 2u^2 \end{vmatrix} = -4uv\mathbf{i} - 2u^2\mathbf{j} + \mathbf{k}$$

points upward, as required. Since $\mathbf{F}(\mathbf{r}(u, v)) = u^2 v\mathbf{i} + uv^3\mathbf{j} + 4u^3 v^2\mathbf{k}$ and $\mathbf{F} \cdot \mathbf{N} = -2u^3 v^3$, it follows that

$$\iint_S (x^2 y\mathbf{i} + xy^3\mathbf{j} + 2xyz\mathbf{k}) \cdot d\mathbf{S} = -2 \iint_{D=[0,1]\times[0,2]} u^3 v^3\, dA$$
$$= -2 \int_0^1 \left(\int_0^2 u^3 v^3 dv \right) du$$
$$= -2 \left(\int_0^1 u^3\, du \right) \left(\int_0^2 v^3\, dv \right)$$
$$= -2 \left(\tfrac{1}{4} u^4 \Big|_0^1 \right) \left(\tfrac{1}{4} v^4 \Big|_0^2 \right) = -2.$$

25. We use the parametrization $\mathbf{r}(\theta, \phi) = a\cos\theta\sin\phi\,\mathbf{i} + a\sin\theta\sin\phi\,\mathbf{j} + a\cos\phi\,\mathbf{k}$, where $0 \le \theta \le 2\pi$ and $0 \le \phi \le \pi$ (we usually use u and v; however, the parameters in this case come from spherical coordinates, and are denoted accordingly). The surface normal $\mathbf{N}$ is computed to be (see Example 7.14 in Section 7.1) $\mathbf{N}(\theta, \phi) = -a\sin\phi\,\mathbf{r}(\theta, \phi)$. Since $0 \le \phi \le \pi$ implies that $\sin\phi \geq 0$, the vector $\mathbf{N}$ points inward, Therefore, we use $-\mathbf{N} = a\sin\phi\,\mathbf{r}(\theta,\phi)$. Recall that (see Section 2.8) $\mathbf{r}(\theta,\phi) = a\mathbf{e}_\rho$; therefore, $-\mathbf{N} = a^2\sin\phi\,\mathbf{e}_\rho$. It follows that

$$\iint_S \mathbf{F} \cdot d\mathbf{S} = \iint_{D=[0,2\pi]\times[0,\pi]} \mathbf{F} \cdot (-\mathbf{N})dA$$

$$= \int_0^{2\pi}\left(\int_0^\pi a^2\sin\phi\mathbf{F}\cdot\mathbf{e}_\rho d\phi\right)d\theta$$

$$= a^2\int_0^{2\pi}\left(\int_0^\pi F_\rho\sin\phi\,d\phi\right)d\theta,$$

where $F_\rho = \mathbf{F}\cdot\mathbf{e}_\rho$ is the component of $\mathbf{F}$ in the direction of $\mathbf{e}_\rho$.

27. The projection of the plane $x + 2y + 8z = 8$ onto the first quadrant in the xy-plane is the triangular region D bounded by $x = 0$, $y = 0$ and $x + 2y = 8$. A unit normal is given by $\bar{\mathbf{n}} = \pm\nabla f/\|\nabla f\|$, where $f = x+2y+8z$; so $\bar{\mathbf{n}} = \pm(1,2,8)/\sqrt{69}$. We need a downward-pointing normal, so we must take $\mathbf{n} = -(1,2,8)/\sqrt{69}$. Thus

$$\iint_S \mathbf{F} \cdot d\mathbf{S} = \iint_D \mathbf{F} \cdot \mathbf{n}\,\frac{dA}{|\mathbf{n}\cdot\mathbf{k}|}$$

$$= -\iint_D ((x+y)\mathbf{i}+\mathbf{j}+z\mathbf{k}) \cdot \frac{\mathbf{i}+2\mathbf{j}+8\mathbf{k}}{\sqrt{69}}\,\frac{dA}{8/\sqrt{69}}$$

$$= -\tfrac{1}{8}\iint_D (x+y+2+8z)\,dA = -\tfrac{1}{8}\iint_D (10-y)\,dA$$

$$= -\tfrac{1}{8}\int_0^4\left(\int_0^{8-2y} (10-y)dx\right)dy$$

$$= -\tfrac{1}{8}\int_0^4 (10-y)(8-2y)\,dy$$

$$= -\tfrac{1}{8}\int_0^4 (80-28y+2y^2)\,dy$$

$$= -\tfrac{1}{8}\left(80y-14y^2+\tfrac{2}{3}y^3\right)\Big|_0^4 = -\tfrac{1}{8}\left(320-224+\tfrac{128}{3}\right) = -\tfrac{52}{3}.$$

29. The projection onto the xy-plane is the disk D given by $x^2 + y^2 \leq 9$. The hemisphere can be viewed as the level surface of $g(x,y,z) = x^2 + y^2 + z^2$ of value 9; so $\nabla g/\|\nabla g\| = (2x,2y,2z)/6 = (x,y,z)/3$ gives a unit normal direction. Since the vector (x,y,z) points away from the origin (and hence away from the sphere), it follows that $\nabla g/\|\nabla g\|$ is the outward unit normal to S. Hence (let $\mathbf{n} = (x\mathbf{i}+y\mathbf{j}+z\mathbf{k})/3$)

$$\iint_S (x^2\mathbf{i}+2z\mathbf{k}) \cdot d\mathbf{S} = \iint_D (x^2\mathbf{i}+2z\mathbf{k}) \cdot \mathbf{n}\,\frac{dA}{|\mathbf{n}\cdot\mathbf{k}|}$$

$$= \iint_D (x^2\mathbf{i}+2z\mathbf{k}) \cdot \tfrac{1}{3}(x\mathbf{i}+y\mathbf{j}+z\mathbf{k}) \frac{dA}{\tfrac{1}{3}|(x\mathbf{i}+y\mathbf{j}+z\mathbf{k})\cdot\mathbf{k}|}$$

$$= \iint_D \frac{x^3+2z^2}{|z|}\,dA$$

$$= \iint_D \left(\frac{x^3}{z}+2z\right)dA,$$

since $z \geq 0$ (by assumption) and so $|z| = z$. Eliminate z and change to polar coordinates:

$$\iint_D \left(\frac{x^3}{z} + 2z \right) dA = \iint_D \left(\frac{x^3}{\sqrt{9 - x^2 - y^2}} + 2\sqrt{9 - x^2 - y^2} \right) dA$$

$$= \int_0^{2\pi} \left(\int_0^3 \left(\frac{r^3 \cos^3 \theta}{\sqrt{9 - r^2}} + 2\sqrt{9 - r^2} \right) r dr \right) d\theta$$

$$= \left(\int_0^{2\pi} \cos^3 \theta \, d\theta \right) \left(\int_0^3 \frac{r^4}{\sqrt{9 - r^2}} dr \right)$$

$$+ \left(\int_0^{2\pi} d\theta \right) \left(\int_0^3 2r\sqrt{9 - r^2} \, dr \right).$$

This is not as bad as it looks. The first summand is zero, since $\int_0^{2\pi} \cos^3 \theta \, d\theta = 0$ (think of the graph of $\cos^3 \theta$ — it looks somewhat like $\cos \theta$ (the precise shape of the graph is not needed here)); what is important is the fact that the area between $\cos^3 \theta$ and the x-axis above the x-axis is the same as the area between $\cos^3 \theta$ and the x-axis below the x-axis (due to the symmetry of $\cos^3 \theta$). Therefore, (integral="area above" – "area below") the integral of $\cos^3 \theta$ over $[0, 2\pi]$ is zero. Since

$$\int_0^3 2r\sqrt{9 - r^2} \, dr = -\tfrac{2}{3}(9 - r^2)^{3/2} \Big|_0^3 = \tfrac{2}{3} 9^{3/2} = 18,$$

it follows that $\iint_S (x^2 \mathbf{i} + 2z\mathbf{k}) \cdot d\mathbf{S} = 2\pi(18) = 36\pi$.

7.5. Integrals: Properties and Applications

1. Since the surface area of a hemisphere of radius a is $2\pi a^2$, it follows that

$$\overline{p} = \frac{1}{2\pi a^2} \iint_S (1 - z^2) \, dS,$$

where S is given by $x^2 + y^2 + z^2 = a^2$, $z \geq 0$. Parametrize it by

$$\mathbf{r}(u, v) = (a \cos v \cos u, a \cos v \sin u, a \sin v), \qquad (u, v) \in D,$$

where D is the rectangle $[0, 2\pi] \times [0, \pi/2]$ in the uv-plane (this parametrization was introduced in Example 7.4). From Example 7.7, $\mathbf{N} = a \cos v \, \mathbf{r}(u, v)$, and so (since $\|\mathbf{r}(u, v)\| = a$)

$$\|\mathbf{N}\| = a|\cos v| \, \|\mathbf{r}(u, v)\| = a^2 \cos v.$$

It follows that

$$\iint_S (1 - z^2) \, dS = \iint_D (1 - a^2 \sin^2 v) a^2 \cos v \, dA$$

$$= a^2 \int_0^{2\pi} \left(\int_0^{\pi/2} \cos v dv \right) du - a^4 \int_0^{2\pi} \left(\int_0^{\pi/2} \sin^2 v \cos v dv \right) du.$$

Now $\int_0^{\pi/2} \cos v \, dv = 1$ and

$$\int_0^{\pi/2} \sin^2 v \cos v \, dv = \tfrac{1}{3} \sin^3 v \Big|_0^{\pi/2} = \tfrac{1}{3}$$

and therefore

$$\bar{p} = \frac{1}{2\pi a^2} \left[a^2(2\pi) - a^4(2\pi)\tfrac{1}{3} \right] = 1 - \tfrac{1}{3}a^2.$$

3. Parametrize the surface S by (see Example 7.3)

$$\mathbf{r}(u, v) = (v\cos u, v\sin u, v), \qquad (u, v) \in D,$$

where D is the rectangle in the uv-plane defined by $0 \le u \le 2\pi$ and $1 \le v \le 4$. The surface normal $\mathbf{N}$ is (see Example 7.5)

$$\mathbf{N} = \mathbf{T}_u \times \mathbf{T}_v = \begin{vmatrix} \mathbf{i} & \mathbf{j} & \mathbf{k} \\ -v\sin u & v\cos u & 0 \\ \cos u & \sin u & 1 \end{vmatrix} = v\cos u\,\mathbf{i} + v\sin u\,\mathbf{j} - v\mathbf{k}.$$

So $\|\mathbf{N}\| = |v|\sqrt{2} = v\sqrt{2}$. The surface area of S is

$$A(S) = \iint_S dS = \iint_D v\sqrt{2}\,dA$$

$$= \int_0^{2\pi} \left(\int_1^4 v\sqrt{2}\,dv \right) du = \left(\int_0^{2\pi} du \right) \sqrt{2}\,\tfrac{1}{2}v^2 \Big|_1^4 = 15\pi\sqrt{2}.$$

The distance from (x, y, z) to the z-axis (i.e., to the point $(0, 0, z)$) is given by $d(x, y, z) = \sqrt{x^2 + y^2}$. So

$$\iint_S d(x, y, z)\,dS = \iint_S \sqrt{x^2 + y^2}\,dS = \iint_D v\,(v\sqrt{2})\,dA$$

$$= \sqrt{2} \int_0^{2\pi} \left(\int_1^4 v^2\,dv \right) du$$

$$= \sqrt{2}(2\pi)\,\tfrac{1}{3}v^3 \Big|_1^4 = 42\pi\sqrt{2}.$$

The average distance is

$$\bar{d} = \frac{1}{A(S)} \iint_S d(x, y, z)\,dS = \frac{42\pi\sqrt{2}}{15\pi\sqrt{2}} = \frac{42}{15}.$$

5. The total charge is given by the triple integral

$$\iiint_V \rho(x, y, z)\,dV = \iiint_V (2x^2 + y)\,dV,$$

where V is the given tetrahedron. The limits of integration are given by $0 \le x \le 1$, $0 \le y \le 3 - 3x$ (the line through $(1, 0)$ and $(0, 3)$ in the xy-plane has the equation $3x + y = 3$), and $0 \le z \le 4 - 4x - 4y/3$ (the plane through $(1, 0, 0)$, $(0, 3, 0)$ and $(0, 0, 4)$ has the equation $x/1 + y/3 + z/4 = 1$). It follows that

$$\iiint_V (2x^2 + y)\,dV = \int_0^1 \left(\int_0^{3-3x} \left(\int_0^{4-4x-4y/3} (2x^2 + y)dz \right) dy \right) dx$$

$$= \int_0^1 \left(\int_0^{3-3x} (2x^2 + y)z \Big|_0^{4-4x-4y/3}\,dy \right) dx$$

$$= \int_0^1 \left(\int_0^{3-3x} -\tfrac{4}{3}(2x^2 + y)(-3 + 3x + y)dy \right) dx$$

$$= \int_0^1 \left(-\tfrac{4}{9}y^3 - \tfrac{2}{3}(2x^2 - 3 + 3x)y^2 - \tfrac{8}{3}x^2(-3 + 3x)y\right)\Big|_0^{3-3x} dx$$

$$= \int_0^1 (6 - 18x + 30x^2 - 30x^3 + 12x^4)\, dx = \tfrac{19}{10}.$$

7. By symmetry, $\bar{x} = 0$; so we need $\bar{y}$ only. The mass m of the wire is (density) $\cdot$ (length) $= \rho\, a\pi$. The moment about the y-axis is

$$M_x = \int_{\mathbf{c}} y\rho ds = \rho \int_{\mathbf{c}} y ds,$$

where $\mathbf{c}$ is a parametrization of the wire. Take $\mathbf{c}(t) = (a\cos t, a\sin t)$, $0 \le t \le \pi$. Then $\mathbf{c}'(t) = (-a\sin t, a\cos t)$ and $\|\mathbf{c}'(t)\| = a$; so

$$M_x = \rho \int_0^{\pi} a\sin t\, a\, dt = 2a^2\rho.$$

Therefore,

$$\bar{y} = \frac{M_x}{m} = \frac{2a^2\rho}{a\pi\rho} = \frac{2a}{\pi}.$$

Hence the coordinates of the center of mass are $(0, 2a/\pi)$.

9. Parametrize the wire using $\mathbf{c}(t) = (a\cos t, a\sin t)$, where $t \in [0, \pi/2]$; in that case, $\mathbf{c}'(t) = (-a\sin t, a\cos t)$ and $\|\mathbf{c}'(t)\| = a$. The mass of the wire is

$$m = \int_{\mathbf{c}} \rho\, ds = \int_{\mathbf{c}} bx\, ds = \int_0^{\pi/2} ab\cos t\, a\, dt = a^2 b\, \sin t\Big|_0^{\pi/2} = a^2 b.$$

The moments are

$$M_x = \int_{\mathbf{c}} y\rho\, ds = \int_0^{\pi/2} a^2 b\sin t\cos t\, a\, dt$$

$$= \tfrac{1}{2}a^3 b \int_0^{\pi/2} \sin 2t\, dt = \tfrac{1}{2}a^3 b \left(-\tfrac{1}{2}\cos 2t\Big|_0^{\pi/2}\right) = \tfrac{1}{2}a^3 b$$

and

$$M_y = \int_{\mathbf{c}} x\rho\, ds = \int_0^{\pi/2} a^2 b\cos^2 t\, a\, dt$$

$$= a^3 b \int_0^{\pi/2} \cos^2 t\, dt$$

$$= a^3 b \left(\tfrac{1}{2}t - \tfrac{1}{4}\sin 2t\right)\Big|_0^{\pi/2} = \tfrac{1}{4}a^3 b\pi$$

Therefore,

$$\bar{x} = \frac{M_y}{m} = \frac{a\pi}{4} \qquad \text{and} \qquad \bar{y} = \frac{M_x}{m} = \frac{a}{2}$$

are the coordinates of the center of mass.

11. Parametrize the paraboloid as $\mathbf{r}(u, v) = (u, v, 2u^2 + 2v^2)$, where $(u, v) \in D$, and D is the disk $x^2 + y^2 \le 4$ (obtained by substituting $z = 8$ into $z = 2(x^2 + y^2)$). Then

$$\mathbf{N} = \mathbf{T}_u \times \mathbf{T}_v = \begin{vmatrix} \mathbf{i} & \mathbf{j} & \mathbf{k} \\ 1 & 0 & 4u \\ 0 & 1 & 4v \end{vmatrix} = -4u\mathbf{i} - 4v\mathbf{j} + \mathbf{k},$$

and $\|\mathbf{N}\| = \sqrt{1 + 4u^2 + 4v^2}$. In the evaluation of the integral we will use polar coordinates $u = r\cos\theta$, $v = r\sin\theta$. The mass is

$$
\begin{aligned}
m &= \iint_S \rho \, dS = \rho \iint_S dS \\
&= \rho \iint_D \sqrt{1 + 4u^2 + 4v^2}\, dA = \rho \int_0^{2\pi} \left(\int_0^2 \sqrt{1 + 4r^2}\, r\,dr \right) d\theta \\
&= \rho \left(\theta \Big|_0^{2\pi} \right) \left(\tfrac{1}{12}(1 + 4r^2)^{3/2} \Big|_0^2 \right) \\
&= \rho\, 2\pi\, \tfrac{1}{12}(17^{3/2} - 1) = \tfrac{1}{6}\pi\rho\,(17^{3/2} - 1).
\end{aligned}
$$

By symmetry, $\bar{x} = 0$ and $\bar{y} = 0$. The moment with respect to the xy-plane is

$$
\begin{aligned}
M_{xy} &= \iint_S z\rho\, dS = \rho \iint_S z\, dS \\
&= \rho \iint_D (2u^2 + 2v^2)\sqrt{1 + 4u^2 + 4v^2}\, dA \\
&= 2\rho \int_0^{2\pi} \left(\int_0^2 r^3 \sqrt{1 + 4r^2}\, dr \right) d\theta
\end{aligned}
$$

(proceed using formulas from a table of integrals)

$$
\begin{aligned}
&= 2\rho(2\pi) \left(\tfrac{1}{20}r^2(1 + 4r^2)^{3/2} - \tfrac{1}{120}(1 + 4r^2)^{3/2} \right) \Big|_0^2 \\
&= 4\pi\rho \left(\left(\tfrac{4}{20}17^{3/2} - \tfrac{1}{120}17^{3/2} \right) - \left(-\tfrac{1}{120} \right) \right) \\
&= \tfrac{1}{30}\pi\rho\,(23(17^{3/2}) + 1).
\end{aligned}
$$

Therefore,

$$
\bar{z} = \frac{M_{xy}}{m} = \frac{\tfrac{1}{30}\pi\rho(23(17^{3/2}) + 1)}{\tfrac{1}{6}\pi\rho(17^{3/2} - 1)} \approx 4.6695.
$$

The moment of inertia about the z-axis is

$$
\begin{aligned}
I_z &= \iint_S (x^2 + y^2)\rho\, dS = \rho \iint_S (x^2 + y^2)\, dS \\
&= \rho \iint_D (u^2 + v^2)\sqrt{1 + 4u^2 + 4v^2}\, dA \\
&= \rho \int_0^{2\pi} \left(\int_0^2 r^3 \sqrt{1 + 4r^2}\, dr \right) d\theta
\end{aligned}
$$

(proceed as in computing M_{xy})

$$
= \tfrac{1}{60}\pi\rho\,(23(17^{3/2}) + 1) \approx 84.4635\rho.
$$

13. The mass of V is ($v(V)$ denotes the volume of V)

$$
m = \iiint_V \rho\, dV = bv(V) = b\tfrac{2}{3}\pi a^3 = \tfrac{2}{3}\pi a^3 b.
$$

By symmetry, $\overline{x} = \overline{y} = 0$. The moment with respect to the xy-plane is $M_{xy} = \iiint_V bz\, dV$. Passing to spherical coordinates $x = \rho \sin \phi \cos \theta$, $y = \rho \sin \phi \sin \theta$ and $z = \rho \cos \phi$, we get

$$M_{xy} = \int_0^{2\pi} \left(\int_0^{\pi/2} \left(\int_0^a b\rho \cos \phi \ \rho^2 \sin \phi \ d\rho \right) d\phi \right) d\theta$$

$$= b \int_0^{2\pi} \left(\int_0^{\pi/2} \sin \phi \cos \phi \ \tfrac{1}{4}\rho^4 \Big|_0^a \, d\phi \right) d\theta$$

$$= \tfrac{1}{4}a^4 b \int_0^{2\pi} \left(\int_0^{\pi/2} \tfrac{1}{2} \sin 2\phi d\phi \right) d\theta$$

$$= \tfrac{1}{4}a^4 b \int_0^{2\pi} \left(-\tfrac{1}{4} \cos 2\phi \right) \Big|_0^{\pi/2} \, d\theta$$

$$= \tfrac{1}{4}a^4 b \, (2\pi) \, \tfrac{1}{2} = \tfrac{1}{4}a^4 b \pi.$$

Therefore,

$$\overline{z} = \frac{M_{xy}}{m} = \frac{a^4 b\pi/4}{2\pi a^3 b/3} = \tfrac{3}{8}a.$$

15. Parametrize the hemisphere by $\mathbf{r}(u, v) = (a \cos u \sin v, a \sin u \sin v, a \cos v)$, where $0 \leq u \leq 2\pi$ and $0 \leq v \leq \pi/2$. Then (see Example 7.14) $\mathbf{N} = a(-\sin v) \cdot \mathbf{r}(u, v)$, and since $\|\mathbf{r}(u, v)\| = a$, it follows that $\|\mathbf{N}\| = a^2 \sin v$. The highest point is $z_0 = a$, and so

$$\iint_S \mu(a - z) \, dS = \mu \int_0^{2\pi} \left(\int_0^{\pi/2} (a - a \cos v)a^2 \sin v dv \right) du$$

$$= a^3 \mu \left(\int_0^{2\pi} du \right) \left(\int_0^{\pi/2} (1 - \cos v) \sin v \, dv \right)$$

$$= 2\pi \mu a^3 \int_0^{\pi/2} (\sin v - \sin v \cos v) \, dv$$

$$= 2\pi \mu a^3 \left(-\cos v + \tfrac{1}{4} \cos 2v \right) \Big|_0^{\pi/2} = \pi \mu a^3.$$

17. Parametrize the cylinder by $\mathbf{r}(u, v) = (2 \cos u, 2 \sin u, v)$, where $(u, v) \in D$, and D is the rectangle $[0, 2\pi] \times [0, 10]$ in the uv-plane. Then

$$\mathbf{N} = \begin{vmatrix} \mathbf{i} & \mathbf{j} & \mathbf{k} \\ -2 \sin u & 2 \cos u & 0 \\ 0 & 0 & 1 \end{vmatrix} = (2 \cos u, 2 \sin u, 0)$$

and $\|\mathbf{N}\| = 2$. The highest point has its z-coordinate equal to 10, and so

$$\iint_S \mu(10 - z) \, dS = \mu \int_0^{2\pi} \left(\int_0^{10} (10 - v)dv \right) du$$

$$= \mu \left(\int_0^{2\pi} du \right) \left(10v - \tfrac{1}{2}v^2 \right) \Big|_0^{10}$$

$$= \mu(2\pi)(50) = 100\pi\mu.$$

19. For example, let $f(x, y, z) = z$ and $\mathbf{c}(t) = (0, 0, t)$, where $-1 \leq t \leq 1$. Then $\|\mathbf{c}'(t)\| = 1$ and

$$\left| \int_{\mathbf{c}} f \, ds \right| = \left| \int_{-1}^{1} t \, dt \right| = 0,$$

whereas

$$\int_{\mathbf{c}} |f| = \int_{-1}^{1} |t| \, dt = 1.$$

21. For points on S, $0 \leq z \leq 1$; it follows that $1 \leq z^2 + 1 \leq 2$, and hence

$$\iint_S 1 \, dS \leq \iint_S (z^2 + 1) \, dS \leq \iint_S 2 \, dS$$

and therefore

$$2\pi \leq \iint_S (z^2 + 1) \, dS \leq 4\pi$$

(since the surface area of a hemisphere of radius 1 is 2π).

23. Since $-1 \leq \sin a \leq 1$ for any real number a, it follows that $e^{-1} \leq e^{\sin(x+y)} \leq e$ and

$$e^{-1} \ell(\mathbf{c}) \leq \int_{\mathbf{c}} e^{\sin(x+y)} \, ds \leq e\ell(\mathbf{c}),$$

where $\ell(\mathbf{c})$ is the length of $\mathbf{c}$. From $\mathbf{c}'(t) = (-2 \sin t, 2 \cos t, 1)$ and $\|\mathbf{c}'(t)\| = \sqrt{5}$, it follows that

$$\ell(\mathbf{c}) = \int_{\pi}^{4\pi} \sqrt{5} \, dt = 3\pi\sqrt{5},$$

and therefore

$$\frac{3\pi\sqrt{5}}{e} \leq \int_{\mathbf{c}} e^{\sin(x+y)} \, ds \leq 3\pi e\sqrt{5}.$$

Chapter Review

True/false Quiz

1. True. Converting to Cartesian coordinates, we get $y = x$, which represents a plane perpendicular to the xy-plane. Or, compute the normal $\mathbf{N} = \mathbf{T}_u \times \mathbf{T}_v = (1, 1, 0) \times (0, 0, 1) = (1, -1, 0)$. Since $\mathbf{N} \cdot \mathbf{k} = 0$, it follows that the plane is perpendicular to the xy-plane.

3. False. $x^2 + z^2 - 1$ is a cylider whose axis of symmetry is the y-axis. Thus, its normal vectors are perpendicular to the y-axis.

5. True. $\iint_S dS$ is the surface area of the sphere of radius 1 (recall that the surface area of the sphere of radius r is $4\pi r^2$).

7. False. $\iint_{S_*} dS^* = 2$, since the surface integral of a real-valued function is independent of the orientation.

9. False. For instance, the flux of $\mathbf{F}$ through any region in the xy-plane is zero. Also, the flux through a surface whose normal makes an angle θ, $\pi/2 \leq \theta \leq \pi$, with respect to $\mathbf{i}$ will be negative.

11. False. The surface integral depends on the way the surface is placed into the flow. Recall that the flux (of a constant vector field; in our case $\mathbf{F} = \mathbf{i}$) is given by $A(S)\|\mathbf{F}\| \cos \theta = A(S) \cos \theta$, where θ is the angle between $\mathbf{F}$ and the surface normal.

13. False. Let $\mathbf{F}(x, y, z) = x\mathbf{k}$ (clearly, $\mathbf{F}$ is not parallel to the xy-plane). Then

$$\iint_S \mathbf{F}\,dS = \iint_D (x\mathbf{k}) \cdot \mathbf{k}\,dA = \iint_D x\,dA = \int_0^{2\pi} \left(\int_0^1 r^2 \cos\theta\,dr \right) d\theta = 0.$$

15. False. $\iiint_W dV$ is the volume of the ball of radius 2, and thus is equal to $(4/3)(2)^3\pi = 32\pi/3$.

Review Exercises and Problems

1. The ellipsoid $x^2/a^2 + y^2/a^2 + z^2/c^2 = 1$ is obtained when the ellipse $y^2/a^2 + z^2/c^2 = 1$ is rotated about the z-axis. We parametrize the ellipse (in the yz-plane) by $(a \cos t, c \sin t)$, $0 \leq t \leq 2\pi$. The ellipsoid can now be parametrized as a surface of revolution (see (7.7) in Section 7.2 and the text surrounding it) $\mathbf{r}(u, t) = (a \cos t \cos u, a \cos t \sin u, c \sin t)$, where $0 \leq t, u \leq 2\pi$ (or $0 \leq u \leq \pi$).

3. For example, let $f(x, y, z) = z$ and let S be the part of the plane $z = x$ defined by $-1 \leq x \leq 1$ and $0 \leq y \leq 1$. Parametrize the plane by $\mathbf{r}(u, v) = (u, v, u)$, $(u, v) \in D = [-1, 1] \times [0, 1]$; then

$$\mathbf{N} = \mathbf{T}_u \times \mathbf{T}_v = \begin{vmatrix} \mathbf{i} & \mathbf{j} & \mathbf{k} \\ 1 & 0 & 1 \\ 0 & 1 & 0 \end{vmatrix} = -\mathbf{i} + \mathbf{k}$$

and $\|\mathbf{N}\| = \sqrt{2}$. It follows that

$$\left| \iint_S f\,dS \right| = \left| \iint_D u\sqrt{2}\,dA \right| = \sqrt{2} \left| \int_0^1 \left(\int_{-1}^1 u\,du \right) dv \right| = 0,$$

since $\int_{-1}^1 u\,du = 0$. However,

$$\iint_S |f|\,dS = \iint_D |u|\sqrt{2}\,dA = \sqrt{2} \int_0^1 \left(\int_{-1}^1 |u|\,du \right) dv$$

$$= \sqrt{2} \int_0^1 \left(-\int_{-1}^0 u\,du + \int_0^1 u\,du \right) dv = \sqrt{2} \int_0^1 dv = \sqrt{2}.$$

5. (a) Parametrize the cylinder by $\mathbf{r}(u, v) = (R \cos u, R \sin u, v)$, where $0 \leq u \leq 2\pi$ and $-H \leq v \leq H$. From $\mathbf{T}_u = (-R \sin u, R \cos u, 0)$ and $\mathbf{T}_v = (0, 0, 1)$ we get $\mathbf{N} = \mathbf{T}_u \times \mathbf{T}_v = R(\cos u, \sin u, 0)$. From

$$\mathbf{E}(\mathbf{r}) = \frac{Q}{4\pi\epsilon_0} \frac{\mathbf{r}}{\|\mathbf{r}\|^3} = \frac{Q}{4\pi\epsilon_0} \frac{(R \cos u, R \sin u, v)}{(R^2 + v^2)^{3/2}}$$

we compute

$$\mathbf{E}(\mathbf{r}) \cdot \mathbf{N} = \frac{Q}{4\pi\epsilon_0} \frac{R^2}{(R^2 + v^2)^{3/2}}$$

and therefore

$$\iint_S \mathbf{E}(\mathbf{r}) \cdot d\mathbf{S} = \iint_{[0,2\pi]\times[-H,H]} \mathbf{E}(\mathbf{r}) \cdot \mathbf{N} dA = \frac{Q}{4\pi\epsilon_0} \int_0^{2\pi} \left(\int_{-H}^H \frac{R^2}{(R^2 + v^2)^{3/2}} dv \right) du$$

$$= \frac{Q}{4\pi\epsilon_0} 2\pi R^2 \int_{-H}^H \frac{1}{(R^2 + v^2)^{3/2}} dv = \frac{QR^2}{2\epsilon_0} \frac{2H}{R^2\sqrt{R^2 + H^2}} = \frac{QH}{\epsilon_0\sqrt{R^2 + H^2}}.$$

Note: from the table of integrals,

$$\int_{-H}^H \frac{1}{(R^2 + v^2)^{3/2}} = \frac{v}{R^2\sqrt{R^2 + v^2}}\bigg|_{-H}^H = \frac{2H}{R^2\sqrt{R^2 + H^2}}$$

(b) Parametrize the upper disk by $\mathbf{r}(u, v) = (u, v, H)$, where (u, v) satisfy $u^2 + v^2 \leq R^2$. Then

$$\mathbf{E}(\mathbf{r}) = \frac{Q}{4\pi\epsilon_0} \frac{(u, v, H)}{(u^2 + v^2 + H^2)^{3/2}}$$

and since $\mathbf{N} = \mathbf{k}$,

$$\mathbf{E}(\mathbf{r}) \cdot \mathbf{N} = \frac{Q}{4\pi\epsilon_0} \frac{H}{(u^2 + v^2 + H^2)^{3/2}}.$$

The surface integral is computed to be (use polar coordinates)

$$\iint_S \mathbf{E}(\mathbf{r}) \cdot d\mathbf{S} = \frac{QH}{4\pi\epsilon_0} \iint_D \frac{1}{(u^2 + v^2 + H^2)^{3/2}} dA = \frac{QH}{4\pi\epsilon_0} \int_0^{2\pi} \left(\int_0^R \frac{r}{(r^2 + H^2)^{3/2}} dr \right) d\theta$$

$$= -\frac{QH}{4\pi\epsilon_0} 2\pi \frac{1}{\sqrt{r^2 + H^2}}\bigg|_0^R = \frac{QH}{2\epsilon_0} \left(\frac{-1}{\sqrt{R^2 + H^2}} + \frac{1}{H} \right).$$

Note that (due to symmetry) the bottom disk gives exactly the same value for the surface integral. Adding up the results in (a) and (b), we get that

$$\frac{QH}{\epsilon_0\sqrt{R^2 + H^2}} + 2\frac{QH}{2\epsilon_0} \left(\frac{-1}{\sqrt{r^2 + H^2}} + \frac{1}{H} \right) = \frac{Q}{\epsilon_0}$$

represents the flux of $\mathbf{E}$ thorugh the closed cylinder.

7. From $x^2/a^2 - y^2/b^2 = z_0/c$ we see that the level curves are hyperbolas (with center at $(0, 0)$ and asymptotes $y = \pm bx/a$ if $z_0 \neq 0$ and a pair of lines $y = \pm bx/a$ if $z_0 = 0$. When $y = 0$, we get $x^2/a^2 = z/c$, i.e., $z = cx^2/a^2$. So, the intersection of the paraboloid with the xz-plane is the parabola $z = cx^2/a^2$. Similarly, we obtain that the intersection of the paraboloid with the yz-plane is the parabola $z = -cy^2/b^2$. A plot of the paraboloid is shown in the figure below.

To obtain a parametrization, we let $x = u$ and $y = v$. Thus, $z = c(x^2/a^2 - y^2/b^2) = c(u^2/a^2 - v^2/b^2)$, and $\mathbf{r}(u, v) = (u, v, cu^2/a^2 - cv^2/b^2)$, where $u, v \in \mathbb{R}$. The parametrization $\mathbf{r}$ is clearly differentiable. Since the normal vector $\mathbf{N} = \mathbf{T}_u \times \mathbf{T}_v = (1, 0, 2cu/a^2) \times (0, 1, -2cv/b^2) = (-2cu/a^2, 2cv/b^2, 1)$ is non-zero for all (u, v), we conclude that $\mathbf{r}$ is smooth.

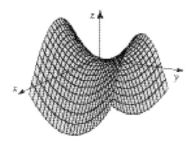

Figure 7

9. The surface in question (call it S) is the part of the plane $x + 2y + 3z = 6$ in the first octant. In other words (write the equation as $x/6 + y/3 + z/2 = 1$), S is the triangle with vertices at $(6, 0, 0)$, $(0, 3, 0)$ and $(0, 0, 2)$.

The required volume is given by $\iint_S \mathbf{F} \cdot d\mathbf{S}$. Parametrize S by $\mathbf{r}(u, v) = (u, v, (6 - u - 2v)/3)$, where $(u, v) \in D$, and D is the triangle in the uv-plane whose vertices are at $(6, 0)$, $(0, 3)$ and $(0, 0)$. Thus, D is bounded by the lines $u = 0$, $v = 0$ and $v = -u/2 + 3$.

The tangent vectors are $\mathbf{T}_u = (1, 0, -1/3)$, $\mathbf{T}_v = (0, 1, -2/3)$, and the surface normal is $\mathbf{N}(u, v) = \mathbf{T}_u \times \mathbf{T}_v = (1/3, 2/3, 1)$. It follows that

$$\mathbf{F}(\mathbf{r}(u, v)) \cdot \mathbf{N} = (2u, 0, v) \cdot (1/3, 2/3, 1) = \frac{2}{3}u + v,$$

and thus

$$\iint_S \mathbf{F} \cdot d\mathbf{S} = \iint_D \mathbf{F} \cdot \mathbf{N}\, dA = \int_0^6 \left(\int_0^{-u/2+3} \left(\frac{2}{3}u + v \right) dv \right) du$$

$$= \int_0^6 \left(\frac{2}{3}uv + \frac{1}{2}v^2 \Big|_0^{-u/2+3} \right) du = \int_0^6 \left(-\frac{5}{24}u^2 + \frac{1}{2}u + \frac{9}{2} \right) du = 21.$$

11. Using the projection onto the xy-plane, we get ($A(S)$ denotes the surface area)

$$A(S) = \iint_S dS = \iint_D \frac{dA}{|\mathbf{n} \cdot \mathbf{k}|},$$

where $\mathbf{n}$ is a unit normal to S. Since the gradient of f is a vector field perpendicular to S, it follows that

$$\mathbf{n} = \pm \frac{\nabla f}{\|\nabla f\|} = \pm \frac{f_x \mathbf{i} + f_y \mathbf{j} + f_z \mathbf{k}}{\sqrt{f_x^2 + f_y^2 + f_z^2}}.$$

In either case,

$$|\mathbf{n} \cdot \mathbf{k}| = \frac{|f_z|}{\sqrt{f_x^2 + f_y^2 + f_z^2}}$$

and

$$A(S) = \iint_D \frac{\sqrt{f_x^2 + f_y^2 + f_z^2}}{|f_z|}\, dA,$$

where dA refers to integration with respect to x and y.

In the case $z = g(x, y)$, $(x, y) \in D$, we let $f(x, y, z) = g(x, y) - z$ and use the formula for $A(S)$ we have just derived; thus $f_x = g_x$, $f_y = g_y$ and $f_z = -1$ and

$$A(S) = \iint_D \frac{\sqrt{g_x^2 + g_y^2 + 1}}{1} \, dA = \iint_D \sqrt{g_x^2 + g_y^2 + 1} \, dA.$$

13. By definition,

$$\iint_S dS = \iint_D \|\mathbf{N}\| \, dA,$$

where (use subscripts for partial derivatives)

$$\mathbf{N} = \mathbf{T}_u \times \mathbf{T}_v = (y_u z_v - y_v z_u, z_u x_v - z_v x_u, x_u y_v - x_v y_u);$$

see formula (7.4) in Section 7.1. Hence

$$\|\mathbf{N}\| = (y_u z_v - y_v z_u)^2 + (z_u x_v - z_v x_u)^2 + (x_u y_v - x_v y_u)^2$$
$$= y_u^2 z_v^2 + y_v^2 z_u^2 - 2 y_u y_v z_u z_v + z_u^2 x_v^2 + z_v^2 x_u^2 - 2 x_u x_v z_u z_v + x_u^2 y_v^2 + x_v^2 y_u^2 - 2 x_u x_v y_u y_v.$$

On the other hand,

$$EG - F^2 = \|\mathbf{r}_u\|^2 \|\mathbf{r}_v\|^2 - (\mathbf{r}_u \cdot \mathbf{r}_v)^2$$
$$= (x_u^2 + y_u^2 + z_u^2)(x_v^2 + y_v^2 + z_v^2) - (x_u x_v + y_u y_v + z_u z_v)^2$$
$$= x_u^2 x_v^2 + x_u^2 y_v^2 + x_u^2 z_v^2 + y_u^2 x_v^2 + y_u^2 y_v^2 + y_u^2 z_v^2 + z_u^2 x_v^2 + z_u^2 y_v^2 + z_u^2 z_v^2$$
$$\quad - x_u^2 x_v^2 - y_u^2 y_v^2 - z_u^2 z_v^2 - 2 x_u x_v y_u y_v - 2 x_u x_v z_u z_v - 2 y_u y_v z_u z_v$$
$$= \|\mathbf{N}\|.$$

Therefore,

$$\iint_S dS = \iint_D \|\mathbf{N}\| \, dA = \iint_D (EG - F^2) \, dA.$$

15. The curve is given by $\mathbf{c}(t) = (t^2, 1, -t^2)$, $t \in [0, 1]$. The moment of inertia with respect to the x-axis is

$$I_x = \int_\mathbf{c} (y^2 + z^2) \rho(x, y, z) ds = \int_0^1 (1 + t^4)(2)(2t\sqrt{2}) \, dt,$$

since $\mathbf{c}'(t) = (2t, 0, -2t)$ and $\|\mathbf{c}'(t)\| = 2|t|\sqrt{2} = 2t\sqrt{2}$, keeping in mind that $t \geq 0$. We conclude that

$$I_x = 4\sqrt{2} \int_0^1 (1 + t^4) t \, dt = 4\sqrt{2} \frac{2}{3} = \frac{8\sqrt{2}}{3}.$$

Similarly,

$$I_y = \int_\mathbf{c} (x^2 + z^2) \rho(x, y, z) ds = \int_0^1 (2t^4)(2)(2t\sqrt{2}) \, dt = 8\sqrt{2} \int_0^1 t^5 \, dt = \frac{4\sqrt{2}}{3}$$

and

$$I_z = \int_\mathbf{c} (x^2 + y^2) \rho(x, y, z) ds = \int_0^1 (t^4 + 1)(2)(2t\sqrt{2}) \, dt = 4\sqrt{2} \int_0^1 (t^4 + 1) t \, dt = \frac{8\sqrt{2}}{3}.$$

8. CLASSICAL INTEGRATION THEOREMS OF VECTOR CALCULUS

8.1. Green's Theorem

1. (a) $\mathbf{F}$ is a constant vector field, so there is no turning (in the sense explained in Section 5.3; see Example 5.21 for details). Alternatively, assume that $\mathbf{c}(t) = (x(t), y(t))$, where $t \in [c, d]$. Then

$$\int_{\mathbf{c}} \mathbf{F} \cdot d\mathbf{s} = \int_c^d (a, b) \cdot (x'(t), y'(t)) \, dt = \int_c^d (ax'(t) + by'(t)) \, dt$$

$$= (ax(t) + by(t)) \Big|_c^d = (ax(d) + by(d)) - (ax(c) + by(c)) = 0,$$

since $\mathbf{c}$ is assumed closed (so that $x(c) = x(d)$ and $y(c) = y(d)$).

(b) Let D be the region enclosed by $\mathbf{c}$ (and assume that it satisfies Assumption 8.1). By Green's Theorem, $\int_{\mathbf{c}} \mathbf{F} \cdot d\mathbf{s} = \iint_D 0 \, dA = 0$.

3. By Green's Theorem,

$$\int_{\mathbf{c}} (-2y\mathbf{i} + x\mathbf{j}) \cdot d\mathbf{s} = \iint_D \left(\frac{\partial}{\partial x}(x) - \frac{\partial}{\partial y}(-2y) \right) dA = \iint_D 3 \, dA,$$

where D is the region enclosed by $\mathbf{c}$; it is an ellipse with semi-axes of length 2 (in the x-direction) and 1 (in the y-direction). It follows that

$$\iint_D 3 \, dA = 3 \operatorname{area}(D) = 6\pi.$$

(Recall that the area of an ellipse with semi-axes a and b is $ab\pi$.)

5. Let $\mathbf{F} = e^{x+y}\mathbf{j} - e^{x-y}\mathbf{i}$; then

$$curl\, \mathbf{F} = \begin{vmatrix} \mathbf{i} & \mathbf{j} & \mathbf{k} \\ \partial/\partial x & \partial/\partial y & \partial/\partial z \\ -e^{x-y} & e^{x+y} & 0 \end{vmatrix} = (e^{x+y} - e^{x-y})\mathbf{k},$$

and, by Green's Theorem,

$$\int_{\mathbf{c}} (e^{x+y}\mathbf{j} - e^{x-y}\mathbf{i}) \cdot d\mathbf{s} = \iint_D (e^{x+y} - e^{x-y})\mathbf{k} \cdot \mathbf{k} \, dA = \iint_D (e^{x+y} - e^{x-y}) \, dA$$

(D is the region in the xy-plane bounded by the lines $x = 0$, $y = 0$ and $y = x$). It follows that

$$\int_{\mathbf{c}} (e^{x+y}\mathbf{j} - e^{x-y}\mathbf{i}) \cdot d\mathbf{s} = \iint_D (e^{x+y} - e^{x-y}) \, dA$$

$$= \int_0^1 \left(\int_0^x (e^{x+y} - e^{x-y}) dy \right) dx$$

$$= \int_0^1 (e^{x+y} + e^{x-y}) \Big|_0^x dx$$

$$= \int_0^1 (e^{2x} + 1 - 2e^x) \, dx$$

$$= \left(\tfrac{1}{2}e^{2x} + x - 2e^x\right)\Big|_0^1 = \tfrac{1}{2}e^2 - 2e + \tfrac{5}{2}.$$

7. Let D be the region in the xy-plane bounded by the graphs of $y = 2x^3$ and $y = 2x$ on $[0, 1]$. By Green's Theorem,

$$\int_{\mathbf{c}} (2x^2y^2\mathbf{i} - x\mathbf{j}) \cdot d\mathbf{s} = \iint_D \left(\frac{\partial}{\partial x}(-x) - \frac{\partial}{\partial y}(2x^2y^2)\right) dA$$

$$= \iint_D (-1 - 4x^2y)\, dA$$

$$= \int_0^1 \left(\int_{2x^3}^{2x} (-1 - 4x^2y) dy\right) dx$$

$$= -\int_0^1 (y + 2x^2y^2)\Big|_{2x^3}^{2x} dx$$

$$= -\int_0^1 (2x + 8x^4 - 2x^3 - 8x^8)\, dx$$

$$= -\left(x^2 + \tfrac{8}{5}x^5 - \tfrac{1}{2}x^4 - \tfrac{8}{9}x^9\right)\Big|_0^1$$

$$= -\left(1 + \tfrac{8}{5} - \tfrac{1}{2} - \tfrac{8}{9}\right) = -\tfrac{109}{90}.$$

9. Using Green's Theorem,

$$\int_{\mathbf{c}} (x^2y^2\mathbf{i} + y^4\mathbf{j}) \cdot d\mathbf{s} = -2 \iint_D yx^2\, dA,$$

where D is the disk $x^2 + y^2 \leq 1$. So

$$\iint_D yx^2\, dA = \int_{-1}^1 \left(\int_{-\sqrt{1-x^2}}^{\sqrt{1-x^2}} yx^2 dy\right) dx = \int_{-1}^1 x^2 \left(\int_{-\sqrt{1-x^2}}^{\sqrt{1-x^2}} y\, dy\right) dx = 0,$$

since $\int_{-\sqrt{1-x^2}}^{\sqrt{1-x^2}} y\, dy = 0$; recall that an integral of an odd function over a symmetric interval (with respect to 0) is zero.

11. By Green's Theorem,

$$\int_{\mathbf{c}} (\cosh y\mathbf{i} + x \sinh y\mathbf{j}) \cdot d\mathbf{s} = \iint_D (\sinh y - \sinh y)\, dA = 0;$$

D is the region bounded by $y = 2x$, $y = 4x$ and $x = 1$.

13. The curve $\mathbf{c}$ encloses the disk D of radius $\sqrt{2}$ centered at the origin. By Green's Theorem,

$$\int_{\mathbf{c}} (\arctan(y/x)\mathbf{i} + \arctan(x/y)\mathbf{j}) \cdot d\mathbf{s} = \iint_D \left(\frac{\partial}{\partial x}\big(\arctan(x/y)\big) - \frac{\partial}{\partial y}\big(\arctan(y/x)\big)\right) dA.$$

The integrand on the right side is equal to

$$\frac{1}{1 + x^2/y^2}\frac{1}{y} - \frac{1}{1 + y^2/x^2}\frac{1}{x} = \frac{y}{x^2 + y^2} - \frac{x}{x^2 + y^2} = \frac{y - x}{x^2 + y^2},$$

and hence (changing to polar coordinates)

$$\int_{\mathbf{c}} (\arctan(y/x)\mathbf{i} + \arctan(x/y)\mathbf{j}) \cdot d\mathbf{s} = \iint_D \frac{y - x}{x^2 + y^2}\, dA$$

$$= \int_0^{2\pi} \left(\int_0^{\sqrt{2}} \frac{r\sin\theta - r\cos\theta}{r^2} \, r dr \right) d\theta$$

$$= \int_0^{2\pi} \left(\int_0^{\sqrt{2}} (\sin\theta - \cos\theta) dr \right) d\theta$$

$$= \sqrt{2} \int_0^{2\pi} (\sin\theta - \cos\theta) \, d\theta = 0,$$

since $\int_0^{2\pi} \sin\theta \, d\theta = \int_0^{2\pi} \cos\theta \, d\theta = 0$.

15. Combining $y = 4x$ and $y = 2x^2$, we get $x^2 = 2x$; i.e., $x = 0$ and $x = 2$. We will use the formula $\int_{\mathbf{c}} x \, dy$ to compute the area of D, where $\mathbf{c}$ consists of the parabolic path $y = 2x^2$ (call it $\mathbf{c}_1$) from $(0,0)$ to $(2,8)$ and the straight line segment $y = 4x$ (call it $\mathbf{c}_2$) from $(2,8)$ back to $(0,0)$. Parametrize $\mathbf{c}_1$ by $\mathbf{c}_1(t) = (t, 2t^2)$, $0 \leq t \leq 2$, so that

$$\int_{\mathbf{c}_1} x \, dy = \int_0^2 (t)(4t) \, dt = \int_0^2 4t^2 \, dt = \frac{4}{3}t^3 \Big|_0^2 = \frac{32}{3}.$$

Parametrize $\mathbf{c}_2$ by $\mathbf{c}_2(t) = (2 - 2t, 8 - 8t)$, $0 \leq t \leq 1$; it follows that

$$\int_{\mathbf{c}_2} x \, dy = \int_0^1 (2 - 2t)(-8) \, dt = -16 \int_0^1 (1 - t) \, dt = -16 \left(t - \frac{1}{2}t^2 \right) \Big|_0^1 = -8.$$

Consequently, the area of D is

$$A(D) = \int_{\mathbf{c}} x \, dy = \int_{\mathbf{c}_1} x \, dy + \int_{\mathbf{c}_2} x \, dy = \frac{32}{3} - 8 = \frac{8}{3}.$$

17. The boundary of D (oriented counterclockwise) consists of the following curves: the straight line segment from $(0,1)$ to $(0,0)$ (call it $\mathbf{c}_1$), straight line segment from $(0,0)$ to $(1,0)$ (call it $\mathbf{c}_2$) and the graph of $x^{2/3} + y^{2/3} = 1$ from $(1,0)$ back to $(0,1)$ (call it $\mathbf{c}_3$).

Parametrize $\mathbf{c}_1$ by $\mathbf{c}_1(t) = (0, 1 - t)$, $t \in [0,1]$; then

$$\int_{\mathbf{c}_1} x \, dy = \int_0^1 0 \, dt = 0.$$

Parametrize $\mathbf{c}_2$ by $\mathbf{c}_2(t) = (t, 0)$, $t \in [0,1]$; then

$$\int_{\mathbf{c}_2} x \, dy = \int_0^1 0 \, dt = 0.$$

Parametrize $\mathbf{c}_3$ by $\mathbf{c}_3(t) = (\cos^3 t, \sin^3 t)$, $t \in [0, \pi/2]$; then

$$\int_{\mathbf{c}_3} x \, dy = \int_0^{\pi/2} (\cos^3 t)(3 \sin^2 t \cos t) \, dt = 3 \int_0^{\pi/2} \cos^4 t \sin^2 t \, dt.$$

Using integration tables, or doing it by hand (expressing $\sin^2 t$ and $\cos^2 t$ using double-angle formulas) we get

$$3 \int_0^{\pi/2} \cos^4 t \sin^2 t \, dt = 3 \left(-\frac{1}{6} \sin t \cos^5 t + \frac{1}{24} \cos^3 t \sin t + \frac{1}{16} \sin t \cos t + \frac{1}{16}t \right) \Big|_0^{\pi/2}$$

$$= \frac{3}{32}\pi.$$

Consequently, the area of D is

$$A(D) = \int_{\mathbf{c}} x\,dy = \int_{\mathbf{c}_1} x\,dy + \int_{\mathbf{c}_2} x\,dy + \int_{\mathbf{c}_3} x\,dy = \tfrac{3}{32}\pi.$$

19. Consider the triangular region D determined by the given points; it is bounded by the lines $y = 0$, $y = 2x$ and $y = 3 - x$. By Green's Theorem, the work of $\mathbf{F}$ is

$$W = \int_{\mathbf{c}} (x\mathbf{i} + (x^2 + 3y^2)\mathbf{j}) \cdot d\mathbf{s} = \iint_D 2x\,dA$$

(combining $y = 2x$ and $y = 3 - x$ we get $x = 1$, and thus $y = 2$)

$$= \int_0^2 \left(\int_{y/2}^{3-y} 2x\,dx \right) dy$$

$$= \int_0^2 x^2 \Big|_{y/2}^{3-y} dy$$

$$= \int_0^2 \left(9 - 6y + y^2 - \tfrac{1}{4}y^2\right) dy = \left(\tfrac{1}{4}y^3 - 3y^2 + 9y\right) \Big|_0^2 = 8.$$

21. The moments of inertia are given by

$$I_x = \iint_D y^2 \rho\,dA = \rho \iint_D y^2\,dA$$

and

$$I_y = \iint_D x^2 \rho\,dA = \rho \iint_D x^2\,dA,$$

since ρ is constant.

We have to find the components of a vector field $\mathbf{F} = P\mathbf{i} + Q\mathbf{j}$ such that

$$\int_{\partial D} \mathbf{F} \cdot d\mathbf{s} = \iint_D \left(\frac{\partial Q}{\partial x} - \frac{\partial P}{\partial y} \right) dA = \rho \iint_D y^2\,dA.$$

There are many possibilities; for example, let $\partial P/\partial y = 0$, and $\partial Q/\partial x = y^2$; in that case, $P = C_1(x)$ and $Q = xy^2 + C_2(y)$. Therefore,

$$I_x = \rho \int_{\mathbf{c}} (C_1(x)\mathbf{i} + (xy^2 + C_2(y))\mathbf{j}) \cdot d\mathbf{s},$$

where $C_1(x)$ is any function of x only and $C_2(y)$ is any function of y only. Similarly,

$$I_y = \rho \int_{\mathbf{c}} (D_1(x)\mathbf{i} + (\tfrac{1}{3}x^3 + D_2(y))\mathbf{j}) \cdot d\mathbf{s},$$

where $D_1(x)$ and $D_2(y)$ are functions of the indicated variables.

23. From $f = e^x \cos y$ we get $f_x = f_{xx} = e^x \cos y$, $f_y = -e^x \sin y$ and $f_{yy} = -e^x \cos y$. It follows that $\Delta f = f_{xx} + f_{yy} = 0$, and therefore $\iint_D \Delta f\,dA = 0$.

Now we compute the right side. The gradient of f is given by $\nabla f = e^x \cos y\,\mathbf{i} - e^x \sin y\,\mathbf{j}$. Parametrize the line segment from $(0,0)$ to $(1,0)$ by $\mathbf{c}_1(t) = (t, 0)$, $t \in [0, 1]$. The outward unit normal is $\mathbf{n}_1 = -\mathbf{j}$, and so

$$D_{\mathbf{n}_1} f = \nabla f \cdot (-\mathbf{j}) = (e^x \cos y\,\mathbf{i} - e^x \sin y\,\mathbf{j}) \cdot (-\mathbf{j}) = e^x \sin y$$

and

$$\int_{\mathbf{c}_1} D_{\mathbf{n}_1} f \, ds = \int_{\mathbf{c}_1} e^x \sin y \, ds = \int_0^1 0 \, dt = 0.$$

Parametrize the line segment from $(1,0)$ to $(1,2)$ by $\mathbf{c}_2(t) = (1,t)$, $t \in [0,2]$. The outward unit normal is $\mathbf{n}_2 = \mathbf{i}$, and so

$$D_{\mathbf{n}_2} f = \nabla f \cdot \mathbf{i} = (e^x \cos y \, \mathbf{i} - e^x \sin y \, \mathbf{j}) \cdot \mathbf{i} = e^x \cos y$$

and

$$\int_{\mathbf{c}_2} D_{\mathbf{n}_2} f \, ds = \int_{\mathbf{c}_2} e^x \cos y \, ds = \int_0^2 e \cos t \, dt = e \sin t \Big|_0^2 = e \sin 2.$$

Parametrize the line segment from $(1,2)$ to $(0,2)$ by $\mathbf{c}_3(t) = (1-t,2)$, $t \in [0,1]$. The outward unit normal is $\mathbf{n}_3 = \mathbf{j}$, and so $D_{\mathbf{n}_3} f = -e^x \sin y$ and

$$\int_{\mathbf{c}_3} D_{\mathbf{n}_3} f \, ds = -\int_{\mathbf{c}_3} e^x \sin y \, ds = -\int_0^1 e^{1-t} \sin 2 \, dt = (\sin 2) e^{1-t} \Big|_0^1 = (1-e) \sin 2.$$

Finally, parametrize the line segment from $(0,2)$ to $(0,0)$ by $\mathbf{c}_4(t) = (0, 2-t)$, $t \in [0,2]$. The outward unit normal is $\mathbf{n}_4 = -\mathbf{i}$, and so $D_{\mathbf{n}_4} f = -e^x \cos y$ and

$$\int_{\mathbf{c}_4} D_{\mathbf{n}_4} f \, ds = -\int_{\mathbf{c}_4} e^x \cos y \, ds = -\int_0^2 \cos(2-t) \, dt = \sin(2-t) \Big|_0^2 = -\sin 2.$$

It follows that

$$\int_{\mathbf{c}} D_{\mathbf{n}} f \, ds = \int_{\mathbf{c}_1} D_{\mathbf{n}_1} f \, ds + \int_{\mathbf{c}_2} D_{\mathbf{n}_2} f \, ds + \int_{\mathbf{c}_3} D_{\mathbf{n}_3} f \, ds + \int_{\mathbf{c}_4} D_{\mathbf{n}_4} f \, ds$$
$$= e \sin 2 + (1-e) \sin 2 - \sin 2 = 0.$$

8.2. The Divergence Theorem

1. (a) From $\mathbf{F} = (F_1, F_2, F_3) = (x(x^2+y^2+z^2)^{-3/2}, y(x^2+y^2+z^2)^{-3/2}, z(x^2+y^2+z^2)^{-3/2})$ we compute $\partial F_1/\partial x = (-2x^2+y^2+z^2)/(x^2+y^2+z^2)^{5/2}$, with similar expressions for $\partial F_2/\partial y$ (interchange x and y) and $\partial F_3/\partial z$ (interchange x and z). Adding up the three partial derivatives, we get $div\,\mathbf{F} = 0$.

(b) We cannot use the Divergence Theorem to calculate $\iint_{S_1} \mathbf{F} \cdot d\mathbf{S}$ since $\mathbf{F}$ is not defined at the origin.

Parametrize S_1 by

$$\mathbf{r}(u,v) = (\cos v \cos u, \cos v \sin u, \sin v), \qquad 0 \leq u \leq 2\pi, \quad -\pi/2 \leq v \leq \pi/2.$$

the surface normal was computed in Example 7.7 to be $\mathbf{N}(u,v) = \cos v \, \mathbf{r}(u,v)$. Since it points away from the origin, it is the outward normal for the sphere. It follows that (note that $\|\mathbf{r}\| = 1$ and $\mathbf{r} \cdot \mathbf{r} = \|\mathbf{r}\|^2 = 1$)

$$\iint_{S_1} \frac{\mathbf{r}}{\|\mathbf{r}\|^3} \cdot d\mathbf{S} = \iint_{[0,2\pi] \times [-\pi/2, \pi/2]} \mathbf{r} \, (\cos v \, \mathbf{r}) \, dA = \int_0^{2\pi} \left(\int_{-\pi/2}^{\pi/2} \cos v \, dv \right) du = 4\pi.$$

(c) We are allowed to use the Divergence Theorem since $\mathbf{F}$ is defined and C^1 on and inside S_2. Denoting by W the three-dimensional solid enclosed by S_2, we get

$$\iint_{S_2} \mathbf{F} \cdot d\mathbf{S} = \iiint_W \operatorname{div} \mathbf{F} \, dV = 0,$$

since $\operatorname{div} \mathbf{F} = 0$.

3. The flux out of the given parallelepiped (name it S) is given by $\iint_S \mathbf{F} \cdot d\mathbf{S}$. By the Divergence Theorem,

$$\iint_S \mathbf{F} \cdot d\mathbf{S} = \iiint_W \operatorname{div} \mathbf{F} \, dV = \iiint_W 3 \, dV = 3 \operatorname{volume}(W) = 90.$$

5. Since

$$\operatorname{div}\left((y^2 + \sin z)\mathbf{i} + (e^{\sin z} + 2)\mathbf{j} + (xy + \ln x)\mathbf{k}\right) = 0,$$

the Divergence Theorem implies that

$$\iint_S \left((y^2 + \sin z)\mathbf{i} + (e^{\sin z} + 2)\mathbf{j} + (xy + \ln x)\mathbf{k}\right) \cdot d\mathbf{S} = \iiint_W 0 \, dV = 0$$

(W is the three-dimensional solid enclosed by S).

7. By the Divergence Theorem,

$$\iint_S \left((x + y^2 + 1)\mathbf{i} + (y + xz)\mathbf{j}\right) \cdot d\mathbf{S} = \iiint_W \operatorname{div}\left((x + y^2 + 1)\mathbf{i} + (y + xz)\mathbf{j}\right) dV$$
$$= \iiint_W 2 \, dV,$$

where W is the part of the cone $z = x^2 + y^2$ between the planes $z = 1$ and $z = 2$. Instead of a somewhat messy integration, we proceed by noticing that

$$\iiint_W 2 \, dV = 2 \operatorname{volume}(W).$$

Since $\operatorname{volume}(W) = $ (volume of the cone of radius 2 and height 2) $-$ (volume of the cone of radius 1 and height 1) $= 8\pi/3 - \pi/3 = 7\pi/3$, it follows that $\iiint_W 2 \, dV = 14\pi/3$.

9. Since

$$\operatorname{div}\left(-e^x \cos y\mathbf{i} + e^x \sin y\mathbf{j} + \mathbf{k}\right) = 0,$$

the Divergence Theorem implies that

$$\iint_S \left(-e^x \cos y\mathbf{i} + e^x \sin y\mathbf{j} + \mathbf{k}\right) \cdot d\mathbf{S} = \iiint_W 0 \, dV = 0$$

(W is the three-dimensional solid enclosed by S).

11. By the Divergence Theorem,

$$\iint_S \mathbf{F} \cdot d\mathbf{S} = \iiint_W \operatorname{div} \mathbf{F} \, dV = 4 \iiint_W x \, dV,$$

where W is the three-dimensional solid inside the hemisphere and above the xy-plane. Changing to polar coordinates, we get

$$\iint_S \mathbf{F} \cdot d\mathbf{S} = 4 \int_0^{2\pi} \left(\int_0^{\pi/2} \left(\int_0^1 \rho \sin\phi \cos\theta \, \rho^2 \sin\phi d\rho \right) d\phi \right) d\theta$$

$$= 4 \left(\int_0^{2\pi} \cos\theta \, d\theta \right) \left(\int_0^{\pi/2} \left(\int_0^1 \rho^3 \sin^2\phi d\rho \right) d\phi \right) = 0,$$

since $\int_0^{2\pi} \cos\theta \, d\theta = 0$.

13. By the Divergence Theorem,

$$\iint_S (ye^z \mathbf{i} + yz\mathbf{k}) \cdot d\mathbf{S} = \iiint_W div \, (ye^z \mathbf{i} + yz\mathbf{k}) \, dV = \iiint_W y \, dV,$$

where W is the tetrahedron bounded by the coordinate planes and the plane $x + 2y + z = 4$. It is defined by $0 \le y \le 2$, $0 \le x \le 4 - 2y$ (to get this, substitute $z = 0$ into $x + 2y + z = 4$) and $0 \le z \le 4 - x - 2y$. Thus

$$\iiint_W y \, dV = \int_0^2 \left(\int_0^{4-2y} \left(\int_0^{4-x-2y} y dz \right) dx \right) dy$$

$$= \int_0^2 \left(\int_0^{4-2y} (4 - x - 2y)y \, dx \right) dy$$

$$= \int_0^2 \left(4xy - \tfrac{1}{2}x^2y - 2y^2x \right) \Big|_0^{4-2y} dy$$

$$= \int_0^2 (2y^3 - 8y^2 + 8y) \, dy$$

$$= \left(\tfrac{1}{2}y^4 - \tfrac{8}{3}y^3 + 4y^2 \right) \Big|_0^2 = \tfrac{8}{3}.$$

15. Using the Divergence Theorem, we compute

$$\iint_S \|\mathbf{r}\|^{-2}\mathbf{r} \cdot d\mathbf{S} = \iiint_W div \, (\|\mathbf{r}\|^{-2}\mathbf{r}) \, dV = \iiint_W \|\mathbf{r}\|^{-2} \, dV,$$

because

$$div \left(\frac{\mathbf{r}}{\|\mathbf{r}\|^2} \right) = div \left(\frac{(x, y, z)}{x^2 + y^2 + z^2} \right) = \frac{-x^2 + y^2 + z^2}{(x^2 + y^2 + z^2)^2} + \frac{-y^2 + x^2 + z^2}{(x^2 + y^2 + z^2)^2}$$

$$+ \frac{-z^2 + x^2 + y^2}{(x^2 + y^2 + z^2)^2} = \frac{1}{x^2 + y^2 + z^2} = \frac{1}{\|\mathbf{r}\|^2}.$$

17. Write $\iint_S xyz \, dS$ as

$$\iint_S xyz \, dS = \iint_S xyz\mathbf{n} \cdot \mathbf{n} \, dS,$$

where $\mathbf{n}$ is the unit outward normal to S. Since S is a sphere centered at the origin, $\mathbf{n}$ has the same direction as the position vector of a point; i.e., $\mathbf{n} = (x\mathbf{i} + y\mathbf{j} + z\mathbf{k})/\sqrt{x^2 + y^2 + z^2} = x\mathbf{i} + y\mathbf{j} + z\mathbf{k}$ (since $x^2 + y^2 + z^2 = 1$). The divergence of $xyz\mathbf{n}$ is computed to be

$$div \, (xyz\mathbf{n}) = div \, xyz(x\mathbf{i} + y\mathbf{j} + z\mathbf{k}) = 6xyz,$$

and, consequently

$$\iint_S xyz\mathbf{n} \cdot \mathbf{n} \, dS = \iint_S xyz\mathbf{n} \cdot d\mathbf{S} = \iiint_W div\,(xyz\mathbf{n}) \, dV = 6 \iiint_W xyz \, dV,$$

where W is the ball $0 \le x^2 + y^2 + z^2 \le 1$. Using spherical coordinates to evaluate the integral, we get

$$6 \int_0^{2\pi} \left(\int_0^\pi \left(\int_0^1 \rho^3 \sin^2 \phi \cos \phi \, \sin \theta \cos \theta \, \rho^2 \sin \phi d\rho \right) d\phi \right) d\theta = 0.$$

The result was obtained without too much integration — factor $\sin \theta \cos \theta$ all the way to the outermost integration and use the fact that $\int_0^{2\pi} \sin \theta \cos \theta \, d\theta = \frac{1}{2} \int_0^{2\pi} \sin 2\theta \, d\theta = 0$.

19. If we show that

$$div\,(f\nabla g) = f\Delta g + \nabla f \cdot \nabla g,$$

then we are done — the statement follows from the Divergence Theorem.

Using a product rule for the divergence (see Section 4.8), we get

$$div\,(f\nabla g) = f div\,(\nabla g) + \nabla f \cdot \nabla g = f\Delta g + \nabla f \cdot \nabla g,$$

since, by definition, $div\,(\nabla g) = \Delta g$.

21. By definition, $D_{\mathbf{n}} f = \nabla f \cdot \mathbf{n}$, where $\mathbf{n}$ is a unit normal to the surface. So

$$\iint_S D_{\mathbf{n}} f \, dS = \iint_S \nabla f \cdot \mathbf{n} \, dS = \iint_S \nabla f \, d\mathbf{S}.$$

Now proceed using the Divergence Theorem and the definition of the Laplace operator:

$$\iint_S \nabla f \, d\mathbf{S} = \iiint_W div\,(\nabla f) \, dV = \iiint_W \Delta f \, dV.$$

23. Since

$$div \left(\frac{y\mathbf{i}}{x^2 + y^2} - \frac{x\mathbf{j}}{x^2 + y^2} \right) = \frac{-2xy}{(x^2 + y^2)^2} - \frac{-2xy}{(x^2 + y^2)^2} = 0,$$

it follows that

$$\iint_{\mathbf{c}} \mathbf{F} \cdot \mathbf{n} \, ds = \iint_R div\,\mathbf{F} \, dA = 0.$$

25. The outward flux of $\mathbf{F}$ is given by

$$\iint_{\mathbf{c}} \mathbf{F} \cdot \mathbf{n} \, ds = \iint_D div\,\mathbf{F} \, dA = \iint_D x \, dA$$

where D is the triangle bounded by the lines $y = x$, $y = 2x$ and $x = 1$. Hence

$$\iint_{\mathbf{c}} \mathbf{F} \cdot \mathbf{n} ds = \int_0^1 \left(\int_x^{2x} x dy \right) dx = \int_0^1 xy \Big|_x^{2x} dx = \int_0^1 x^2 \, dx = \frac{1}{3}.$$

The counterclockwise circulation of $\mathbf{F}$ is computed using Green's Theorem:

$$\iint_{\mathbf{c}} \mathbf{F} \cdot ds = \iint_D curl\,\mathbf{F} \cdot \mathbf{k} \, dA = \iint_D (-3y) \, dA$$

$$= -3 \int_0^1 \left(\int_x^{2x} y dy \right) dx = -\frac{3}{2} \int_0^1 (y^2) \Big|_x^{2x} dx = -\frac{3}{2} \int_0^1 3x^2 \, dx = -\frac{3}{2}.$$

27. Since $div\,\mathbf{F} = e^x e^y + 2e^y = e^y(e^x + 2)$, the outward flux of $\mathbf{F}$ is given by

$$\iint_{\mathbf{c}} \mathbf{F} \cdot \mathbf{n}\, ds = \iint_R div\,\mathbf{F}\, dA = \iint_R e^y(e^x + 2)\, dA.$$

From Fubini's Theorem, it follows that

$$\iint_R e^y(e^x + 2)\, dA = \int_0^3 \left(\int_0^4 e^y(e^x + 2) dy \right) dx$$

$$= (e^4 - 1) \int_0^3 (e^x + 2)\, dx = (e^4 - 1)(e^x + 2x) \Big|_0^3 = (e^4 - 1)(e^3 + 5).$$

Since

$$curl\,\mathbf{F} = \begin{vmatrix} \mathbf{i} & \mathbf{j} & \mathbf{k} \\ \partial/\partial x & \partial/\partial y & \partial/\partial z \\ e^x e^y & 2e^y & 0 \end{vmatrix} = -e^x e^y \mathbf{k},$$

it follows that the counterclockwise circulation of $\mathbf{F}$ is (by Green's Theorem)

$$\iint_{\mathbf{c}} \mathbf{F} \cdot d\mathbf{s} = \iint_R curl\,\mathbf{F} \cdot \mathbf{k}\, dA = - \iint_R e^x e^y\, dA$$

$$= - \int_0^4 \left(\int_0^3 e^x e^y dx \right) dy$$

$$= -(e^3 - 1) \int_0^4 e^y\, dy = -(e^3 - 1)(e^4 - 1).$$

29. (a) Parametrize the bottom square by $\mathbf{r}(u, v) = (u, v, 0), 0 \le u, v \le 1$. Then $\mathbf{T}_u = (1, 0, 0)$, $\mathbf{T}_v = (0, 1, 0)$, $\mathbf{N} = (0, 0, 1)$, $\mathbf{F}(\mathbf{r}(u, v)) \cdot \mathbf{N}(u, v) = (0, 0, e^{-u^2}) \cdot (0, 0, 1) = e^{-u^2}$, and so

$$\iint_S \mathbf{F} \cdot d\mathbf{S} = \iint_{[0,1] \times [0,1]} e^{-u^2}\, dudv.$$

It is not possible to evaluate $\int e^{-u^2} du$ as a compact formula (i.e., without using power series).

Note: the surface integrals along the top and the bottom surfaces cancel each other. Since the integrals along the four vertical sides are zero, it follows that $\iint_S \mathbf{F} \cdot d\mathbf{S} = 0$.

(b) Using the Divergence Theorem, we get

$$\iint_S \mathbf{F} \cdot d\mathbf{S} = \iiint_W div\,\mathbf{F}\, dV = 0,$$

since $div\,\mathbf{F} = 0$.

8.3. Stokes' Theorem

1. (a) Since

$$curl\,\mathbf{F} = \begin{vmatrix} \mathbf{i} & \mathbf{j} & \mathbf{k} \\ \partial/\partial x & \partial/\partial y & \partial/\partial z \\ -y & x & 0 \end{vmatrix} = 2\mathbf{k},$$

Stokes' Theorem gives

$$\int_{\mathbf{c}} \mathbf{F} \cdot d\mathbf{s} = \iint_S 2\mathbf{k} \cdot d\mathbf{S},$$

where S is the square $[0,1] \times [0,1]$ in the plane $z = 1$. Parametrizing S by $\mathbf{r}(u,v) = (u,v,1)$, $0 \leq u,v \leq 1$, we get $\mathbf{N} = \mathbf{T}_u \times \mathbf{T}_v = (0,0,1) = \mathbf{k}$. Note that $\mathbf{c}$ is positively oriented, as required by the theorem. Hence

$$\iint_S 2\mathbf{k} \cdot d\mathbf{S} = \iint_{[0,1]\times[0,1]} (2\mathbf{k}) \cdot \mathbf{k} \, dA = 2.$$

(b) Since $curl\,\mathbf{F} \cdot \mathbf{n} = (2\mathbf{k}) \cdot \mathbf{j} = 0$, Stokes' Theorem implies that $\int_{\mathbf{c}} \mathbf{F} \cdot d\mathbf{s} = 0$.

3. Complete the square in $x^2 + 2x + y^2 = 3$ to get $(x+1)^2 + y^2 = 4$; parametrize the intersection $\mathbf{c}$ of the cylinder $x^2 + 2x + y^2 = 3$ and the plane $z = x$ as

$$\mathbf{c}(t) = (2\cos t - 1, 2\sin t, 2\cos t - 1), \qquad t \in [0, 2\pi].$$

Then $\mathbf{c}'(t) = (-2\sin t, 2\cos t, -2\sin t)$ and

$$\iint_{\mathbf{c}} \mathbf{F} \cdot d\mathbf{s} = \int_0^{2\pi} (4\cos^2 t, 0, -(2\cos t - 1)^2) \cdot (-2\sin t, 2\cos t, -2\sin t)\, dt$$

$$= \int_0^{2\pi} (-8\cos t \sin t + 2\sin t)\, dt$$

$$= \int_0^{2\pi} (-4\sin 2t + 2\sin t)\, dt = 0,$$

since $\int_0^{2\pi} \sin t \, dt = 0$ and $\int_0^{2\pi} \sin 2t \, dt = 0$.

Parametrize the part of the plane inside the cylinder (call it S) by $\mathbf{r}(u,v) = (u,v,u)$, where $(u+1)^2 + v^2 \leq 4$. The surface normal vector is $\mathbf{N} = (-1,0,1)$ and, since $curl\,\mathbf{F} = 2x\mathbf{j}$, it follows that

$$\iint_S curl\,\mathbf{F} \cdot \mathbf{N}\, dS = \iint_S (0, 2u, 0) \cdot (-1,0,1)\, dS = \iint_S 0\, dS = 0.$$

5. The $curl$ of $\mathbf{F}$ is

$$curl\,\mathbf{F} = \begin{vmatrix} \mathbf{i} & \mathbf{j} & \mathbf{k} \\ \partial/\partial x & \partial/\partial y & \partial/\partial z \\ 2x+y & 0 & -3x+y+z \end{vmatrix} = \mathbf{i} + 3\mathbf{j} - \mathbf{k};$$

so, by Stokes' Theorem,

$$\int_{\mathbf{c}} \mathbf{F} \cdot d\mathbf{s} = \iint_S curl\,\mathbf{F} \cdot d\mathbf{S} = \iint_S (\mathbf{i} + 3\mathbf{j} - \mathbf{k}) \cdot d\mathbf{S},$$

where S is the part of the plane $x + 4y + 3z = 1$ in the first octant, oriented by a normal that points away from the side that does not face the origin; i.e., that normal has positive $\mathbf{k}$ component. Parametrize the surface of the plane by

$$\mathbf{r}(u,v) = (u, v, (1-u-4v)/3), \qquad (u,v) \in D,$$

where D is the region in the uv-plane defined by $u = 0$, $v = 0$ and $u + 4v = 1$. Then

$$\mathbf{N} = \begin{vmatrix} \mathbf{i} & \mathbf{j} & \mathbf{k} \\ 1 & 0 & -1/3 \\ 0 & 1 & -4/3 \end{vmatrix} = \tfrac{1}{3}\mathbf{i} + \tfrac{4}{3}\mathbf{j} + \mathbf{k}$$

is a normal with the desired direction. It follows that

$$\iint_S (\mathbf{i} + 3\mathbf{j} - \mathbf{k}) \cdot d\mathbf{S} = \iint_D (\mathbf{i} + 3\mathbf{j} - \mathbf{k}) \cdot (\tfrac{1}{3}\mathbf{i} + \tfrac{4}{3}\mathbf{j} + \mathbf{k}) \, dA$$

$$= \iint_D \tfrac{10}{3} \, dA = \tfrac{10}{3} \tfrac{1}{8} = \tfrac{5}{12},$$

since $\iint_D dA = \text{area}(D) = 1/8$ (D is a right triangle with sides 1 and 1/4).

7. The *curl* of $\mathbf{F}$ is computed to be

$$curl\,\mathbf{F} = \begin{vmatrix} \mathbf{i} & \mathbf{j} & \mathbf{k} \\ \partial/\partial x & \partial/\partial y & \partial/\partial z \\ x^2 + z^2 & y^2 z^2 & 0 \end{vmatrix} = -2y^2 z \mathbf{i} + 2z\mathbf{j};$$

by Stokes' Theorem,

$$\int_{\mathbf{c}} \mathbf{F} \cdot d\mathbf{s} = \iint_S curl\,\mathbf{F} \cdot d\mathbf{S} = \iint_S (-2y^2 z \mathbf{i} + 2z\mathbf{j}) \cdot d\mathbf{S},$$

where S is the part of the plane $z = y$ defined by $1 \leq x \leq 2$ and $0 \leq y \leq 4$. Parametrize S by

$$\mathbf{r}(u, v) = (u, v, v), \qquad (u, v) \in D = [1, 2] \times [0, 4].$$

The surface normal $\mathbf{N}$ is

$$\mathbf{N} = \begin{vmatrix} \mathbf{i} & \mathbf{j} & \mathbf{k} \\ 1 & 0 & 0 \\ 0 & 1 & 1 \end{vmatrix} = -\mathbf{j} + \mathbf{k}$$

(notice that the orientation requirement is satisfied); therefore,

$$\iint_S (-2y^2 z \mathbf{i} + 2z\mathbf{j}) \cdot d\mathbf{S} = \iint_D (-2v^3 \mathbf{i} + 2v\mathbf{j}) \cdot (-\mathbf{j} + \mathbf{k}) \, dA$$

$$= -\iint_D 2v \, dA = -\int_1^2 \left(\int_0^4 2v\,dv \right) du$$

$$= -\int_1^2 v^2 \Big|_0^4 \, du = -\int_1^2 16 \, du = -16.$$

9. Since

$$curl\,\mathbf{F} = \begin{vmatrix} \mathbf{i} & \mathbf{j} & \mathbf{k} \\ \partial/\partial x & \partial/\partial y & \partial/\partial z \\ y^2 & y^2 & y^2 \end{vmatrix} = 2y\mathbf{i} - 2y\mathbf{k},$$

Stokes' Theorem implies that

$$\iint_{\mathbf{c}} y^2 (\mathbf{i} + \mathbf{j} + \mathbf{k}) \cdot ds = \iint_{\overline{S}} (2y\mathbf{i} - 2y\mathbf{k}) \cdot d\mathbf{S},$$

where $\overline{S}$ is a surface whose boundary is $\mathbf{c}$. Instead of taking $\overline{S}$ to be the part of the sphere (and thus going through a messy parametrization and evaluation of an integral), we use the plane instead. In other words, parametrize the part of the plane $z = 1/2$ (call it S) whose boundary is $\mathbf{c}$ by $\mathbf{r}(u, v) = (u, v, 1/2)$, where $0 \leq u^2 + v^2 \leq \frac{3}{4}$. The surface normal is $\mathbf{N} = \mathbf{k}$;

that forces the boundary curve to be oriented counterclockwise as seen from above (which is precisely what the assumption states). Hence

$$\iint_S (2y\mathbf{i} - 2y\mathbf{k}) \cdot d\mathbf{S} = \iint_{\{0 \le u^2 + v^2 \le 3/4\}} (2y\mathbf{i} - 2y\mathbf{k}) \cdot \mathbf{k}\, dS$$

$$= -2 \int_0^{2\pi} \left(\int_0^{\sqrt{3}/2} r \sin\theta\, r dr \right) d\theta = 0,$$

since $\int_0^{2\pi} \sin\theta\, d\theta = 0$.

11. By Stokes' Theorem,

$$\int_{\mathbf{c}} (\mathbf{F} \times \mathbf{r}) \cdot d\mathbf{s} = \iint_S curl\,(\mathbf{F} \times \mathbf{r}) \cdot d\mathbf{S}.$$

Let $\mathbf{F} = (F_1, F_2, F_3)$ (since $\mathbf{F}$ is a constant vector field, its components are constants). So

$$\mathbf{F} \times \mathbf{r} = \begin{vmatrix} \mathbf{i} & \mathbf{j} & \mathbf{k} \\ F_1 & F_2 & F_3 \\ x & y & z \end{vmatrix} = (F_2 z - F_3 y)\mathbf{i} + (F_3 x - F_1 z)\mathbf{j} + (F_1 y - F_2 x)\mathbf{k}$$

and

$$curl\,(\mathbf{F} \times \mathbf{r}) = \begin{vmatrix} \mathbf{i} & \mathbf{j} & \mathbf{k} \\ \partial/\partial x & \partial/\partial y & \partial/\partial z \\ F_2 z - F_3 y & F_3 x - F_1 z & F_1 y - F_2 x \end{vmatrix}$$

$$= (F_1 + F_1)\mathbf{i} - (-F_2 - F_2)\mathbf{j} + (F_3 + F_3)\mathbf{k} = 2\mathbf{F}.$$

It follows that

$$\int_{\mathbf{c}} (\mathbf{F} \times \mathbf{r}) \cdot d\mathbf{s} = \iint_S curl\,(\mathbf{F} \times \mathbf{r}) \cdot d\mathbf{S} = 2 \iint_S \mathbf{F} \cdot d\mathbf{S}.$$

13. Parametrize the circle by $\mathbf{c}(t) = \cos t\,\mathbf{i} + \sin t\,\mathbf{j}$, $t \in [0, 2\pi]$. Then $\mathbf{F}(\mathbf{c}(t)) = -2\sin t\,\mathbf{i} + 2\cos t\,\mathbf{j}$, $\mathbf{c}'(t) = -\sin t\,\mathbf{i} + \cos t\,\mathbf{j}$ and

$$\int_{\mathbf{c}} \mathbf{F} \cdot d\mathbf{s} = \int_0^{2\pi} (-2\sin t\,\mathbf{i} + 2\cos t\,\mathbf{j}) \cdot (-\sin t\,\mathbf{i} + \cos t\,\mathbf{j})\, dt = \int_0^{2\pi} 2\, dt = 4\pi.$$

On the other hand, since $curl\,\mathbf{F} = \mathbf{0}$, it follows that

$$\iint_S curl\,\mathbf{F}\, d\mathbf{S} = \iint_S \mathbf{0}\, d\mathbf{S} = 0.$$

The vector field $\mathbf{F}$ is not defined at $(0,0)$, and therefore Stokes' Theorem does not apply.

15. The $curl$ of $\mathbf{F}$ is

$$curl\,\mathbf{F} = \begin{vmatrix} \mathbf{i} & \mathbf{j} & \mathbf{k} \\ \partial/\partial x & \partial/\partial y & \partial/\partial z \\ 2x/(x^2 + y) & 1/(x^2 + y) & 0 \end{vmatrix}$$

$$= \left(-\frac{2x}{(x^2 + y)^2} + \frac{2x}{(x^2 + y)^2} \right) \mathbf{k} = \mathbf{0}.$$

The domain of $\mathbf{F}$ is $\mathbb{R}^2$, with the points on the parabola $y = -x^2$ removed — so it is not simply-connected. However, if we restrict its domain as shown in Figure 15, then it becomes simply-connected. By Theorem 5.8 in Section 5.4, the circulation is zero.

Alternatively, notice that $\mathbf{F} = \nabla \ln(x^2 + y)$; proceed by using the Fundamental Theorem of Calculus (see Theorem 5.5).

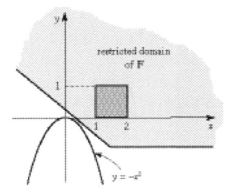

Figure 15

17. Since
$$curl\,\mathbf{F} = \begin{vmatrix} \mathbf{i} & \mathbf{j} & \mathbf{k} \\ \partial/\partial x & \partial/\partial y & \partial/\partial z \\ y & 2z & 3x \end{vmatrix} = -2\mathbf{i} - 3\mathbf{j} - \mathbf{k},$$

Stokes' Theorem implies that
$$\int_{\mathbf{c}} \mathbf{F} \cdot d\mathbf{s} = \iint_S (-2\mathbf{i} - 3\mathbf{j} - \mathbf{k}) \cdot d\mathbf{S},$$

where S is any surface (with the appropriate orientation) whose boundary is the intersection of $x^2 + y^2 = 1$ and $z = y$. Take the plane and use the parametrization $\mathbf{r}(u, v) = (u, v, v)$, where $(u, v) \in D$, and D is the disk $x^2 + y^2 \le 1$. Then
$$\mathbf{N} = \begin{vmatrix} \mathbf{i} & \mathbf{j} & \mathbf{k} \\ 1 & 0 & 0 \\ 0 & 1 & 1 \end{vmatrix} = -\mathbf{j} + \mathbf{k},$$

and so
$$\iint_S \mathbf{F} \cdot d\mathbf{S} = \iint_D (-2\mathbf{i} - 3\mathbf{j} - \mathbf{k}) \cdot (-\mathbf{j} + \mathbf{k})\, dA$$
$$= \iint_D 2\, dA = 2\,\text{area}(D) = 2\pi.$$

19. We will use Stokes' Theorem. Since
$$curl\,\mathbf{F} = \begin{vmatrix} \mathbf{i} & \mathbf{j} & \mathbf{k} \\ \partial/\partial x & \partial/\partial y & \partial/\partial z \\ x & -yz & 1 \end{vmatrix} = y\mathbf{i},$$

it follows that

$$\int_{\mathbf{c}} \mathbf{F} \cdot d\mathbf{s} = \iint_{S} y\mathbf{i} \cdot d\mathbf{S},$$

where S is any surface whose boundary is $\mathbf{c}$. Take S to be the part of the plane inside the paraboloid — parametrize it by $\mathbf{r}(u, v) = (u, v, 2v)$, where $(u, v) \in D$, and D is the disk $u^2 + (v - 1)^2 \leq 1$ (this has been obtained by combining $z = x^2 + y^2$ and $z = 2y$). Then

$$\mathbf{N} = \begin{vmatrix} \mathbf{i} & \mathbf{j} & \mathbf{k} \\ 1 & 0 & 0 \\ 0 & 1 & 2 \end{vmatrix} = -2\mathbf{j} + \mathbf{k},$$

and so

$$\iint_{S} y\mathbf{i} \cdot d\mathbf{S} = \iint_{D} y\mathbf{i} \cdot (-2\mathbf{j} + \mathbf{k}) \, dA = \iint_{D} 0 \, dA = 0.$$

21. We are not able to use Stokes' Theorem, since in that case we would have to find a surface whose boundary is $\mathbf{c}$ (such surfaces exist, but their parametrizations are quite complicated).

We proceed by computing the path integrals directly (actually we do not have much choice, because $curl\, \mathbf{F} = -2\mathbf{k} \neq \mathbf{0}$ and we cannot use the Fundamental Theorem). From $\mathbf{c}_1(t) = (\cos t, \sin t, t)$ we get $\mathbf{c}_1'(t) = (-\sin t, \cos t, 1)$ and thus

$$\int_{\mathbf{c}_1} \mathbf{F} \cdot d\mathbf{s} = \int_0^{3\pi} (2\cos t + \sin t, 2\sin t - \cos t, 0) \cdot (-\sin t, \cos t, 1) \, dt = \int_0^{3\pi} (-1) \, dt = -3\pi.$$

Parametrize the line segment by

$$\mathbf{c}_2(t) = (-1, 0, 3\pi) + t(2, 0, -3\pi) = (-1 + 2t, 0, 3\pi - 3\pi t),$$

where $t \in [0, 1]$. Then $\mathbf{c}_2'(t) = (2, 0, -3\pi)$ and

$$\int_{\mathbf{c}_2} \mathbf{F} \cdot d\mathbf{s} = \int_0^1 (-2 + 4t, 1 - 2t, 0) \cdot (2, 0, -3\pi) \, dt = \int_0^1 (-4 + 8t) \, dt = (-4t + 4t^2)\Big|_0^1 = 0.$$

It follows that

$$\int_{\mathbf{c}} \mathbf{F} \cdot d\mathbf{s} = \int_{\mathbf{c}_1} \mathbf{F} \cdot d\mathbf{s} + \int_{\mathbf{c}_2} \mathbf{F} \cdot d\mathbf{s} = -3\pi.$$

23. By Stokes' Theorem,

$$\int_{\mathbf{c}} f\nabla f \cdot d\mathbf{s} = \iint_{S} curl\,(f\nabla f) \cdot d\mathbf{S}.$$

Using a product rule (see Section 4.8) we get

$$curl\,(f\nabla f) = f curl\,(\nabla f) + \nabla f \times \nabla f = \mathbf{0},$$

since both summands are $\mathbf{0}$. It follows that $\int_{\mathbf{c}} f\nabla f \cdot d\mathbf{s} = 0$.

8.4. Differential Forms and Classical Integration Theorems

1. The expression $f\beta$ is a 1-form, as a product of a scalar function and a 1-form. Its value is $f\beta = e^{x+y}(x^2 + y^2)dz$.

3. The expression $\alpha \wedge \beta$ is the wedge product of two 1-forms. It is equal to the 2-form

$$\alpha \wedge \beta = (3xdx + yzdy) \wedge (x^2 + y^2)dz = 3x(x^2 + y^2)dxdz + yz(x^2 + y^2)dydz.$$

5. Since β is a 1-form and μ is a 2-form their difference $\beta - \mu$ is not defined.

7. The wedge product $\beta \wedge \gamma$ is a 3-form, and so the difference $\beta \wedge \gamma - \nu$ is defined. From

$$\beta \wedge \gamma = (x^2 + y^2)dz \wedge (2dxdy - x \cos y \, dydz)$$
$$= 2(x^2 + y^2)dzdxdy - x(x^2 + y^2) \cos y \, dzdydz = 2(x^2 + y^2)dxdydz$$

it follows that

$$(\beta \wedge \gamma) - \nu = 2(x^2 + y^2)dxdydz - \sin x \, dxdydz = (2x^2 + 2y^2 - \sin x)dxdydz.$$

9. Since ν is a 3-form and α is a 1-form, their wedge product is a 4-form. In $\mathbb{R}^3$ there are no non-zero 4-forms, and so $\nu \wedge \alpha = 0$ (0 denotes a zero 4-form).

Alternatively, notice that each of the two terms in $\nu \wedge \alpha$ contains a wedge product of four basic 1-forms with a repeated factor (and so each term is zero).

11. The differential $d\alpha$ is the 1-form

$$d\alpha = \frac{\partial}{\partial x}(e^{xyz})dx + \frac{\partial}{\partial y}(e^{xyz})dy + \frac{\partial}{\partial z}(e^{xyz})dz$$
$$= yze^{xyz}dx + xze^{xyz}dy + xye^{xyz}dz = e^{xyz}(yzdx + xzdy + xydz).$$

13. From $\alpha = x^2(dxdy + dydz)$ we get

$$d\alpha = d(x^2) \wedge dxdy + d(x^2) \wedge dydz = 2xdx \wedge dxdy + 2xdx \wedge dydz = 2xdxdydz.$$

15. By definition,

$$d(\arctan x) = \frac{\partial}{\partial x}(\arctan x) \, dx + \frac{\partial}{\partial y}(\arctan x) \, dy = \frac{1}{1 + x^2}dx.$$

17. Since

$$d\left(\frac{x}{x^2 + y^2}\right) = \frac{\partial}{\partial x}\left(\frac{x}{x^2 + y^2}\right)dx + \frac{\partial}{\partial y}\left(\frac{x}{x^2 + y^2}\right)dy$$
$$= \frac{x^2 + y^2 - x(2x)}{(x^2 + y^2)^2}dx + \frac{x(-2y)}{(x^2 + y^2)^2}dy = \frac{y^2 - x^2}{(x^2 + y^2)^2}dx - \frac{2xy}{(x^2 + y^2)^2}dy,$$

and (computed similarly)

$$d\left(\frac{y}{x^2 + y^2}\right) = \frac{-2xy}{(x^2 + y^2)^2}dx + \frac{x^2 - y^2}{(x^2 + y^2)^2}dy,$$

it follows that

$$d\alpha = d\left(\frac{x}{x^2 + y^2}\right) \wedge dx - d\left(\frac{y}{x^2 + y^2}\right) \wedge dy$$
$$= \frac{y^2 - x^2}{(x^2 + y^2)^2}dx \wedge dx - \frac{2xy}{(x^2 + y^2)^2}dy \wedge dx - \left(\frac{-2xy}{(x^2 + y^2)^2}dx \wedge dy + \frac{x^2 - y^2}{(x^2 + y^2)^2}dy \wedge dy\right)$$

$$= \frac{4xy}{(x^2 + y^2)^2} dx dy.$$

19. Since

$$d\alpha = d(x^3 dy - 2y dx + z dz) = 3x^2 dx \wedge dy - 2dy \wedge dx + dz \wedge dz = (3x^2 + 2)dx dy,$$

it follows that

$$d(d\alpha) = d((3x^2 + 2)dx dy) = d(3x^2 + 2) \wedge dx dy = 6x dx \wedge dx dy = 0.$$

21. By definition,

$$df = \frac{2x}{x^2 + y^2 + z^2 + 1} dx + \frac{2y}{x^2 + y^2 + z^2 + 1} dy + \frac{2z}{x^2 + y^2 + z^2 + 1} dz.$$

Since

$$\nabla f = \left(\frac{2x}{x^2 + y^2 + z^2 + 1}, \frac{2y}{x^2 + y^2 + z^2 + 1}, \frac{2z}{x^2 + y^2 + z^2 + 1} \right),$$

the corresponding components are equal. The identity $d(df) = 0$ represents $curl\,(\nabla f) = \mathbf{0}$.

23. From $\mathbf{c}(t) = (\sqrt{t}, t^2, t^3)$ it follows that $x = \sqrt{t}$, $y = t^2$ and $z = t^3$, and therefore

$$\int_{\mathbf{c}} (y dy - xyz dz) = \int_0^1 \left(y \frac{dy}{dt} - xyz \frac{dz}{dt} \right) dt$$

$$= \int_0^1 (2t^3 - 3t^{15/2}) dt = \left(\frac{1}{2}t^4 - \frac{6}{17}t^{17/2} \right) \Big|_0^1 = \frac{1}{2} - \frac{6}{17} = \frac{5}{34}.$$

25. (a) Using the definition of the differential, we get

$$d\alpha = d(y e^{xy} dx + x e^{xy} dy) = (e^{xy} + xy e^{xy})dy \wedge dx + (e^{xy} + xy e^{xy})dx \wedge dy = 0.$$

Consequently, α is closed. By Theorem 8.6 α is also exact (it is defined on $\mathbb{R}^3$, which is a simply-connected set).

(b) Since $d\alpha = d(x dx + y dy + z dz) = dx \wedge dx + dy \wedge dy + dz \wedge dz = 0$, we conclude that α is closed. Proceeding as in (a), we can show that it is also exact. Alternatively, $d\left(\frac{1}{2}(x^2 + y^2 + z^2) \right) = x dx + y dy + z dz = \alpha$ implies that α is exact.

(c) From

$$d\alpha = d(-y dx + x dy + z dz) = -dy \wedge dx + dx \wedge dy = 2dx \wedge dy,$$

we conclude that α is not closed. Consequently, it cannot be exact either (recall that "exact" implies "closed").

(d) Since

$$d\alpha = d(-2y dx dy - y dz dy + z dx dz) = -2dy \wedge dx dy - dy \wedge dz dy + dz \wedge dx dz = 0,$$

it follows that α is closed. By Theorem 8.6 α is also exact (it is defined on $\mathbb{R}^3$, which is a star-shaped set).

(e) Since

$$d\alpha = d(\sin x\, dxdy + (\sin y + \cos z)dydz) = \cos x dx \wedge dxdy + (\cos y dy - \sin z dz) \wedge dydz = 0,$$

we conclude that α is closed. By Theorem 8.6 α is also exact (it is defined on $\mathbb{R}^3$, which is a star-shaped set).

(f) From α is a 3-form, $d\alpha = 0$ by definition. Thus α is closed. From

$$d\left(-\frac{1}{yz^2}\cos(xyz^2)dydz\right) = \alpha$$

(this was done by guess-and-check), we conclude that α is exact. Notice that we do not have a version of Theorem 8.6 for 3-forms.

27. We have to find a 0-form (i.e., a real-valued function) β such that $d\beta = \alpha$. Since $d\beta = \beta_x dx + \beta_y dy + \beta_z dz$, we get $\beta_x = y$, $\beta_y = x$ and $\beta_z = 4$. Proceeding as in any other exercise involving finding a potential function, we get $\beta = xy + 4z + C$. Thus, using Exercise 26,

$$\int_c \alpha = \int_c d\beta = \beta(0,4,1) - \beta(0,0,0) = 4.$$

29. Parametrize S by $\mathbf{r}(u,v) = (\cos u, \sin u, v)$, $(u,v) \in D$, where D is the rectangle in the uv-plane given by $0 \leq u \leq 2\pi$, $0 \leq v \leq 3$. By definition,

$$\int_S (xdxdy + ydydz + zdzdx) = \iint_D \left(x\frac{\partial(x,y)}{\partial(u,v)} + y\frac{\partial(y,z)}{\partial(u,v)} + z\frac{\partial(z,x)}{\partial(u,v)}\right)dA,$$

where

$$\frac{\partial(x,y)}{\partial(u,v)} = \begin{vmatrix} -\sin u & 0 \\ \cos u & 0 \end{vmatrix} = 0,$$

$$\frac{\partial(y,z)}{\partial(u,v)} = \begin{vmatrix} \cos u & 0 \\ 0 & 1 \end{vmatrix} = \cos u$$

and

$$\frac{\partial(z,x)}{\partial(u,v)} = \begin{vmatrix} 0 & 1 \\ -\sin u & 0 \end{vmatrix} = \sin u;$$

It follows that

$$\int_S xdxdy + ydydz + zdzdx = \iint_D (\sin u \cos u + v \sin u)\, dA$$

$$= \int_0^{2\pi}\left(\int_0^3 \tfrac{1}{2}\sin 2u\, dv\right)du + \int_0^{2\pi}\left(\int_0^3 v\sin u dv\right)du = 0,$$

since $\int_0^{2\pi}\sin 2u\, du = 0$ and $\int_0^{2\pi}\sin u\, du = 0$.

31. Parametrize S by $\mathbf{r}(u,v) = (u,v,u^2+v^2)$, $(u,v) \in D$, where D is the disk $x^2 + y^2 \leq 4$. Then

$$\int_S (dxdy + zdzdx) = \iint_D \left(\frac{\partial(x,y)}{\partial(u,v)} + z\frac{\partial(z,x)}{\partial(u,v)}\right)dA,$$

where

$$\frac{\partial(x,y)}{\partial(u,v)} = \begin{vmatrix} 1 & 0 \\ 0 & 1 \end{vmatrix} = 1$$

and

$$\frac{\partial(z,x)}{\partial(u,v)} = \begin{vmatrix} 2u & 2v \\ 1 & 0 \end{vmatrix} = -2v.$$

So

$$\int_S (dxdy + zdzdx) = \iint_D \left(1 + (u^2 + v^2)(-2v)\right) dA$$

$$= \iint_D 1\, dA - 2 \iint_D (u^2 v + v^3)\, dA.$$

Changing to polar coordinates $u = r\cos\theta$, $v = r\sin\theta$, we get

$$\iint_D (u^2 v + v^3)\, dA = \int_0^{2\pi} \left(\int_0^2 (r^3 \cos^2\theta \sin\theta + r^3 \sin^3\theta) r\, dr \right) d\theta$$

$$= \left(\int_0^{2\pi} \cos^2\theta \sin\theta\, d\theta \right) \left(\int_0^2 r^4\, dr \right) + \left(\int_0^{2\pi} \sin^3\theta\, d\theta \right) \left(\int_0^2 r^4\, dr \right)$$

$$= \left(-\tfrac{1}{3} \cos^3\theta \Big|_0^{2\pi} \right) \left(\tfrac{1}{5} r^5 \Big|_0^2 \right) + 0$$

$$= \left(-\tfrac{1}{3} - (-\tfrac{1}{3}) \right) \tfrac{32}{5} = 0,$$

since $\int_0^{2\pi} \sin^3\theta\, d\theta = 0$ (use integration tables or symmetry). It follows that

$$\int_S (dxdy + zdzdx) = \iint_D 1\, dA = 4\pi.$$

33. Let D be the region bounded by the semicircle and the line segment and let **c** be its positively oriented boundary. By Green's Theorem,

$$\int_{\mathbf{c}} xdx + xdy = \iint_D d(xdx + xdy) = \iint_D dxdy = \iint_D dA = \text{area}(D) = \tfrac{\pi}{2}.$$

35. Let $D = [0,1] \times [-2,2]$; by assumption, **c** is its positively oriented boundary. Green's Theorem implies that

$$\int_{\mathbf{c}} (2x + 2y^2)dx - 3xdy = \int_D d((2x + 2y^2)dx - 3xdy)$$

$$= \int_D (4ydydx - 3dxdy) = \int_D (-3 - 4y)dxdy$$

(this is an integral of a 2-form; interpret it as an iterated integral)

$$= \int_{-2}^2 \left(\int_0^1 (-3 - 4y)dx \right) dy = \int_{-2}^2 (-3 - 4y)\, dx$$

$$= \left(-3y - 2y^2 \right) \Big|_{-2}^2 = -12.$$

37. Since

$$d\alpha = d(ydydz + xzdzdx + zdxdy) = dxdydz,$$

it follows that

$$\int_S ydydz + xzdzdx + zdxdy = \int_W dxdydz = \iiint_W dV,$$

where W is the three-dimensional solid region bounded by S. Using cylindrical coordinates, we get

$$\iiint_W dV = \int_0^{2\pi} \left(\int_0^1 \left(\int_0^{r^2} rdz \right) dr \right) d\theta$$

$$= \int_0^{2\pi} \left(\int_0^1 r^3 dr \right) d\theta = \tfrac{1}{4}2\pi = \tfrac{1}{2}\pi.$$

39. Since

$$d\alpha = d(xdydz + ydzdx + zdxdy) = 3dxdydz,$$

the Divergence Theorem implies that (W is the solid region between the spheres of radius 1 and 2)

$$\int_S xdydz + ydzdx + zdxdy = 3 \int_W dxdydz$$

$$= 3 \text{ volume}(W) = 3 \left(\tfrac{4}{3}\pi\,2^3 - \tfrac{4}{3}\pi\,1^3 \right) = 28\pi.$$

8.5. Vector Calculus in Electromagnetism

1. There are infinitely many answers. For example,

$$\mathbf{A}_0 = (2xz^2 + f(x))\mathbf{i} + (xy + g(y))\mathbf{j} + (yz + h(z))\mathbf{k},$$

where f, g and h are differentiable functions of variables indicated.

In general, $curl\,\mathbf{A}_0 = curl\,\mathbf{A}$ implies $curl\,(\mathbf{A}_0 - \mathbf{A}) = 0$; i.e., $\mathbf{A}_0 - \mathbf{A} = grad\,f$ (this is due to Theorem 5.4 in Section 5.4; the domain of $\mathbf{A}$ and $\mathbf{A}_0$ is $\mathbb{R}^3$, which is simply-connected). It follows that $\mathbf{A}_0 = \mathbf{A} + grad\,f$, where f is any differentiable function.

3. Formula (8.25) states that

$$\mathbf{J} = \frac{1}{\mu_0} curl\,\mathbf{B} - \epsilon_0 \frac{\partial \mathbf{E}}{\partial t}.$$

Using the linearity of div and switching div and $\partial/\partial t$ we get

$$div\,\mathbf{J} = div \left(\frac{1}{\mu_0} curl\,\mathbf{B} - \epsilon_0 \frac{\partial \mathbf{E}}{\partial t} \right)$$

$$= \frac{1}{\mu_0} div\,(curl\,\mathbf{B}) - \epsilon_0 div \left(\frac{\partial \mathbf{E}}{\partial t} \right) = -\epsilon_0 \frac{\partial}{\partial t}(div\,\mathbf{E}),$$

since $div\,(curl\,\mathbf{B}) = 0$. From (8.23) it follows that

$$div\,\mathbf{J} = -\epsilon_0 \frac{\partial}{\partial t} \left(\frac{\rho}{\epsilon_0} \right) = -\frac{\partial \rho}{\partial t}.$$

5. Since

$$div\,\mathbf{E} = div\,\left(e^t(x+y)\mathbf{i} + e^t(y+z)\mathbf{j} + e^t z\mathbf{k}\right) = 3e^t$$

and $3e^t = \rho/\epsilon_0$ (by assumption), equation (8.23) is satisfied. From

$$curl\,\mathbf{E} = e^t \begin{vmatrix} \mathbf{i} & \mathbf{j} & \mathbf{k} \\ \partial/\partial x & \partial/\partial y & \partial/\partial z \\ x+y & y+z & z \end{vmatrix} = e^t(-\mathbf{i}-\mathbf{k})$$

and

$$-\frac{\partial \mathbf{B}}{\partial t} = -(e^t\mathbf{i} + e^t\mathbf{k})$$

it follows that equation (8.24) is satisfied. Now

$$curl\,\mathbf{B} = \begin{vmatrix} \mathbf{i} & \mathbf{j} & \mathbf{k} \\ \partial/\partial x & \partial/\partial y & \partial/\partial z \\ e^t & x^2 - z^2 & e^t + 1 \end{vmatrix} = 2z\mathbf{i} + 2x\mathbf{k}$$

and

$$\mu_0\left(\mathbf{J} + \epsilon_0\frac{\partial \mathbf{E}}{\partial t}\right) = \mu_0\mathbf{J} + \mu_0\epsilon_0\frac{\partial \mathbf{E}}{\partial t}$$
$$= \left(2z - \mu_0\epsilon_0 e^t(x+y)\right)\mathbf{i} - \mu_0\epsilon_0 e^t(y+z)\mathbf{j}$$
$$+ \left(2x - \mu_0\epsilon_0 e^t z\right)\mathbf{k} + \mu_0\epsilon_0 e^t\left((x+y)\mathbf{i} + (y+z)\mathbf{j} + z\mathbf{k}\right)$$
$$= 2z\mathbf{i} + 2x\mathbf{k}$$

imply that equation (8.25) is satisfied. Clearly,

$$div\,\mathbf{B} = div\,(e^t\mathbf{i} + (x^2 - z^2)\mathbf{j} + (e^t + 1)\mathbf{k}) = 0,$$

so (8.26) is satisfied, and we are done.

7. Let E_r, E_θ and E_z be the components of $\mathbf{E}$, and let B_r, B_θ and B_z be the components of $\mathbf{B}$ in cylindrical coordinates. From the formula in Section 4.8 for *curl* in cylindrical coordinates, we get

$$curl\,\mathbf{E} = \frac{1}{r}\begin{vmatrix} \mathbf{e}_r & r\mathbf{e}_\theta & \mathbf{e}_z \\ \frac{\partial}{\partial r} & \frac{\partial}{\partial \theta} & \frac{\partial}{\partial z} \\ E_r & rE_\theta & E_z \end{vmatrix}$$

and so the formula $curl\,\mathbf{E} = -\partial\mathbf{B}/\partial t$ reads (compare components of $\mathbf{e}_r$, $\mathbf{e}_\theta$ and $\mathbf{e}_z$)

$$\frac{1}{r}\left(\frac{\partial E_z}{\partial \theta} - \frac{\partial(rE_\theta)}{\partial z}\right) = \frac{1}{r}\left(\frac{\partial E_z}{\partial \theta} - r\frac{\partial E_\theta}{\partial z}\right) = -\frac{\partial B_r}{\partial t}$$
$$\frac{1}{r}\left(-\frac{\partial E_z}{\partial r} + \frac{\partial E_r}{\partial z}\right) = -\frac{\partial B_\theta}{\partial t}$$
$$\frac{1}{r}\left(\frac{\partial(rE_\theta)}{\partial r} - \frac{\partial E_r}{\partial \theta}\right) = \frac{1}{r}\left(E_\theta + r\frac{\partial E_\theta}{\partial r} - \frac{\partial E_r}{\partial \theta}\right) = -\frac{\partial B_z}{\partial t}.$$

9. Since $\mathbf{E} = \sin x \sin t\mathbf{j}$, it follows that $div\,\mathbf{E} = (\partial/\partial y)(\sin x \sin t) = 0$. From $\mathbf{B} = \sin x \cos t\mathbf{k}$ it follows that $div\,\mathbf{B} = (\partial/\partial z)(\sin x \cos t) = 0$. From

$$curl\,\mathbf{E} = \begin{vmatrix} \mathbf{i} & \mathbf{j} & \mathbf{k} \\ \partial/\partial x & \partial/\partial y & \partial/\partial z \\ 0 & \sin x \sin t & 0 \end{vmatrix} = \cos x \sin t\mathbf{k}$$

and

$$-\frac{\partial \mathbf{B}}{\partial t} = -\sin x(-\sin t)\mathbf{k}$$

it follows that (8.28) is satisfied. However,

$$curl\,\mathbf{B} = \begin{vmatrix} \mathbf{i} & \mathbf{j} & \mathbf{k} \\ \partial/\partial x & \partial/\partial y & \partial/\partial z \\ 0 & 0 & \sin x \cos t \end{vmatrix} = -\cos x \cos t\mathbf{j},$$

whereas

$$\frac{\partial \mathbf{E}}{\partial t} = \sin x \cos t\mathbf{j};$$

so (8.29) does not hold.

8.6. Vector Calculus in Fluid Flow

1. Let $\mathbf{v} = (v_1, v_2, v_3)$. Then

$$div\,(\rho\,\mathbf{v}) = \frac{\partial}{\partial x}(\rho v_1) + \frac{\partial}{\partial y}(\rho v_2) + \frac{\partial}{\partial z}(\rho v_3)$$

$$= \frac{\partial \rho}{\partial x}v_1 + \rho\frac{\partial v_1}{\partial x} + \frac{\partial \rho}{\partial y}v_2 + \rho\frac{\partial v_2}{\partial y} + \frac{\partial \rho}{\partial z}v_3 + \rho\frac{\partial v_3}{\partial z}$$

$$= \left(\frac{\partial \rho}{\partial x}, \frac{\partial \rho}{\partial y}, \frac{\partial \rho}{\partial z}\right) \cdot (v_1, v_2, v_3) + \rho\left(\frac{\partial v_1}{\partial x} + \frac{\partial v_2}{\partial y} + \frac{\partial v_3}{\partial z}\right) = \nabla\rho \cdot \mathbf{v} + \rho\,div\,\mathbf{v}.$$

3. Assume that $\mathbf{v} = (v_1, v_2, v_3)$, $\mathbf{w} = (w_1, w_2, w_3)$, and that $\mathbf{v} \cdot \mathbf{a} = \mathbf{w} \cdot \mathbf{a}$ holds for all vectors $\mathbf{a}$ in $\mathbb{R}^3$. Substituting $\mathbf{a} = \mathbf{i} = (1, 0, 0)$ into $\mathbf{v} \cdot \mathbf{a} = \mathbf{w} \cdot \mathbf{a}$, we get $v_1 = w_1$. Similarly, substituting $\mathbf{a} = \mathbf{j}$ and then $\mathbf{a} = \mathbf{k}$ we show that $v_2 = w_2$ and $v_3 = w_3$. Thus, $\mathbf{v} = \mathbf{w}$.

Substituting $\mathbf{a} = \mathbf{i}$ into $\mathbf{v} \cdot \mathbf{a} = \iiint_W \mathbf{w} \cdot \mathbf{a}\, dV$ we get

$$v_1 = \iiint_W w_1\, dV.$$

Similarly, substituting $\mathbf{a} = \mathbf{j}$ and then $\mathbf{a} = \mathbf{k}$ we show that $v_2 = \iiint_W w_2\, dV$ and $v_3 = \iiint_W w_3\, dV$, i.e., all components of $\mathbf{v}$ and $\iiint_W \mathbf{w}\, dV$ are equal.

5. By definition, $\mathbf{v} \cdot \nabla\mathbf{v}$ is a vector whose components are $\mathbf{v} \cdot \nabla v_1$, $\mathbf{v} \cdot \nabla v_2$ and $\mathbf{v} \cdot \nabla v_3$. From $\mathbf{v} = (x+3t, 2y+2t, 3z+t)$ we get $\nabla v_1 = (1, 0, 0)$, $\nabla v_2 = (0, 2, 0)$, $\nabla v_3 = (0, 0, 3)$, and $\mathbf{v} \cdot \nabla v_1 = x+3t$, $\mathbf{v} \cdot \nabla v_2 = 2(2y+2t)$, and $\mathbf{v} \cdot \nabla v_3 = 3(3z+t)$. Thus, $\mathbf{v} \cdot \nabla\mathbf{v} = (x+3t, 4y+4t, 9z+3t)$.

7. We compare $\mathbf{i}$ components. Let $\mathbf{v} = (v_1, v_2, v_3)$ and $\mathbf{w} = (w_1, w_2, w_3)$. The $\mathbf{i}$ component of $\nabla(\mathbf{v} \cdot \mathbf{w})$ is

$$\frac{\partial}{\partial x}(v_1 w_1 + v_2 w_2 + v_3 w_3) = \frac{\partial v_1}{\partial x} w_1 + v_1 \frac{\partial w_1}{\partial x} + \frac{\partial v_2}{\partial x} w_2 + v_2 \frac{\partial w_2}{\partial x} + \frac{\partial v_3}{\partial x} w_3 + v_3 \frac{\partial w_3}{\partial x}.$$

On the right side, the $\mathbf{i}$ component of $\mathbf{v} \cdot \nabla \mathbf{w} + \mathbf{w} \cdot \nabla \mathbf{v}$ is

$$\mathbf{v} \cdot \nabla \mathbf{w}_1 + \mathbf{w} \cdot \nabla \mathbf{v}_1 = v_1 \frac{\partial w_1}{\partial x} + v_2 \frac{\partial w_1}{\partial y} + v_3 \frac{\partial w_1}{\partial z} + w_1 \frac{\partial v_1}{\partial x} + w_2 \frac{\partial v_1}{\partial y} + w_3 \frac{\partial v_1}{\partial z}.$$

The $\mathbf{i}$ component of

$$\mathbf{v} \times curl\, \mathbf{w} = \begin{vmatrix} \mathbf{i} & \mathbf{j} & \mathbf{k} \\ v_1 & v_2 & v_3 \\ \frac{\partial w_3}{\partial y} - \frac{\partial w_2}{\partial z} & \frac{\partial w_1}{\partial z} - \frac{\partial w_3}{\partial x} & \frac{\partial w_2}{\partial x} - \frac{\partial w_1}{\partial y} \end{vmatrix}$$

is equal to

$$v_2 \left(\frac{\partial w_2}{\partial x} - \frac{\partial w_1}{\partial y} \right) - v_3 \left(\frac{\partial w_1}{\partial z} - \frac{\partial w_3}{\partial x} \right).$$

Similarly, the $\mathbf{i}$ component of $\mathbf{w} \times curl\, \mathbf{v}$ is

$$w_2 \left(\frac{\partial v_2}{\partial x} - \frac{\partial v_1}{\partial y} \right) - w_3 \left(\frac{\partial v_1}{\partial z} - \frac{\partial v_3}{\partial x} \right).$$

Thus, we verified that the $\mathbf{i}$ components of $\nabla(\mathbf{v} \cdot \mathbf{w})$ and $\mathbf{v} \cdot \nabla \mathbf{w} + \mathbf{w} \cdot \nabla \mathbf{v} + \mathbf{v} \times curl\, \mathbf{w} + \mathbf{w} \times curl\, \mathbf{v}$ are equal. In a similar way, we prove the identity for $\mathbf{j}$ and $\mathbf{k}$ components.

Substituting $\mathbf{w} = \mathbf{v}$ into $\nabla(\mathbf{v} \cdot \mathbf{w}) = \mathbf{v} \cdot \nabla \mathbf{w} + \mathbf{w} \cdot \nabla \mathbf{v} + \mathbf{v} \times curl\, \mathbf{w} + \mathbf{w} \times curl\, \mathbf{v}$ we get $\nabla(\mathbf{v} \cdot \mathbf{v}) = 2\mathbf{v} \cdot \nabla \mathbf{v} + 2\mathbf{v} \times curl\, \mathbf{v}$, and thus $\mathbf{v} \cdot \nabla \mathbf{v} = \frac{1}{2} \nabla(\|\mathbf{v}\|^2) - \mathbf{v} \times curl\, \mathbf{v}$.

9. To prove $\mathbf{v} \cdot \nabla(\rho \mathbf{v}) = \rho \mathbf{v} \cdot \nabla \mathbf{v} + div\,(\rho \mathbf{v})\mathbf{v}$ we have to show that the $\mathbf{i}$, $\mathbf{j}$ and $\mathbf{k}$ components of vectors on both sides coincide. We prove the equality of $\mathbf{i}$ components (others are done in the same way). Let $\mathbf{v} = (v_1, v_2, v_3)$. The $\mathbf{i}$ component of $\mathbf{v} \cdot \nabla(\rho \mathbf{v})$ is

$$\mathbf{v} \cdot \nabla(\rho v_1) = v_1 \frac{\partial}{\partial x}(\rho v_1) + v_2 \frac{\partial}{\partial y}(\rho v_1) + v_3 \frac{\partial}{\partial z}(\rho v_1).$$

The $\mathbf{i}$ component of $\rho \mathbf{v} \cdot \nabla \mathbf{v} + div\,(\rho \mathbf{v})\mathbf{v}$ is

$$\rho \mathbf{v} \cdot \nabla v_1 + \left(\frac{\partial}{\partial x}(\rho v_1) + \frac{\partial}{\partial y}(\rho v_2) + \frac{\partial}{\partial z}(\rho v_3) \right) v_1$$

$$= \rho \left(v_1 \frac{\partial v_1}{\partial x} + v_2 \frac{\partial v_1}{\partial y} + v_3 \frac{\partial v_1}{\partial z} \right)$$

$$+ \left(\frac{\partial \rho}{\partial x} v_1 + \rho \frac{\partial v_1}{\partial x} + \frac{\partial \rho}{\partial y} v_2 + \rho \frac{\partial v_2}{\partial y} + \frac{\partial \rho}{\partial z} v_3 + \rho \frac{\partial v_3}{\partial z} \right) v_1$$

$$= v_1 \left(\rho \frac{\partial v_1}{\partial x} + \frac{\partial \rho}{\partial x} v_1 \right) + v_2 \left(\rho \frac{\partial v_1}{\partial y} + \frac{\partial \rho}{\partial y} v_1 \right) + v_3 \left(\rho \frac{\partial v_1}{\partial z} + \frac{\partial \rho}{\partial z} v_1 \right)$$

$$+ \rho v_1 \left(\frac{\partial v_1}{\partial x} + \frac{\partial v_2}{\partial y} + \frac{\partial v_3}{\partial z} \right).$$

The last term is zero, since $div\, \mathbf{v} = 0$, and so we are done.

Chapter Review

True/false Quiz

1. True. Take any simple closed curve **c**. By Green's Theorem, $\int_{\mathbf{c}} \mathbf{F} \cdot ds = \iint_S curl\,\mathbf{F} \cdot d\mathbf{S} = 0$, since $curl\,\mathbf{F} = \mathbf{0}$. The conclusion now follows from Theorem 5.8. Note: There is actually no need to compute the path integral. Since $curl\,\mathbf{F} = \mathbf{0}$, we use Theorem 5.8 directly to show that **F** is path-independent.

3. False. It works only if the domain of **F** is a simply-connected set (see Theorem 5.8).

5. True. Follows from the Divergence Theorem.

7. True. By the Divergence Theorem, $\iint_S curl\,\mathbf{F} \cdot d\mathbf{S} = \iiint_W div\,(curl\,\mathbf{F})\,dV = 0$, since $div\,(curl\,\mathbf{F}) = 0$.

9. True. In order to compute the differential, we need to calculate partial derivatives of the coefficient of $dxdy$ (which is 1). Since they are all zero, we conclude that $d(dxdy) = 0$.

11. True. This is equation (8.55) in Section 8.6

13. True. Follows from (8.16) in Section 8.4.

Review Exercises and Problems

1. Let S be a sphere of radius R. Recall that $\iint_S \mathbf{F} \cdot d\mathbf{S} = \iint_S \mathbf{F} \cdot \mathbf{n}\,dS$, where **n** is a unit normal vector to S. Since S is centred at the origin, the direction vector $\mathbf{r} = x\mathbf{i} + y\mathbf{j} + z\mathbf{k}$ is normal to S. Thus, $\mathbf{n} = \mathbf{r}/\|\mathbf{r}\| = \mathbf{r}/R$ is the outward-pointing normal vector field on S. Since

$$\mathbf{F} \cdot \mathbf{n} = -GM\frac{\mathbf{r}}{\|\mathbf{r}\|^3} \cdot \frac{\mathbf{r}}{R} = -GM\frac{\|\mathbf{r}\|^2}{R^4} = -GM\frac{1}{R^2},$$

we conclude that

$$\iint_S \mathbf{F} \cdot d\mathbf{S} = -\iint_S \frac{GM}{R^2}dS = -\frac{GM}{R^2}4\pi R^2 = -4\pi GM.$$

Alternatively, we could imitate the calculation done in the first part of the derivation of Gauss' Law (see Example 8.12).

3. There are several ways of doing this. We use one of the area formulas derived from Green's theorem. Parametrize the ellipse by $\mathbf{c}(t) = (a\cos t, b\sin t)$, $t \in [0, 2\pi]$. Then

$$\int_{\mathbf{c}} x\,dy = \int_0^{2\pi} (a\cos t)(b\cos t)\,dt = ab\int_0^{2\pi} \cos^2 t\,dt = ab\pi.$$

5. Using Green's Theorem, we get

$$\iint_{\partial D} \phi(f_x dy - f_y dx) = \iint_{\partial D} (\phi f_x dy - \phi f_y dx)$$
$$= \int_D d(\phi f_x dy - \phi f_y dx)$$

$$= \int_D (\phi_x f_x + \phi f_{xx}) dx dy - (\phi_y f_y + \phi f_{yy}) dy dx$$

$$= \int_D (\phi_x f_x + \phi_y f_y + \phi(f_{xx} + f_{yy})) dx dy$$

$$= \int_D (\phi_x f_x + \phi_y f_y) dx dy,$$

since, by assumption, $f_{xx} + f_{yy} = 0$.

7. Parametrize the segment from $(0,0)$ to $(2,0)$ by $\mathbf{c}_1(t) = (2t, 0)$, $0 \le t \le 1$. Then

$$\int_{\mathbf{c}_1} \alpha = \int_{\mathbf{c}_1} x^2 dx + xy \, dy = \int_0^1 (4t^2)(2t) \, dt = \left. \tfrac{8}{3} t^3 \right|_0^1 = \tfrac{8}{3}.$$

Parametrize the line segment from $(2,0)$ to $(0,1)$ by $\mathbf{c}_2(t) = (2,0) + t(-2,1) = (2 - 2t, t)$, $0 \le t \le 1$. It follows that

$$\int_{\mathbf{c}_2} \alpha = \int_{\mathbf{c}_2} x^2 dx + xy \, dy$$

$$= \int_0^1 \left((2 - 2t)^2(-2) + (2 - 2t)t \right) dt = \int_0^1 (-10t^2 + 18t - 8) \, dt = -\tfrac{7}{3}.$$

Finally, parametrize the segment from $(0,1)$ to $(0,0)$ by $\mathbf{c}_3(t) = (0, 1 - t)$, $0 \le t \le 1$. Then

$$\int_{\mathbf{c}_3} \alpha = \int_{\mathbf{c}_3} x^2 dx + xy \, dy = \int_0^1 0 \, dt = 0,$$

and thus $\int_{\mathbf{c}} \alpha = \tfrac{8}{3} - \tfrac{7}{3} + 0 = \tfrac{1}{3}$. Next, we use Green's Theorem. From

$$d\alpha = d(x^2 dx + xy \, dy) = (2x \, dx) \wedge dx + (y \, dx + x \, dy) \wedge dy = y \, dy \, dx$$

it follows that (D is the region inside $\mathbf{c}$)

$$\int_{\mathbf{c}} \alpha = \iint_D y \, dx \, dy = \int_0^1 \left(\int_0^{2 - 2y} y \, dx \right) dy$$

(the equation of the line through $(0,1)$ and $(2,0)$ is $x + 2y = 2$; hence the upper limit from the inner integration). Thus

$$\int_{\mathbf{c}} \alpha = \int_0^1 \left. yx \right|_0^{2 - 2y} dy = \int_0^1 (2y - 2y^2) \, dy = \left. (y^2 - \tfrac{2}{3} y^3) \right|_0^1 = \tfrac{1}{3}.$$

9. Using the identity $\operatorname{div}(f \nabla g - g \nabla f) = f \Delta g - g \Delta f$ (that was proved in Section 4.8), we obtain

$$\iint_S (f \nabla g - g \nabla f) \cdot d\mathbf{S} = \iiint_W \operatorname{div}(f \nabla g - g \nabla f) \, dV = \iiint_W (f \Delta g - g \Delta f) \, dV,$$

by Gauss' Divergence Theorem.

11. The differential of α is computed to be

$$d\alpha = d((x^2 + 2xy) dy \, dz + (2y + x^2 z) dz \, dx + 4x^2 y^3 dx \, dy)$$

$$= d(x^2 + 2xy) \wedge dy \, dz + d(2y + x^2 z) \wedge dz \, dx + d(4x^2 y^3) \wedge dx \, dy$$

$$= ((2x + 2y) dx + 2y \, dy) \wedge dy \, dz + (2 \, dy + 2xz \, dx + x^2 dz) \wedge dz \, dx$$

$$+ (8xy^3dx + 12x^2y^2) \wedge dxdy$$
$$= (2x + 2y + 2)dxdydz.$$

By Gauss' Divergence Theorem (change to cylindrical coordinates),

$$\int_S \alpha = \int_W d\alpha = 2\int_W (x + y + 1)dxdydz$$
$$= 2\int_0^{\pi/2} \left(\int_0^3 \left(\int_0^1 (r(\cos\theta + \sin\theta) + 1)rdz \right) dr \right) d\theta$$
$$= 2\int_0^{\pi/2} \left(\int_0^3 (r^2(\cos\theta + \sin\theta) + r)dr \right) d\theta$$
$$= 2\int_0^{\pi/2} \left(9(\cos\theta + \sin\theta) + \tfrac{9}{2} \right) d\theta$$
$$= (18(\sin\theta - \cos\theta) + 9\theta) \Big|_0^{\pi/2} = 36 + \tfrac{9}{2}\pi.$$

13. By Stokes' Theorem, $\iint_S curl\,\mathbf{F} \cdot d\mathbf{S} = \int_\mathbf{c} \mathbf{F} \cdot d\mathbf{s}$. The path integral on the right side is the work of $\mathbf{F}$ along $\mathbf{c}$. Since $\mathbf{F}$ is perpendicular to $\mathbf{c}$, it follows that the work is zero.

Alternatively, parametrize $\mathbf{c}$ by a function $\mathbf{c}(t)$, $t \in [a, b]$. Then, by assumption, $\mathbf{F}(\mathbf{c}(t))$ is perpendicular to $\mathbf{c}'(t)$, so that

$$\int_\mathbf{c} \mathbf{F} \cdot d\mathbf{s} = \int_a^b \mathbf{F}(\mathbf{c}(t)) \cdot \mathbf{c}'(t)\, dt = \int_a^b 0\, dt = 0.$$

15. (a) By definition,

$$d(\mathcal{B}) = d(B_x dtdx + B_y dtdy + B_z dtdz + c^{-2}(E_x dydz + E_y dzdx + E_z dxdy))$$
$$= \frac{\partial B_x}{\partial y}dtdxdy - \frac{\partial B_x}{\partial z}dtdzdx - \frac{\partial B_y}{\partial x}dtdxdy + \frac{\partial B_y}{\partial z}dtdydz$$
$$+ \frac{\partial B_z}{\partial x}dtdzdx - \frac{\partial B_z}{\partial y}dtdydz + c^{-2}\left(\frac{\partial E_x}{\partial t}dtdydz + \frac{\partial E_x}{\partial x}dxdydz \right.$$
$$\left. + \frac{\partial E_y}{\partial t}dtdzdx + \frac{\partial E_y}{\partial y}dxdydz + \frac{\partial E_z}{\partial t}dtdxdy + \frac{\partial E_z}{\partial z}dxdydz \right).$$

It follows that

$$d\mathcal{B} + \mu_0 \mathcal{D} = \left(\frac{\partial B_x}{\partial y} - \frac{\partial B_y}{\partial x} + c^{-2}\frac{\partial E_z}{\partial t} + \mu_0 \mathbf{J}_z \right) dtdxdy$$
$$+ \left(\frac{\partial B_z}{\partial x} - \frac{\partial B_x}{\partial z} + c^{-2}\frac{\partial E_y}{\partial t} + \mu_0 \mathbf{J}_y \right) dtdzdx + \left(\frac{\partial B_y}{\partial z} - \frac{\partial B_z}{\partial y} + c^{-2}\frac{\partial E_x}{\partial t} + \mu_0 \mathbf{J}_x \right) dtdydz$$
$$+ \left(c^{-2}\frac{\partial E_x}{\partial x} + c^{-2}\frac{\partial E_y}{\partial y} + c^{-2}\frac{\partial E_z}{\partial z} - \mu_0 \rho \right) dxdydz$$
$$= \left(-(curl\,\mathbf{B})_z + c^{-2}\frac{\partial E_z}{\partial t} + \mu_0 \mathbf{J}_z \right) dtdxdy + \left(-(curl\,\mathbf{B})_y + c^{-2}\frac{\partial E_y}{\partial t} + \mu_0 \mathbf{J}_y \right) dtdzdx$$
$$+ \left(-(curl\,\mathbf{B})_x + c^{-2}\frac{\partial E_x}{\partial t} + \mu_0 \mathbf{J}_x \right) dtdydz + (c^{-2}div\,\mathbf{E} - \mu_0 \rho)\, dxdydz.$$

Maxwell's equation

$$curl\,\mathbf{B}(x, y, z, t) = \mu_0\left(\mathbf{J}(x, y, z, t) + \epsilon_0\frac{\partial\mathbf{E}(x, y, z, t)}{\partial t}\right)$$

(see (8.25)) is the reason why the first three components of $d\mathcal{B} + \mu_0\mathcal{D}$ are zero. The fact that the $dxdydz$ component is zero follows from the equation (8.23).

(b) Computing the differential of $d\mathcal{B} + \mu_0\mathcal{D} = 0$ we get $d(d\mathcal{B}) + \mu_0 d\mathcal{D} = 0$; i.e., $\mu_0 d\mathcal{D} = 0$, since the differential applied twice is zero. Thus $d\mathcal{D} = 0$.

(c) From $\mathcal{D} = J_x\,dtdydz + J_y\,dtdzdx + J_z\,dtdxdy - \rho dxdydz$ it follows that

$$d\mathcal{D} = \left(-\frac{\partial J_x}{\partial x} - \frac{\partial J_y}{\partial y} - \frac{\partial J_z}{\partial z} - \frac{\partial\rho}{\partial t}\right)dtdxdydz.$$

The equation $d\mathcal{D} = 0$ implies that $-div\,\mathbf{J} - \partial\rho/\partial t = 0$; ie, $div\,\mathbf{J} = -\partial\rho/\partial t$ (this is the continuity equation for $\mathbf{J}$ and ρ).

(d) From $d\mathcal{B} + \mu_0\mathcal{D} = 0$ it follows that $\mu_0\mathcal{D} = -d\mathcal{B}$ and

$$\mathcal{D} = -\frac{1}{\mu_0}d\mathcal{B} = d\left(-\frac{1}{\mu_0}\mathcal{B}\right).$$

Thus, if $\alpha = -\mathcal{B}/\mu_0$, then $d\alpha = \mathcal{D}$.

17. (a) We get two Maxwell's equations: $curl\,\mathbf{E} = -\partial\mathbf{B}/\partial t$ and $div\,\mathbf{B} = 0$. The computation is performed at the end of Section 8.5.

(b) We get

$$*\mathcal{E} = *(-E_x\,dtdx - E_y\,dtdy - E_z\,dtdz + B_z\,dxdy + B_x\,dydz + B_y\,dzdx)$$
$$= c^{-2}(E_x\,dydz + E_y\,dzdx + E_z\,dxdy) + B_z\,dtdz + B_x\,dtdx + B_y\,dtdy).$$

This is the form $\mathcal{B}$ from the text of the Exercise 15.

(c) Computing the differential of $*\mathcal{E}$, we get

$$d(*\mathcal{E}) = c^{-2}\left(\frac{\partial E_x}{\partial x} + \frac{\partial E_y}{\partial y} + \frac{\partial E_z}{\partial z}\right)dxdydz+$$
$$+ \left(\frac{\partial B_z}{\partial x} - \frac{\partial B_x}{\partial z}\right)dtdzdx + \left(\frac{\partial B_x}{\partial y} - \frac{\partial B_y}{\partial x}\right)dtdxdy + \left(\frac{\partial B_y}{\partial z} - \frac{\partial B_z}{\partial y}\right)dtdydz$$
$$= c^{-2}div\,\mathbf{E}\,dxdydz - (curl\,\mathbf{B})_y\,dtdzdx - (curl\,\mathbf{B})_z\,dtdxdy - (curl\,\mathbf{B})_x\,dtdydz.$$

The equation $d(*\mathcal{E}) = 0$ represent the equations $div\,\mathbf{E} = 0$ and $curl\,\mathbf{B} = \mathbf{0}$ (these are Maxwell's equations in the special case when there are no charges and no currents present).

19. By definition,

$$d\alpha = d(f(x, y)dx + g(y, z)dy + h(x, z)dz)$$
$$= (f_x(x, y)dx + f_y(x, y)dy)\wedge dx + (g_y(y, z)dy + g_z(y, z)dz)\wedge dy$$
$$+ (h_x(x, z)dx + h_z(x, z)dz)\wedge dz$$
$$= f_y(x, y)dy\wedge dx + g_z(y, z)dz\wedge dy + h_x(x, z)dx\wedge dz$$
$$= -(f_y(x, y)dxdy + g_z(y, z)dydz + h_x(x, z)dzdx).$$

Thus

$$d(d\alpha) = -d(f_y(x,y)dxdy + g_z(y,z)dydz + h_x(x,z)dzdx)$$
$$= -((f_{yx}(x,y)dx + f_{yy}(x,y)dy) \wedge dxdy + (g_{zy}(y,z)dy + g_{zz}(y,z)dz) \wedge dydz$$
$$+ (h_{xx}(x,z)dx + h_{xz}(x,z)dz) \wedge dzdx = 0,$$

since (when multiplied out) each term contains a repeated factor. In order for this identity to hold, we have to assume that f, g and h are twice continuously differentiable functions.

Printed and bound by CPI Group (UK) Ltd, Croydon, CR0 4YY

20/10/2024

14576725-0003